Evening News

MATERIAL TEXTS

EVENING NEWS

Optics, Astronomy,
and Journalism
in Early Modern Europe

EILEEN REEVES

PENN

UNIVERSITY OF PENNSYLVANIA PRESS

PHILADELPHIA

Published by
University of Pennsylvania Press
Philadelphia, Pennsylvania 19104-4112
www.upenn.edu/pennpress

Printed in the United States of America on acid-free paper
2 4 6 8 10 9 7 5 3 1

Library of Congress Cataloging-in-Publication Data
Reeves, Eileen Adair.
  Evening news : optics, astronomy, and journalism in early
modern Europe / Eileen Reeves — 1st ed.
    p.   cm. — (Material texts)
  ISBN 978-0-8122-4574-5 (hardcover : alk. paper)
  Includes bibliographical references and index.
  1. Journalism—Europe—History—17th century.
2. Newspaper publishing—Effect of technological innovations
on—Europe—History—17th century. 3. Optics—Social
aspects—Europe—History—17th century.
4. Astronomy—Social aspects—Europe—History—17th
century. 5. Europe—Intellectual life—17th century. I. Title.
II. Series: Material texts.
PN5110.R44   2014
070.9'032   2013036502

CONTENTS

Introduction

Evening News concerns journalism's entanglement with astronomy and optics in the first decades of the seventeenth century. Part of this story, of course, involves the emergence of new technologies of communication and vision, but it is not merely the triumphal tale of the successful diffusion of scientific news in the popular and learned press. My concern is as much with what was distorted in those transmissions of information as it is with the more obvious successes of the medium. Thus while a series of ready-made comparisons between the telescope and the newssheet or newsletter exists—both, for instance, had antecedents elsewhere but emerged as viable commercial commodities from the Dutch Republic, and both purported to offer miniaturized views of remote regions, but less of local value—this work will concentrate above all on the productive shortcomings of those two conduits of information.[1]

While Galileo Galilei figures as an important protagonist of this narrative, *Evening News* is not specifically devoted to the development of journalism in Tuscany, where he lived as the grand duke's mathematician and philosopher subsequent to his 1610 discovery of the "Medici Stars" or Jupiter's satellites. Printed serial news came relatively late to Florence: sometime between 1636 and 1641, two different firms, seemingly rivals, received permission to produce and sell brief pamphlets reporting the latest events in and beyond Italy.[2] Both businesses maintained a Galileian connection of sorts by using the Medici Stars as a sometime printer's mark on the books they published, this discreet image suggesting Florentine loyalty to the astronomer both before and after his abjuration and condemnation in 1633.[3] Matters were otherwise, however, in the case of their gazettes; generally originating and composed elsewhere, the news reflected little in the way of a Florentine perspective, apart from the costly burden of reporting on Medici ceremonies. While it is difficult, therefore, to draw a robust connection between Galileo's activities and the particulars of Florentine journalism, it *is* the case that early

modern discussions of news often involved gratuitous gestures to the astronomer, and that texts concerning the heavens, conversely, are occasionally informed by reportage of earthly events. This book investigates the logic of
these allusions.

Direct Your Curiosity to Heavenly Things

The apparent confluence of optics, astronomy, and journalism in the early
seventeenth century was familiar before the actual emergence of either the
Dutch telescope in September 1608 or the publication of the first periodical
news in this same decade. To choose a celebrated example from antiquity,
when in his *Moralia* Plutarch addressed that most common of human fail-
ings, curiosity, he figured the soul of the inquisitive man as an insalubrious
camera obscura best remedied with the occasional shaft of light or fresh
breeze, and he presented astronomical research as "a great spectacle," and the
only respectable form of newsmongering.[4] "Apply your curiosity to the sun,"
he urged his readers, "where does it set and whence does it rise? Inquire into
the changes in the moon, as you would into those of a human being. . . .
These are secrets of Nature, yet Nature is not vexed with those who find
them out."[5]

While Plutarch conveniently ignored the fact that the periodicity of the
sun's apparent movement or of the moon's phases made the secrets that
characterized them, once discovered, predictable and therefore wholly inadequate as substitutes for news, his advice would have enjoyed a renewed hearing in the early seventeenth century, when interest either in the very basic
question of the sun's stasis or movement or in the more refined one of the
lunar light peaked, and when both issues were part of the debate over Copernicanism. Plutarch's pious observation that he who relied least on sensory
information was often involved in the greatest use of the undistracted mind,
and his approval of the spirit, if not the particulars, of an alarming tale about
Democritus—"that he deliberately destroyed his sight by fixing his eyes on a
red-hot mirror . . . so that his eyes might not repeatedly summon his intellect
outside and disturb it"[6]—would also have met with real understanding in
the early days of sunspot study. There, the roles of the senses and of optical
instruments came under great scrutiny, and those who indulged in direct
observation of the unfamiliar solar phenomena suffered irreversible damage
to their eyesight.

For many early modern readers, the abstemious regime of astronomical study proposed by Plutarch would have hardly served as a surrogate for news, and they would have much preferred an important forerunner of and companion to the newsletter, the almanac. Though it dates to antiquity, the almanac flourished in the print revolution of the late fifteenth century, and because it offered specific sorts of information about the near future—when solar and lunar eclipses would occur, when weather patterns were expected to change, when major religious holidays and feast days were to be celebrated, and when bloodletting, baths, and haircuts were best undertaken—it was presumably of immediate use in many aspects of civil and domestic life.[7] The predictive value of both almanacs and news collections, their cumulative ability to show long-term trends, is what the English statesman Dudley Carleton emphasized in 1616, perhaps somewhat self-consciously, when he reminded his friend John Chamberlain to save what others professed to use as fish wrap:

> Since you can be content with the gazettas you shall have them by every messenger and you need not send them back, though I will desire you to keep them, in that I make (and so have done a long time) a collection of them as some do of almanacs to know the certainty of the weather, yet not with so great diligence as the duke of Urbino, who hath them in process of time from great antiquity.[8]

But the single prognostications of the almanac, rather than the picture it presented over time, had already acquired a pronounced political and confessional character, a trait shared with early newswriting, and one at odds with the Roman Catholic Church's emphasis on free will and its hostility to astrological determinism. Matteo Franzesi, a Florentine who served as a secretary in the Roman Curia in the late 1530s and 1540s, explicitly connected prognostication with the thirst for news: "Let he who guesses / On the basis of conjectures and discussions / Astrologize," he advised his readers in a lighthearted poem on the emergent craze.[9]

While Franzesi's "On News" appeared in 1555 alongside other poems devoted to trivial objects—carrots, toothpicks, sausage—the situation of newsreaders and newswriters given to foretelling future events soon appeared more serious. A papal bull issued by Pius V in mid-March 1572, less than two weeks before his death, was directed against all "those who write, dictate, keep, transmit, and fail to destroy libelous pamphlets and those things called 'newsletters'" and warned against the short-term predictions, not necessarily astrological in origin, associated with such writings.[10] Under a year later his successor Gregory XIII threatened "those scandal-seekers known as *menanti*, and those who accept their writings, and those copying and transmitting libelous pamphlets," noting that such rumormongers dealt not just in a promiscuous and irrelevant blend of tales about the recent past, but also that "they foolishly foretell what will come to be."[11] By 1586, Sixtus V issued a bull against all who practiced judicial astrology and other divinatory arts, and the consumers of their books,[12] and though the specificity of the injunction suggests that almanac makers existed apart from newsvendors, their association remained close in the popular mind, likely because at least some copyists trafficked in both sorts of texts.[13] Tellingly, the format of the *Stupendous and Marvelous Prognostication and Almanac of the Present Year*, a posthumous work of the prolific satirical writer Giulio Cesare Croce, is that of contemporaneous printed newssheets.[14]

The tangled relationship of astrological predictions about the near future with news of the recent past is nicely emblematized in the exchanges between Gloucester and his bastard son Edmund and his true heir Edgar in *King Lear*, first performed in 1606. When Edmund seeks to persuade his father of Edgar's perfidious nature, he does so by showily concealing a fake private letter plotting the old man's death just as Gloucester makes his entry. Gloucester's assumption that such missives necessarily contain faithful accounts of current events—"Edmund, how now, what news?"—is typical of the early modern thirst for the sort of information conveyed in letters from unnamed "gentlemen of quality," and it predisposes him to believe that Edgar is both the writer and his would-be killer.[15] What most convinces Gloucester of the truth of the preposterous letter, the impact of which others, nonplussed, refer to as "strange news,"[16] is a set of prognostications concerning "these late eclipses in the sun and the moon,"[17] an allusion that would have had particular resonance, if not persuasive force, after the several eclipses of fall 1605. While Edmund has a bitter laugh over such astrological determinism in a subsequent soliloquy, he makes a show of being absorbed in reading not a letter but an almanac when approached by the guileless Edgar, to whom he then relates both a number of predictions "and I know not what,"[18] that will come true by the tragedy's end, and the accurate and timely news of Gloucester's rage.

The mutual involvement of news and prognostications appears to have increased after the first decade of the seventeenth century with the rise of journalism and the rapid development of astronomy. While licensers originally assigned the title *Astronomica Denuntiatio ad Astrologos* or *Astronomical Announcement to Students of the Heavens* to Galileo's *Starry Messenger*, it contained little of immediate use for prognosticators, and even seemed a menace to the entire enterprise to some early onlookers.[19] By 1613, however, when Sienese scholar Adriano Politi published his *Tuscan Dictionary*, an abridgment of the *Dictionary of the Italian Language* issued in 1612 by a Florentine academy to which Galileo belonged, he added the novel term *gazzetta*, likely derived from the name of a small Venetian coin, and defined it in a manner that suggested a genuine similarity between civic news and astrological prognostication: "a *gazzetta* is composed of sheets of reports from newsmongers; in Latin one might phrase it 'brief commentary on public events,' or 'ephemerides.'"[20] Jesuit writer François Garasse sought, less subtly, in 1619 to use Galileo's work in a counterattack on the dire prophecies of a Huguenot rival: "Since you want to play [the prophet] Nostradamus against the Roman Catholic Church," he wrote, "what with your predictions, your chronologies, and your preposterous prognostications, allow those who have read Galileo Galilei, and of the discovery of his new stars and unheard-of planets, to send you a reformed almanac," before going on to say what the spyglass revealed to him about his "lunatic" rivals.[21] Garasse's brazen appropriation of the *Starry Messenger* is part of a pattern; in 1630, Galileo's mere presence in Rome inclined a newswriter to allege, very improbably, that the famous "astronomer and astrologer had let people know" his predictions about the gender of an unborn baby in the Barberini family, about the approaching deaths of the baby's father and of his close kinsman Pope Urban VIII, and about the strong prospects for peace that summer in Northern Italy.[22]

The popular obsession with news seemingly generated by an astronomer was explicitly mocked in the first book-length study of the emergent social phenomenon of journalism, Caspar Stieler's *Zeitungs Lust und Nutz* of 1695. In his discussion of the familiar figure of the newsmonger, Stieler characterized information about remote events such as the current states of Mount Etna or Vesuvius, or the safe arrival of Dutch and English ships, as of little relevance to German audiences as "knowing whether or not men or spirits live on the moon or whether there [is] to be found nothing but wildernesses and deserts there."[23] Curiously, idle speculation about an earth-like moon populated with inhabitants and the bold claim to have seen them with the telescope were often attributed to the Frisian David Fabricius, who used the brand-new instrument and the *camera obscura* to observe the spotted sun with his son Johannes in March 1611. Though Johannes went almost entirely unnoticed as the first to publish the surprising news about the sun later that spring, in this typical expropriation of the figure of the astronomer, David Fabricius assumed a long and weird afterlife in technical literature, a history of astronomy, a philosophical dialogue, and even a novel as the purveyor of an irrelevant lunar bulletin.[24]

The anonymous stargazer whose pronouncements gained immediate popular attention for their apparent political and social impact fared somewhat better, and he still flourished as a credible news source in the Romantic era. In 1816, "the year without a summer," an epoch characterized by social unrest, high unemployment, crop failure, and a pale, spotted sun, the prediction of the impending death of the solar body and the fiery end of the world was ascribed by the Parisian *Moniteur Universel* to an astronomer from the observatory in Bologna who had allegedly offered an exact day, July 18, for the disaster. The prediction, probably a misinterpretation of a longitudinal study by a Jesuit astronomer at the Brera Observatory in Milan showing increased rainfall and lowered temperatures in Lombardy over the course of fifty years, was subsequently reported in the English, Belgian, and German press; actors, composers, and satirists eventually capitalized on the nonevent.[25]

An Emergent Industry

Thus far I have treated early modern news as a single entity because the symbiosis between different genres seems to me of crucial importance, but it would be unwise to neglect the real distinctions that obtained here. In what follows I will offer first a very brief overview of the emergent industry, and then a discussion of the relationship of elite newsreaders to a larger public, before returning to the particular meshing of news and optics in the post-telescopic era. Though both manuscript newsletter services for privileged subscribers and printed news accounts of sensational events flourished, more or less independently of each other, in the early sixteenth century, and though the *Mercurius Gallobelgicus* had been in existence since 1594, it was not until the early seventeenth century that printed newsbooks or newssheets, widely available in certain urban areas and perhaps well beyond,[26] costing relatively little to those who purchased them, and nothing to those who heard them read aloud or received them secondhand, emerged as periodical publications.[27] The chronology, though subject to some debate, typically places the first such works in Strasbourg in 1605, Wolfenbüttel in 1609, Basel in 1610, Frankfurt and Vienna in 1615, Berlin in 1617, Hamburg and Amsterdam in 1618, Antwerp in 1620, London in 1621, Delft in 1623, Paris in 1631, The Hague in 1635, Florence in 1636, and Genoa in 1639.[28]

These early printed newsletters and newsbooks usually emerged every week or every fortnight, and offered an assortment of relatively brief reports from other cities, rather than the narration of a single extraordinary event. Produced on poor quality paper and with modest attention to typographical and stylistic presentation, the newsletters or corantos ran from one to twelve sheets, and generally had some sort of consistent title indicating the place and date of production.[29] Conventions about the order of the individual news items vary, but the progression seems mainly to have been from those reports sent from the most distant points—and thus the stalest—to those nearer in place and time.[30] The last paragraphs of a newsletter, often more miscellaneous in character, reflected the latest additions to the text.[31]

Bibliographers and historians of journalism have found it useful to distinguish between these printed items and the many manuscript newsletters that flourished alongside them throughout the seventeenth century, and in many circumstances such distinctions were and are crucial. Manuscript newsletters, even those produced in many copies or widely circulated, would have enjoyed less propagation than their printed analogues. They were sometimes destined for a small list of subscribers who had the requisite capital, power, energy, and ambition to inform themselves more fully, if not more accurately, than those who bought or merely listened to printed newssheets, or who availed themselves of gazettes enjoying broader circulation.[32] In Rome such newswriters were often remarkably well informed—though not infallible— about the most minute developments within the Inquisition, despite that institution's aim for absolute discretion in its trials.[33] Newswriters directing their information to a restricted audience distinguished themselves occasionally from these gazettes and their readers; shortly after the assassination of Henri IV in 1610, for example, a *menante* or newswriter reporting to the duke of Urbino from philo-Protestant and anti-papal Venice wrote that "some fine minds say that the King of England [James I] has been killed too, and the gazettes here are full of this, but few believe it, and to season this sorrow they add another, that the Pope [Paul V] is either dead, or dying."[34]

The writer of manuscript news, especially in Venice and Rome where printed serial news was late to be established, operated for the most part with the connivance and even the cooperation of the authorities, many of whom relied on the restricted individual circuits of information provided by the *menanti*.[35] In England, where the crown made sporadic efforts to control printed news, manuscript newsletters containing more volatile information seem to have been less regulated, though the extent of and reasons for this apparent tolerance are by no means clear.[36] In France, a number of prominent literary figures including François Malherbe, Jean-Louis Guez de Balzac, and Jean Chapelain offered with varying degrees of formality an insider's news to a number of correspondents, often with explicit reference to the inferiority of printed accounts.[37] And from the mid-sixteenth through the mid-seventeenth century, manuscript newsletters were especially important for monarchs, scholars, merchants, Protestant clergymen, and political elites in central and southeastern Europe.[38]

As important as all these sorts of distinctions are, a close look at both production and consumption of early modern news militates, however, against always treating the printed serial and the manuscript newsletter in isolation or apart from neighboring genres of pamphlets, broadsides, separates, and private letters, or greatly distancing the elite newsreader from his down-market counterpart.[39] Manuscript newsletters were typically the source or basis of their printed analogues, the latter often conveying, for reasons of time, space, and political expediency, a blander or cruder version of the original report.[40] At least one observer of the two sorts of news alleged that the stricter regulation of printed accounts in Rome drove up the prices of the manuscript versions of original documents held by copyists and *menanti*, but prevented the "most grave scandals" that accompanied greater diffusion.[41] The passage from one circuit to another was sometimes emphasized or even fabricated: printed accounts in weeklies and in separates were frequently presented as the private letters of "gentlemen of quality."

The direction of the flow of information—from manuscript to print and from a restricted diffusion to a wider audience—appears on occasion to have been reversed: in private correspondence between elite consumers of news, for instance, information taken from printed sources often served simply as a marker of the distance between the sorts of things the writers knew or believed, and the cruder fictions available to the masses. Elsewhere, however, such information retained most or all of its original value, as when French letter writers copied printed *canards* into their personal correspondence, or when English-language corantos originally published in the Netherlands were handwritten by entrepreneurs eager to retail them in England, or when a *menante* in Prague in the employ of the duke of Urbino commented briefly on the contents and tone of a pamphlet written by a "simple rustic in Czech" and circulating in printed form in that agitated city.[42]

To this volatile mix of manuscript and printed news I would like to add, finally, two rather peripheral sorts of information about current events, the newsletters of Jesuit missionaries and the satirical poems routinely attached to the statues of Pasquino and Marforio in Rome. These conduits of news function as antitypes: the Jesuit letters from the society's missions, translated, printed, and circulated throughout Europe, were earnest and lachrymose expressions of piety and of recent miracles in remote settings,[43] while the handwritten and printed texts left at the bases of these statues were subversive commentaries on current events in Rome, the opinions of disgruntled humanists and members of the Curia, but often disguised as the spontaneous outcry of an amorphous and ill-defined populace.[44] The Jesuit *avvisi*, while directed to Rome, their spiritual epicenter, were emphatic in their exoticism, if not always informative in their details; the pasquinades posted on the lumpish, truncated Pasquino and the lolling giant Marforio, though they were soon exported elsewhere as a more mobile satirical tradition, focused mostly upon the ample scandals of that city alone.

The façades of these two kinds of news—heroic piety, miracles, and sacrifice abroad, and a folksy and finally inconsequential popular sentiment in Rome—are especially interesting, for they attempt to mask precisely the issues so objectionable to the enemies of the Society of Jesus, on the one hand, and to the ecclesiastical authorities on the other. While writers of anti-Jesuit tracts routinely mocked the newsletters from the society's missionaries, they were much more conscious of the order's alleged machinations at home, and especially of the enormous network of couriers and messengers necessary to sustain what they saw as the political undertakings of this singular corporate body.[45] Pseudoperiodicals such as the polemical *Mercure Iesuite*, published in Geneva and edited by the prominent Calvinist Jacques Godefroy, emerged with predictable speed, purporting to offer readers a more accurate chronicle of the society. As for the pasquinades, their booming foolery, anonymity, ephemeral quality, and relatively modest circulation were also meant to mask their status as more serious challenges to authority. Because of their sometime utility to various factions within the Curia, the satires stuck to these statues enjoyed a certain measure of tolerance, but neither the authorities nor readers appear to have been taken in by what the French scholar Gabriel Naudé portrayed in 1620 as a kind of malicious and transparent civic spectacle:

[These pamphlets] would better be burned than exposed to view, these centos, colloquies, *avvisi*, letters, "Echoes," "Harangues," "Remonstrations," and others of this sort, these things which people pull from their pockets, pass only to their friends, and sell in secret. They cost quite a lot, and are worth nothing, and are poorly made, as if coming from the hands of a crude, ignorant, and unpolished populace, one which has let itself be led, like "wooden puppets whose strings are moved by others."[46]

What Other People Read

It is worth noting that the interest of the elite in the kind of information available to a more general public was marked by a certain cultural anxiety, typically an overemphasis on classical learning in the midst of a request for current news. Given the strong connection between manuscript newsletters for the elite and their education as humanists, this development might signal either an awareness that the gap between elite and common newsreaders was narrowing, or a genuine interest in the popular perception of particular events.[47] Thus Peter Paul Rubens, writing from Antwerp in 1626 and renowned as a diplomat and painter, sought from Pierre Dupuy in Paris, "copies of public newssheets, but of the better sort, and at my expense," before adding,

I am sorry not to have the same commodity here where news-writers are not used, such that everyone informs himself as best he can, though there is no shortage of storytellers and charlatans who publish in print certain reports that are not worthy of being seen by gentlemen. I will do what I can to inform myself not with such bagatelles, *sed summa sequar fastigia rerum* [but following the traces of the chief events].[48]

Rubens's request is somewhat puzzling—for at that point Antwerp had printed periodical news, and Paris did not—but it is clear that his interest lay with what the public read.[49] His choice to end his stiffly worded request for the newssheets with a line from the *Aeneid* doubtless reflects a wish to distinguish himself in cultural and intellectual terms from this larger audience of newsreaders, even those Latin-readers who could tolerate the "harsh and uncouth stile" of the *Mercurius Gallobelgicus*.⁵⁰ When deploring rumormongering a year later he resorted again to the polished Latin verse of the satirist Juvenal, telling Dupuy that "here they speak of nothing except the siege of Grol, and lots of false rumors are spread daily. Every day it has been lost, it has been won, it is being helped, it has been lost again, and there are skirmishes and battles. The crafty lie 'kills whomsoever the mob bids,' according to the desires of each individual."⁵¹

Rubens's citation is drawn from a portrait of the vulgar and successful huckster in the third of Juvenal's *Satires*, and his expectation—surely correct—was that his correspondent knew all about the temporary ascendancy of "these men [who] were once horn blowers, who went the round of every provincial show, and whose puffed-out cheeks were known in every village."⁵² What such allusions do, in addition to marking off Rubens from that larger mass of literate, but not literary, news consumers, is to imply that a sort of continuum stretched from the world of the Virgilian epic where the goddess Venus, reporting on Dido's past, promised "to follow the traces of the chief events," to Juvenal's satirical sketch of those conjurors Fortune had once capriciously favored in Rome, and to their heirs in the early modern news industry.

An analogous touch marks the embarrassed request that Lorenzo Pignoria, canon of the Cathedral of Padua, made in 1608 of his friend Monsignor Paolo Gualdo, for more details about an ambassador then in Rome: "if you find out anything else, I'd appreciate two lines on him, because this will serve as food for the curiosity of these news devourers of ours, who are happy to hear about *Lamiae turres & pectines Solis* [the witch's castles and the sun's combs,] if nothing better is around."⁵³ While it is probable that Pignoria, like many others, prized news for its fungibility, exchanging one piece of information for another of greater interest, it is also the case that in drawing on church father Tertullian's scathing reference to old wives' tales and bedtime stories, he hoped to distinguish between his own very pronounced taste for current events, and the insatiable appetites of others, the latter being assimilated to a childish and heterodox craving deplored some fourteen centuries earlier.⁵⁴ Such attitudes were durable. The late seventeenth-century *Discourse on the Modern Use and Abuse of the Stories called News Tidings* insisted on the classical and biblical antecedents of the contemporary newsmonger—the vigorous gossiper of Theophrastus's *Characters*, or the rumor-hungry Athenians in Acts of the Apostles XVII—but identified print as particular accelerant.⁵⁵

While the remarks of Rubens and Pignoria betray a certain ambivalence, other elite newsreaders depicted their avid consumption of what was available to the populace as a sin of the past, a peccadillo that could readily be renounced. In 1653, Jean-Louis Guez de Balzac, who had written letters widely celebrated for their stylish communication of current events in the 1620s and 1630s, had seen them reprinted and translated in numerous editions in the subsequent decades, and had defended the Parisian *Gazette* in 1636 from mocking verses from the Netherlands comparing the *Bureau d'Adresse* to the Palace of Fame in Ovid's *Metamorphoses*,[56] offered at last a meditation on a kind of retreat from all news, a position which naturally involved an explicit rejection of the French journal's founder, the late Théophraste Renaudot. In "A Preface to the History of the Next Month, or the Pleasures of the Retired Life," Balzac began his fantastical reverie by recalling that an ambassador of his acquaintance had once spoken of an enchanted castle whose residents had neither calendars, nor clocks nor chimes, no workdays and no holidays, and never heard either good news or bad.[57] Balzac assured his correspondent that he could not do without bells and their regulation of civic and religious life, but that he could indeed dispense with news.

> Truly, if you wanted to make me happy, you would ban everything
> that goes by the name of *relation, gazette, ordinaire, extraordinaire*,
> and so forth. I have the greatest admiration for Monsieur Renaudot,
> and I praised him long ago. But since Plato chased Homer out of
> his republic after having crowned him with flowers and anointed
> him with perfumes, Monsieur Renaudot will not be offended if, for
> conditions favorable to our peace of mind, we close the door to him
> with all sorts of civilities, after having said of him that he is the most
> eloquent of modern historians, that France in some way owes him
> her great reputation, and that if she had lost him, she would have
> been hard put to find someone to replace him.[58]

The apparent ease with which Balzac dismissed Renaudot, founder of French journalism, depended upon his several allusions to the potential availability of information restricted to a more elite public: in this exercise of reverse snobbism, what passed for timely news in the *relation* and *gazette* was beneath the notice of ambassadors and courtiers, whose thorough familiarity with such information allowed them to transfer their attention to the *retardataire* world of enchanted castles where time itself has ceased to exist. The image is also a casual revision of both the conventional literary depiction of the *entrepôt* of sensitive information destined for the intelligentsia, those castles where rumors good and bad and true and false converge,[59] and of the historical fact that the individuals most likely to find themselves in such settings, members of European courts, were among the earliest subscribers to manuscript newsletters, as well as to printed news.[60]

Despite the heavy irony in his comparison of Homer to Renaudot—and by extension, the enduring splendor of the epic to the ephemeral weekly smudge that was the *Gazette*—Balzac's anxiety about the place of serial news within the state is genuine. Just as Provençal scholar Nicolas-Claude Fabri de Peiresc had taken pains to supply a scholar in Genoa with the *Mercure Français*, just as Dutch humanist Constantijn Huygens would regard news and rumors as the raw material of official chronicles, so Balzac, with less enthusiasm, recognized that newsprint, because of its widespread dissemination and relentless chronicling of the quotidian, offered a novel basis for historical writing. To the extent that simplistic, erroneous, mendacious, or politically motivated accounts of the recent past had any impact and efficacy—Balzac evidently believed this to be the case—France did indeed in some way owe her great reputation to Renaudot.

While the postures variously adopted by Pignoria in 1608, by Rubens in the late 1620s, by Peiresc less than a decade later, and by Balzac in 1653 to distinguish themselves from a larger audience of newsreaders do not mean that a public sphere—an autonomous arena for rational debate to which all private citizens had access—yet existed, these elite readers' apparent familiarity with and covert interest in the popular consumption of news is worth note, for it suggests a dawning apprehension that something like that sort of forum was under development. It is clear that there was an increasing interest, on the part of well-informed news consumers, in the impressions an amorphous public had about current events, in the manner in which such people came by their information, in the impact of wide availability upon the solitary work of interpretation, and in the alarming porosity of public and private discourses.[61]

Fellow Horn Blowers

The newsreader's awareness of the various ways in which information about current events reached him is complemented by a similar attentiveness to these distinctions on the part of the newswriters themselves, who evidently did not feel that they had emerged from a common past as "men who were once horn-blowers." Some of the attention devoted to other purveyors of news is simply due to professional rivalry. In 1615, for instance, the *Mercurius Gallobelgicus* gestured to "all those pamphlets, some libelous, some ridiculous" that had threatened the Twelve Years' Truce in the United Provinces in 1608, the *Mercure François* alluded to the imprisonment and death of Noël Léon Morgard, "master almanac maker," in 1614 and inveighed against prognostications, and in 1633 the *Swedish Intelligencer* exhibited a new diffidence about the publications upon which it sometimes depended, telling the reader in the preface that

> I have (as little as might be, and especially in the Kings story) trusted to no *written Relations*; unlesse received from a *knowne hand*, or confirmed by personall *eye*, or *eare-witnesse*. No, I have not singly relyed, so much as upon that diligent amasser of the *Dutch currant-oes*, the *Gallobelgicus*, and the *Arma Suecica; le Soldat Suedois*, I meane by it: upon whose single credit, I have no where written anything; except in those slighter encounters about *Norimberg*. And yet even there, had I beene in the same boxe, with him, and before him; the *High Dutch Relations*.[62]

It is also the case that those who produced manuscript newsletters for more restricted audiences alluded to others in the same trade who apparently performed different functions in the propagation of information about current events. Thus a newsletter written in March 1609 from Rome and destined for the duke of Urbino closed with a hasty scrawl, "The *novelista* [newsmonger] came here to tell me that the Venetians had brought bricks and other materials to make a tower on the sluice they have made on the Po, and that his Holiness [Pope Paul V] was not pleased with the papal legate to Ferrara, who had written nothing of this."[63] Though the emphasis here would seem to be on the subterfuges and silence of the papal legate, rather than on any access to privileged information enjoyed by the *novelista*, in other circumstances the newswriters in the employ of the duke of Urbino referred to the well-placed informants on whom they depended. In a letter sent from Neustadt in Austria in 1614, the newswriter assured the duke that he was much loved by the Belgian Jesuit Bartholomew Viller, confessor to the Habsburg archduke Ferdinand II, and that he received frequent confidences from him.[64] It is by no means clear that this newswriter was actually privy to inside information, but the motif of the political meddling of Jesuit confessors and other members of the order working at courts was both a familiar one in Protestant polemical literature, and an ongoing concern to the superior general of the society in Rome.[65]

The newswriters employed by the duke of Urbino told him with some frequency about the dangers and deaths of those who gathered and disseminated false or inappropriate information about current events, and the rhetorical point of such disclosures was the shoddy work of other journalists. Thus in late February 1608 the duke was told that

> On Monday eight of those who engage in the profession of *avvisi* were imprisoned for having written in their pages that Our Lord had granted a dispensation to the Duke of Savoy so that parties could be given, weddings performed, and meat eaten, and things like that during this Lenten season, which is wholly false, and a deliberate lie. On Thursday morning, the three of them who confessed to writing this were sent to the galleys for five years, and the others, judged innocent, were freed the night before.[66]

The duke's newswriters also relayed dispassionate reports of the professional irresponsibility, imprisonment, and death of the almanac maker Morgard and of his printer in February 1614,[67] and in August of that year, those in Rome sent three accounts of the decapitation of a pamphleteer from Rimini. Chilling in their brevity, these reports differ slightly in details about the display of the decapitated torso. All recorded the victim's name and birthplace, and agreed that the body bore the inscription "for libelous pamphlets and *pasquinate*," but one mentioned that the deceased was "placed for public view in the usual place" on the ponte Sant'Angelo, and added that two bundles of the fatal works were attached to the display, another noted that the killing had taken place in the jail of the Torre di Nona, and a third related that two burning torches illuminated the postmortem scene.[68]

It is difficult to judge if these accounts offer any degree of sympathy for or identification with the pamphleteer, but those of 1611 concerning the execution of humanist and ex-Huguenot Guillaume Reboul in Rome are less ambiguous. The manuscript newsletters sent to the duke of Urbino describe the victim's decapitation in the Torre di Nona and the display of his torso near the ponte Sant'Angelo, relate that he was charged with "having composed and published libelous pamphlets about the most powerful lords and ministers of the French crown," and conclude by emphasizing both this writer's great political knowledge and his most impolitic works.[69]

The duke and other subscribers to manuscript newsletters received the occasional reminder of the danger which these *menanti* themselves faced for making too explicit unpopular information. Most such allusions would simply reside in the cryptic brevity characteristic of these texts, though at least one writer, seeking to enliven the endless tedium of reporting on the pomp, festivities, and delays leading up to a princely marriage in 1608 with something more like fear, intrigue, and stealth, gamely resorted to a kind of bombastic timorousness. In a notice beginning "from someone else from Mantua, who knows something," he wrote,

> I want to say something about this marriage [of Francesco Gonzaga to Margherita of Savoy], but my lips are sealed. I will tell you only that there has been a delay, which, for those who know, is not at all what they believe it to be in the piazza. I don't know what to say: I am afraid, and I am afraid, and I could say it a thousand times, I am afraid, because it is not just a question of the wedding, but of something worse. This Gordian knot can't fail to come undone, though in my view, I believe it has by now been sliced through. When I can tell you something, I will.[70]

Remote Landscapes

I would like to return to the particular convergence of optics and news in the early modern period. That news should have been routinely imagined in visual terms is in some fashion puzzling, for it is certain that consumers often heard rather than read about current events.[71] Nor were these auditors necessarily either those unable to read or those so incurious as to be satisfied with short oral recapitulations. A poem by Michelangelo Buonarroti the Younger, written around 1632 and thus just a few years before the establishment of printed serial news in Florence, implies that in this city, as in many others, gazettes were routinely read aloud to a larger section of the public.

> And until another star takes me
> From these narrow city walls where the limits

Of my life are so tightly bound,
I will sit in a circle, listening to gazettes,
That now Mainz has been taken, now Cologne,
That Sweden is putting the bit on Germany,
And whether that king wants the scepter in Poland,
Or just to hold the Empire in check.[72]

Gabriello Chiabrera's "To Signor Giovanni Francesco Geri," written several decades earlier, likewise depicts a crowd composed of both nobility and merchants gathered in Florence waiting for "fresh letters" brought by the courier from France, and eventually hearing, rather than reading, the news from Lyon, Paris, and the Netherlands.[73] And even those who spent their time reading and writing the news, as for example the poet and sometime diplomat Fulvio Testi, understood the distillate of those lengthy written exchanges as commodities within an oral culture; in a letter sent from Turin in 1630 when much of Northern Italy had become, and would continue to be, a battleground for Spain and France, he noted that "*Rumors of quiet* are resounding everywhere, and the hopes of this exhausted region are increased by the consensus among the *avvisi*."[74]

The emphasis on the aural propagation of news, however, does not erase either a consumer's awareness of the texts from which resounding "rumors of quiet" emerged, or the conventional fiction that closed the gap between well-written words and vivid images of remote scenes. It is possible that individual writers who presented the epistolary exchange as akin to the telescope—letters and lenses alike offering a picture of sorts of a distant person or object—found the motif particularly appropriate when they did *not* have much news to share, for the emphasis in the examples below sometimes seems to fall on the pellucid medium and the candor of the correspondence, rather than on vivid realism of the scene described, as in the case of commercial newsletters. Thus in a letter written decades before the invention of the Dutch telescope, humanist translator Francesco Angelo Coccio imagined he was abandoning his pen for *occhiali da veder lontano* [glasses for seeing far], which allowed him "without getting up from the table, [to] see Venice, [some fifteen miles away], in the way that one sees certain distant cities represented in the landscapes of Flemish painters."[75] Curiously, though Coccio began by envisioning Venice, he ended by imagining his own likeness at his friend's side in that city, as if the visual medium were simultaneously available to both correspondents, and had more, finally, to do with the spirit of their friendship for each other than with the particulars conveyed by any of their letters.

The advent of the Dutch telescope did not mean that this sort of ambiguity vanished: the suggestion that remote correspondents were both looking from afar *and* transported to that distant landscape reemerges in the early seventeenth century in a letter of Angelo Grillo, a Benedictine priest, poet, and enthusiastic supporter of Galileo, to a friend in Antwerp. "Because of the naturalness of your style, it seemed to me that in reading the letter that I myself had accompanied you to Flanders, and undergone all those perils with you, and had not so much experienced them because of the affection I have for you, as *seen* them through the news bulletin that you give me."[76]

The association of optical instruments with the exchange of news between acquaintances was perhaps already a tired trope even in the year before the invention of the Dutch telescope when the Florentine poet Alessandro Allegri sought, in contrarian fashion, to distinguish rather than to compare the work of lenses and texts. Not surprisingly, in that writer's viewpoint, the thing that texts did best, and where they routinely outstripped their instrumental counterparts, was to convey what passed for news to distant friends. Put differently, the quality of information transmitted in epistolary fashion far surpassed that allegedly provided by perspective glasses, which allowed observers to recognize individuals and to read their lips.[77] Thus in a publication of 1607, as a prelude to a short poem about a recurrent (as well as current) event—food shortages—Allegri undid the traditional comparison, writing that if

> glasses make you see well, texts let you understand better; if the
> former let you look at many things, with the latter you may learn
> an infinite number; if the former reveal distant things, the latter
> show you bygone ones; if the former expose details in objects, the
> latter bring out subtlety in minds. With glasses we see no more than
> present things, but with texts, we can even learn the future; with the
> former we maintain eyesight even in old age, but with the latter, one
> gains eternal Fame. Lenses encased in silver and planted on the
> bridge of the nose, or tied to the ears, make us take many idiots for
> scholars, while alphabetic letters, if they are truly imprinted, as in
> the best of studies, and not merely stuck to fine gold with a lick and
> promise, allow those in the know to distinguish men of value. But

of the many and such sought-after advantages of the alphabet, the one that seems to me to trump all is that it allows true friends to know each others' doings, however many thousands of miles apart we might be, and not just our own affairs, but those of others, and not just particular matters, but also universal ones. For this reason, it would seem to me a great injustice, both to me and to you, if I did not give you detailed report of the silent suffering, and the emotion, and the sorrow of poor people in seeing that this year the Fair of Saint Martin was just about ruined—something not mentioned without tears there—because of the dearth of apples, ordinarily the principal purpose of that market. But because I know how much this eats away at you, so as not to dishearten you, I will tell you in rhyme.[78]

For all his efforts to discard the conventional association of lenses with the relaying of news, Allegri reverted to a traditional posture: the "detailed report" he sent about the fair would have provided, in a year of famine, little satisfaction to any correspondent whose primary interest was current events. What the letter conveys, precisely through its lack of news and its redeployment of the literary trope about optical instruments, is nothing but steady affection for his correspondent, what he would have termed, with a pun on its visual connotations, "regard."

The trope approached its logical limits, it would seem, in 1635 in the hands of Thomas Hobbes. In a letter to Sir Gervase Clifton, whose son he was tutoring in Paris, Hobbes admitted that he would write more often if he had either "any curiosity towards newes, or luck in writing them,"[79] and he began and closed another newsless missive in this fashion:

No newes, nor service I can do you, but the ambition to remayne in your memory hath produced this Letter, which is an acknowledgment of your favors, a confession under my hand of my great obligation to you. . . . If you believe this, my letter will have the effect of a perspective glasse, which shewes you not onely a towre afarre off in grosse, but also the battlements and windowes and other principall partes distinctly, as this shewes you my principall quality, gratitude, by which only I pretend to your good opinion, and the honor to be esteemed.[80]

The most striking efforts to dissociate written news and the telescope emerge in descriptions of the technology transfer that resulted in the Galileian instrument. While optical devices both before and after the invention of the telescope were routinely evaluated in terms of the access they gave to remote texts—letters lying open in distant fields and the like—those who sought to differentiate Galileo's "invention" from its Dutch prototype, beginning with Galileo himself, insisted on the oral character of the news from The Hague. They did so in order to suggest that he had encountered something imprecise, malleable, and unsketched, rather than a fixed description, a helpful drawing, or the object itself.[81] "About ten months ago a rumor came to our ears," Galileo claimed in the *Starry Messenger* in 1610; decades later, a disciple who described the simultaneous "invention" of the telescope, sight unseen, in Rome likewise referred to the impetus provided by "so many rumors from the Netherlands."[82] An early biographical sketch of Galileo stated that "having gone [from Padua] to Venice, he heard, among other news read aloud from the gazette from Flanders, that a lens maker had presented a glass that made faraway things seem near to Maurice, Prince of Orange." Unsurprisingly, the fortuitous oral propagation of the stale Netherlandish news story contrasts strongly with the sequel, distinguished by the energy and purpose of codified knowledge: "that very night [Galileo] invented one of his own ingenuity, making a gift of it to the Venetian Senate."[83]

The telescope was occasionally mentioned in news destined for a larger audience, generally in connection with its military potential. Shortly after its development in The Hague, for example, the spyglass figured in a printed separate originating in that city, and eventually in the *Mercure François*; once it was in Galileo's hands, references appear in the manuscript newsletter destined for the duke of Urbino,[84] as well as in printed form in Strasbourg.[85] An early printed description of the telescope, published in Milan in 1615 in a Jesuit writer's treatise on divine providence, had nothing at all to say about its utility in astronomy, but portrayed a twenty-powered device as a means of foreseeing and averting threats from an unspecified enemy:

Thus if you take two spectacle lenses—and let them be well-polished—one for a man of forty years, and the other with a curvature of seven *punti*, and you put them in a tube, the one at the top, and the other at the bottom, you will see things ten miles off that you would not have been able to see from a half-mile. . . . Thus with these lenses and telescopes you will see a cavalry troop, and a company of soldiers from a great distance, and you will know in time to guard against their attack.[86]

The instrument seems sometimes to have been assigned an implicit role in the gathering of news for a particular military event, as a few examples will suggest. In a manuscript newsletter of late May 1615 from the Urbino collection, for instance, the writer reported that "from Milan we learn that . . . with the invention of the Galileian perspective trunk one can see from Castello di Annone what is being done [ten kilometers away] in Asti and beyond, in the way of fortification and trench building on both sides of the Tanaro River," and three separate accounts of skirmishes near Asti follow on the subsequent folios as if to bear out such an assertion.[87] In an installment written in mid-June the newswriter made no more mention of the device, but presented events in Asti in terms consistent with earlier news accounts emphasizing the usefulness of the telescope in detecting fraud of all sorts.

From Milan on the 7th of this month we learn that after this skirmish . . . that the Duke of Savoy had discovered that a bombardier of his within the stronghold [of Asti] had been firing without ammunition because of a secret arrangement he had with the Spanish who had given him a great handful of money, such that they had been getting closer and closer. His Highness had him quietly removed, and placed others there, with reinforcements of artillery, and thus gave a good salvo to these Spaniards, who, in accordance with their agreement, first advanced and then retreated, finding themselves poorly treated. And then [the duke of Savoy] had the bombardier hung on the city walls, with a red purse about his neck.[88]

While the effect of such passages, especially the final emphasis on color and clothing,[89] is one of vivid proximity and immediacy—a relatively rare instance of such in the terse and spare prose of the manuscript newsletters—in subsequent decades the telescope's association with news conveyed something more like an idle detachment from the events at hand. Guez de Balzac, ever ready to assure his correspondents and his readers of the great distance he desired to place between himself and the world whose news he so clearly craved, advised a friend after the siege of Saint-Jean-d'Angély in 1621—in which

Louis XIII himself participated—that he "should do well never more to behold [war], but with *Flanders* spectacles."[90] The telescope seems likewise the instrument of a somewhat craven spectatorship, in a different sort of text, the *Swedish Intelligencer* of 1636, where, masked with a kind of ceremonial languor, it is associated with the impotence of the French, the Genoese, and the Tuscans in the face of the Spanish Armada's seizure and occupation of two small islands just off the coast of Cannes. "The navies of the great Duke of *Florence*, and the Duke of *Tursy*, sent from *Ligorne* under the command of *Melchior Borgia*, to vittaile the Islands of *Honoria* and *Margarita*, with their perspectives beholding the gallantry of the French fleet."[91]

As the range of connotations suggested by this brief survey implies, the motif of the telescope was easily adopted by those who wanted variously to emphasize the candor and transparency of their correspondence, the magnanimous perspective of friendship, the immediacy and detail now available to all newsreaders, and the absolute remoteness of such events. This is perhaps to say no more than that telescopes did not (and still do not) yield identical results for all observers, but this variety also makes the device an ideal image of a functional mendacity, as Galileo's friend Michelangelo Buonarroti the Younger recognized in his *Fiera* of 1619. In this play, the allegorical figure of Falsehood carried a perspective glass. Such instruments, hawked by story-selling and almanac-bearing mountebanks, allowed soldiers to see remote events in the city, but were finally of little appeal to them, since they had encountered better ones elsewhere.[92]

Medium and Message

Flourishing alongside fictions of the transparent communication of information was the notion that the medium, whether textual or optical, so strongly shaped the message that their separation was difficult. The recognition that a significant portion of the news received by early modern audiences often originated in alien linguistic, cultural, and confessional circumstances would have undercut or at least compromised the fantasy of the neutral transfer of such information, but the very fact that these transfers took place—that coherent images of distant regions were routinely conveyed to consumers a world away—suggested that some faith in the validity of the medium was clearly warranted. Consider in this connection a late seventeenth-century print which insists on the complex and ongoing relationship of the optical

FIGURE 1. Giuseppe Maria Mitelli, *The Courier in the Distance*, 1692.

Courtesy of the British Museum, © Trustees of the British Museum.

medium and the journalistic message, Giuseppe Maria Mitelli's *The Courier in the Distance Awaited by Those Crazed with War*. This etching shows eight men pointing to two horses on the horizon of Bologna, and commenting on what sort of information they expect to obtain, and on what bets they have placed; three of them observe the approach of the horsemen through spyglasses. The first of these declares, "It seems like a German courier to me," while the second replies, "To me he looks French," and the third retorts, "It's surely the courier of Spain."[93]

If this print of 1692 depicts a specific historical moment, as satirical works often do, the context would have been the series of battles waged in Piedmont and the Dauphiné by Louis XIV against the coalition of Piemontese troops and their Spanish and Austrian allies. Such attention would indeed have become an obsession in Bologna, for the commander in chief of the Austrian forces was Enea Silvio Caprara, of a prominent Bolognese military family.[94] The early stages of this campaign, though far less important than Louis' contemporaneous engagements with the English and Dutch in the Spanish Netherlands, were clearly of interest well beyond the region; the letters of French commander in Piedmont, Nicolas Catinat, show his irritation in 1691 when details of a failed siege were reported in a French-language gazette printed in the United Provinces.[95]

The differences evoked by the telescopic observers seem above all significant. None of these three men—indeed none in this unkempt and thuggish group—appears the embodiment of reason, vigor, or acuity: all idle in that featureless plain in the frenzied passivity conventionally associated with newsmongers. The slouching trio equipped with spyglasses, in particular, seems a comic version of the Austrian, French, and Spanish soldiery whose exploits they follow, as if the cheap optical devices and the wasteland surrounding them were crucial common denominators shared with the militia, both sides having pursued a scorched-earth policy in Piedmont and in the Dauphiné.[96] Couriers arriving in Bologna in 1692 would have had a variety of news to report: the French would have trumpeted Louis' victory over Dutch and English forces in the Spanish Netherlands, while the Spanish and Austrians would have emphasized their invasion of the Dauphiné. Mitelli's suggestion that each observer saw what he pleased, a variant on one of his prints of 1690, where a Frenchman and a Spaniard come to blows while listening to news being read publicly,[97] may well have corresponded to the actual situation in Bologna in the spring and summer campaigns of 1692.

The spyglasses the trio uses, though adequate to their purpose, appear of a relatively simple type, composed of three or four sections, and not nearly as long or as unwieldy as either astronomical or terrestrial telescopes.[98] Their reliance on these glasses to determine the nationality of the courier is especially interesting, for to recognize the horseman's country of origin and by extension, something of the tenor of the news he carried, is, in a sense, to perform an extremely cursory reading of the incoming information. The familiar analogy between reading and using these optical devices, moreover, finds a notable development in the very materials of which late seventeenth-century Italian telescopes and books were both composed. The tubes of astronomical and terrestrial telescopes and of spyglasses were often made of pasteboard—individual sheets of paper that had been glued together and compressed to yield a somewhat stiffer surface—which were then sometimes covered with leather; the tooling used to decorate the finer instruments compares to that adopted in this period by bookbinders in Italy.[99]

Better telescopes also had other material characteristics that they shared with the finest books: the eye piece and objective lenses, for example, were sometimes protected with caps made of vellum and lined with marbled paper, while an aperture disk would also have been composed of vellum, and the outsides of some tubes were decorated with paste paper, used in seventeenth- and eighteenth-century book binding.[100] One mid-seventeenth-century writer, Count Carlo Antonio Manzini, went so far as to maintain that he adapted such materials to refine his lenses: the final polishing of the glass took place on a piece of pasteboard previously softened by the bookbinder's hammer, on top of which he pasted fine paper typically reserved for deluxe editions, and a minimum of abrasive powder.[101]

While a rough correlation could probably be drawn between the size and overall quality of the telescope and those same aspects of books of the same period, my interest is in the more modest materials of which the spyglasses in Mitelli's print were composed. Just as they appear to lack the features associated with the preservation of finer books—leather tooling, vellum, dec- orations of colored end papers and paste papers—so the pasteboard of which these tubes were made was much more likely to have already served another previous purpose as the surface on which a text had once been written.[102] Put differently, in the case of less expensive instruments such as those depicted here, old and discarded manuscript or printed news offered the physical means of spying out the latest installment of current information.[103]

Mitelli's image suggests, then, that the ephemeral nature of news makes visible and legible more recent reports. In his earlier prints devoted to this subject, strangely exhausted consumers offer histrionic protests to the vendor—"I don't want to hear any more news, no, no, no!" or "We're sick of it, beat it, we're fine without it!"—and a bonfire seems the best means to end the obsessive interest in the Ottoman Empire.[104] But this print offers the most important meditation on the function of those short-lived documents: the newest news can only be seen with the telescope, and the telescope itself may be composed, in a literal sense, of old news. As this volume is designed to show, the late seventeenth-century conflation of the optical medium and the journalistic message embodied in *The Courier in the Distance* was a long time coming, but always visible on the horizon.

And Now the News

Evening News is chronological in structure, each chapter taking as the point of departure particular junctures of early modern journalism, astronomy, and

optics between 1610 and the late 1630s. That said, this study insists throughout on the distorted temporal relationships between reportage and referent:
news not infrequently precedes the event, recycles a remote past, or, when it
arrives too quickly, recounts nothing at all. Chapter 1 addresses the sole topical detail in Galileo's *Starry Messenger*—his peculiar depiction of the lunar
crater Albategnius as "Bohemia" in this publication of 1610—and argues that
this touch was engendered by news from Prague regarding the parlous state
of the Holy Roman Empire under the wavering rule of Rudolf II. What
makes Galileo's remark worth note is the response given it by Johannes
Kepler in his *Discussion with the Starry Messenger*, where legitimate astronomical observation and speculation about the moon's topography merge with
scraps of information taken from popular pamphlets attacking the Jesuits of
that city. The rumor that the ambitions of the Society of Jesus extended to
the lunar globe was to undergo no real revision but rather a series of generic
transformations, emerging in English poetry, polemics, and drama over the
next fifteen years.

Chapter 2 examines the almost simultaneous emergence of what Galileo
named the "Medici satellites" and the Medici regency in France after the
assassination of Henri IV in spring 1610. It shows that various chroniclers
fused these unrelated occurrences into a single narrative, typically by insisting
on the unknown means that made reports of the slaying in Paris available to
Netherlanders and Venetians some two weeks *before* the fact, and by portraying this access to news about remote landscapes in terms consonant with
the recently developed optical instrument. While the association of uncanny
foreknowledge of current events with the telescope is most striking in discussions of the royal assassination, it also appeared in at least one argument over
scientific priority. In a bid to establish that he, rather than his archrival Galileo, was the first to make telescopic observations of sunspots, Jesuit astronomer Christoph Scheiner recalled that in March 1611 he had been motivated by
a spontaneous desire to scrutinize that remote body, rather than by "advance
rumors" of the solar phenomena.[105] The puzzling status of such rumors, often
known as *vox populi*, and their sometime function as doubles of telescopic
vision, are the broader concern of this chapter.

Astronomers and journalists appear in another guise in Chapter 3, "Galileo *Gazzettante*." The pairing is both counterintuitive and counterfactual:
Galileo's extant personal correspondence contains almost no information
about current events, and from around 1613, he had very little to report in

the way of celestial novelties. The contemporary association of Galileo with his friend Traiano Boccalini, author of the caustic and extremely popular *News from Parnassus*, was based less on the individual assertions they offered, than on their relentless presentations of a particular critical perspective strongly associated with Tacitus and Machiavelli. Perspective, outlook, point of view: the language deployed by Boccalini, Galileo, and early seventeenth-century onlookers suggests a connection between a meticulous attention to novelties of the natural world and political perspicacity, as if to insist that the initial acquisition of piecemeal revelations would inevitably lead to an ungovernable tendency to promote new world systems.

The onset of the Thirty Years' War in 1618, the simultaneous development of widespread press networks, and the considerable increase in both astronomical activity and optical applications meant that these journalistic products and the telescope were now more widely available, subject to greater rather than less use, and more crucial to the public's sense of the information it required. Predictably, this need appeared most acute to diplomatic personnel, whose traffic in official reports and covert information was at times balanced by trade in and talk about optical gadgets. Chapter 4 presents the *camera obscura* as a fantastical corrective to the excesses of the telescope, and as an idealized correlate to foreign reportage, in part because projection was described in Latin and in most vernaculars as a kind of printing, the latter an economic activity separate from publishing, and thus an index of legibility, but not necessarily of public disclosure.[106] English ambassador Henry Wotton's enthusiasm for Johannes Kepler's instrument in the otherwise melancholy landscape of defeat after the Battle of White Mountain rests on an understanding of the canvas in the dark room as a passive and objective receptor of neutrally conveyed information to be shared among a political elite. The tidy picture of cameras that don't lie would be further adapted as an emblem of the miniscule proportions to which printed foreign news was reduced during the second wave of press censorship in England.

Chapter 5 is devoted, by contrast, to cameras that do lie, or at the very least, are greatly distant from the fantasy of the rapid projection of objective information to qualified observers. It examines the very polemical presentation of the *camera obscura* in Christoph Scheiner's treatise on the eye at the beginning of the Thirty Years' War, the association of the dark room with a strangely quiescent Amsterdam in the work of René Descartes, the importation of rumor and popular opinion into the *camera obscura* of Constantijn

Huygens's household, and the bizarre conflation of retinal experimentation and news gathering in the correspondence of the French scholar Nicolas-Claude Fabri de Peiresc. These representations of the optical apparatus within the realm of the social simultaneously erode and confirm the emergence of something like the public sphere: the issues of projection, illusion, and the disembodied and disinterested viewer concern them all.

The sixth and final chapter puts aside the sedentary consumer of foreign reports, and is thus concerned not with the hyperbolic treatment of optical devices, but with the emergence and the afterlife of a high-speed vehicle, the amphibious *zeilwagen* or "sailing chariot." Here, as in Chapter 3, my emphasis is on the state of newslessness. When the vehicle itself, rather than the onrushing rumor of some distant object or event, is the focal point, there's nothing to report. This is not to say, however, that the *zeilwagen* simply tore through an unaltered terrain, for the most arresting aspect of descriptions of that contraption was what the speed differential suggests about an evolving audience of news consumers. While the *zeilwagen* or its late analogue, the English mail-coach, signaled a single and static message of authority, the tardy and discrepant tidings in their wake were the tokens of a recent past still under negotiation, a future of sorts.

CHAPTER I

Jesuits on the Moon

Though his astronomical activity generated so much news, Galileo manifested little overt interest in current events. The correspondence of some of his closest friends, particularly Paolo Sarpi, Daniello Antonini, Gianfrancesco Sagredo, and Paolo Gualdo, by contrast, offers a varied budget of news ranging from international events, Venetian politics, and the endless skirmishes within clerical, university, and literary circles. Whether Galileo's apparent silence is an index of circumspection or indifference is difficult to gauge, and it would be in any case unwise to assume that what remains to us of his correspondence is representative, especially in an era where news enjoyed oral as well as printed and manuscript circulation.[1]

An important exception to this seeming distance from current events, and one that survived the transition from private letter to published treatise, was Galileo's curious comparison of an oversized lunar crater, generally identified as Albategnius, with Bohemia.[2] The association first arose in his letter of January 7, 1610, best known not for its lengthy synopsis of his telescopic observations of the moon, but rather for its casual reference to phenomena newly encountered only that evening, the moons of Jupiter.[3] Scholars then and now have rightly been more concerned with this latter and wholly unexpected discovery, for the detection of those satellites was an unrivaled novelty, unlike the observations of the moon's rough surface and secondary light, or of the numberless new stars in particular constellations and in the Milky Way, or even of the eventual phases of Venus or the sunspots, all of which had been either seen or hypothesized prior to the invention of the telescope.

In terms of the popular imagination, however, the impact of telescopic observation of the moon was without parallel. A traditional symbol of the Church, its new depiction as an opaque and rugged sphere was startling, and was significantly complicated by the fact that Christendom itself was then such a fractious entity. To describe a large and centrally located crater as an "enormous round amphitheater," and compare it to Bohemia, as Galileo did in his letter of January 7, 1610, and a few months later in the *Starry Messenger*, was to name the epicenter of confessional tension, and the eventual theater of war. As anticipated by most onlookers, the conflict that would become the Thirty Years' War began in Bohemia.[4]

Galileo's passing gesture to current events found considerable amplification in Johannes Kepler's *Discussion with the Starry Messenger* in May 1610. That Kepler elaborated remarks of religious and political relevance is not surprising: he was living the complicated life of the Protestant imperial astronomer at the Catholic court in Rudolphine Prague; he had long been attracted to the notion of presenting a "moon state," a commentary combining sociopolitical considerations with discussion of the physical features of our satellite; and while he inevitably came to privilege the latter alternative over the former in his posthumously published *Somnium*, it is clear that he could not always resist certain allusions to terrestrial rather than lunar situations. His references to satirical models such as Lucian of Samosata's descriptions of imaginary voyages to the moon—narratives poised midway between sheer fancy and rigorous technical detail, and full of precise observations about the distant earth—in the context of the *Discussion with the Starry Messenger* confirms that he was ready to use the lunar globe as a vantage point for judging its terrestrial neighbor.[5]

There is much to suggest that Kepler intended to convey in certain details of his treatise a commentary on contemporary affairs, especially those involving confessional conflict. Such details are not, however, so substantive that they either articulate a full-fledged meditation on Bohemia's crisis or displace the primary focus of the *Discussion*, an evaluation of Galileo's telescopic arguments; it is rather that the insistence on current events in and near Prague complements the astronomical news borne by the *Starry Messenger*, one part of which had clearly involved the "enormous round amphitheater" of Bohemia. Given that Kepler was uncharacteristically restrained about religious matters, what he offered by way of commentary on confessional issues cannot necessarily be ascribed to him, but rather to a nebulous *vox populi* and to the news pamphlets that seemingly contributed to and reflected such views.

Apart from an outburst about the excesses of the Counter-Reformation in a letter of June 1607, Kepler's criticisms of Catholic policy are few and mild.[6] When writing in April 1607 to his former teacher, Michael Maestlin, who admired neither the Jesuit astronomer Christoph Clavius nor the Society of Jesus, he referred wryly to that priest's description of Tübingen as "the city of the heretics in Germany,"[7] and he received several letters from Protestant friends outside Bohemia inquiring darkly about his safety and his confessional liberties.[8] "My most beloved Lord Kepler," began Lutheran pastor and astronomer David Fabricius in fall 1608 before launching into one of his lengthy missives,

> because of your long and unusual silence, I believe you have either died or left [Prague] on account of the Bohemian uprisings, which you may have ominously attracted through your endless Archimedean speculations about the martial planet; as the saying goes, "Imagination creates the event." If you're dead, this letter is my final farewell; if you're living, it's a warning. Break your silence, and whatever is bothering you, spit it out.[9]

Kepler does not appear to have followed Fabricius's bidding in this or in many other matters, replying mildly that "the causes of my silence are the idleness of age. You're an astrologer; you should have seen my sun besieged by Saturn, so I ought to say nothing of public problems."[10] When he described himself as one who lived "like a private individual on the world's stage," and who "served not Caesar, but all men and posterity," in a letter of 1605 to Maestlin, the point seems to have been to diminish his mentor's envy over his appointment as imperial astronomer, but it also suggests the kind of self-effacement and separation from his particular historical circumstances upon which Kepler insisted in most matters concerning his religious liberties.[11] In an age obsessed with questions of the appropriation of roles and the proper degree of display—one where Sarpi implied that he, his Jesuit enemies, the emperor Rudolph, and all Italians went about in their various disguises, and where, more trivially, the imperial ambassador to Venice believed that the suggestion that certain elements of the *Starry Messenger* were not new somehow "snatched the mask" from Galileo's face—Kepler claimed under certain circumstances to have no real part to play.[12] As I will argue below, however, this silence becomes in certain passages of the *Discussion with the Starry Messenger* a kind of ventriloquism, for what one hears is the raucous *vox populi* associated with the press. To adopt the humble terms of David Fabricius's proposal, Kepler's solution was not to "spit out whatever was bothering him," but rather to speak more covertly "from the gut," in the voice of another, that provided by tabloid journalism.

Kepler made clear, both in the extraordinary rapidity of his response to Galileo and in the breathless tone of the work itself, that he took seriously the Italian astronomer's presentation of his observations as news, and imagined his answer to fit within this genre. His posture is in some way at odds with the reactions of other readers, who tended to note in Galileo's claims to primacy the clear echo of efforts already made by others; Kepler, for his part, appeared content to point out, when possible, those instances in which Galileo's telescopic observations confirmed what he himself had offered as conjecture years before.[13] But this is not to say that the *Starry Messenger* seemed to Kepler definitive in its disclosures; indeed, part of its newsworthy aspect lay in the new debates to which it gave rise. These coming struggles, Kepler suggested in an astonishing passage in his "Address to the Reader," would surely be ardent, but strictly philosophical in tenor:

> I imagine, for intellectual pleasure, a quarrel between adversaries, then struggles, the victor's triumph, the awful threats and torment of the vanquished party, his shame, chains, imprisonment, and exile, all of which promise a certain seriousness: both are fighting for their views as if "for altar and hearth."[14] And it is hardly necessary to remind academics what it means "to defend one's position"—and others in any event can imagine it—for when one does this, he adopts as his own not just true and received opinions, but even absurd and false notions, and indeed, within academia, impious, pernicious, and blasphemous views.[15]

Though presented in parodic fashion, and as a sketch of the absurdities that characterized the academic debate, the sort of exchange Kepler described was a crucial one, and its excesses of some consequence. First of all, a letter Kepler had written in 1598 to Maestlin suggests that he viewed the tactic of defending all manner of nonsense—some version of those "absurd and false notions"—as a Jesuit specialty, albeit one which others were free to adopt, and which he chose studiously to avoid.[16] As the phrase *pro aris et focis*, "for altar and hearth," conveys, the focus of these quarrels often shifted from matters of no confessional import to questions of worship and of private property, the loss of one's home or the dramatic decrease in its value being a

means frequently used by Catholic authorities to punish or persuade recalci-
trant Protestants.[17] At the same time, the Catholic renewal in the Habsburg
lands in the mid-sixteenth century had seen the transfer of large amounts of
ecclesiastical properties lost during the Reformation, formerly held by neigh-
borhood committees or entirely without maintenance or ownership, back
into the hands of the Church, where it was available for resale, lease, or
commercial development; it is no accident that during the Prague Uprising
of February 1611, the most vicious attacks were reserved for the altars, hearths,
and personnel of two cloisters that had been renewed in just this fashion.[18]

The fact that ambiguities in Emperor Rudolph II's Letter of Majesty of
1609 had further facilitated the transfer of crown lands to Catholic prelates,
were vigorously protested, and eventually led to the Defenestration of Prague
would leave little doubt about the importance and inevitability of defending
one's position in matters of altar and hearth. Predictably, the period immediately after the Battle of White Mountain in November 1620 was characterized
by a certain imperial clemency regarding the "life and honor" of the defeated
Protestant faction, but also by a massive confiscation of their property.[19] It is
worth noting that Kepler's odd conflation of academic quarrels with confessional differences and with the eventual loss of property was emblematized
by an actual Jesuit with whom he had frequent correspondence, the mathematician Johannes Reinhard Ziegler. It was Ziegler who had edited one of
Father Christoph Clavius's several responses to Joseph Scaliger's attack on
the Gregorian calendar in 1609, who was overseeing the edition of Clavius's
opus in 1611–1612, and who by the mid-1620s would become the originator
and a prime force behind the movement for an even larger restoration of *all*
property acquired by the Protestants since 1555.[20]

It is also clear that Kepler saw a connection between inflammatory academic rhetoric, confessional antagonisms, and the loss of property, as that
had been his experience in the Styrian city of Graz.[21] He had complained, in
fact, in a private letter of 1607 that the perilous position of non-Catholic
subjects in Styria, where he lived from 1594 until he and his fellow non-
Catholics were expelled in 1600, had been exacerbated because there was no
one to show "the young people in the universities the manner and way to
behave in such places without offending one's conscience and with the necessary 'cleverness of serpents' so that rulers who are of a different faith may not
be disquieted."[22] The statement itself may be either laden with serpentine
cleverness or overburdened with the mildness of doves, for Kepler does not
mention that the Styrian ruler of a different faith who was so disquieted by

his non-Catholic subjects, Archduke Ferdinand II, was strongly influenced by his own Jesuit education in Ingolstadt, and especially by his subsequent contact with members of the society at the college in Graz.[23]

Kepler's preoccupations about the role of education in fomenting civil unrest were if anything greater in Prague.[24] From 1599 to 1600, and again from 1603 to 1612, his close friend Martin Bachazek served as rector of Charles University, a difficult position that another of the astronomer's associates, Johannes Jessenius, would occupy from 1617 until his execution in 1621.[25] The rivalry between Charles University, or the Carolinum, and the Jesuit college at Prague, the Clementinum, was pronounced; Catholics associated the former institution, particularly under Jessenius, with sedition, a charge also made, with reason, against the more distant and radical Lutheran universities of Jena and Altdorf.[26] The Clementinum, for its part, benefited from the aid of several very powerful patrons, and had grown rapidly from the dilapidated monastery previously associated with the Dominicans and destroyed during the Hussite Revolution; by 1600, the Jesuit compound, or *insula*, was equipped with a theater and a library, offered upper and lower schools where philosophy, rhetoric, and humanities were taught to Catholic and non-Catholic students from wealthy and modest backgrounds, and included, in addition to the increasingly well-endowed church of Saint Clement where sermons had been preached in Czech for almost a decade, and the ongoing construction of Saint Salvator, the beautiful newly consecrated Italian Chapel.[27] The Jesuits, or *patres*, as they were called, also worked closely with the Italian Congregation of Prague to aid the poor, the infirm, and the dying; one of the society's historians claimed, for example, that in 1608 the two organizations "led two or three hundred souls, especially the Calixtine Hussites, out of their errors and into Christ's flock."[28] The extraordinary ambitions and success of the society produced not just alarm and hostility among non-Catholics, but also envy among coreligionists; the Prague Jesuits' desire to absorb or suppress the Carolinum throughout the 1620s would fail in large measure because of the resistance of Catholics not associated with the order to that society's perceived monopoly on education.[29]

Many of the charges brought against the society appear today greatly exaggerated, but what is relevant here is the fact that "news" concerning Jesuit education focused precisely on the two colleges Kepler knew best, those of Graz and of Prague. The archetype of such pamphlets was Johannes Cambilhom's *Discoverie of the most secret and subtile practises of the Iesuites*, published in Augsburg in Latin in 1608 and soon thereafter translated to German,

French, Italian, Dutch, and English, cited at some length in the *Mercurius Gallobelgicus* of 1609, incorporated by Lutheran preachers in Augsburg into anti-Jesuit sermons and writings in 1609 and 1610, presented in an anthology of similar pamphlets in 1610, vigorously refuted both in polemical responses authored by Jacob Gretser S.J. in 1609, 1610, and 1612, and in the *Annual Letter* of the Society of Jesus in 1610, and reissued from Basel in an expanded format in 1627. The *Discoverie* included riveting inventions about a telescopic mirror employed by Father Pierre Coton in Paris; as these false rumors emerged on the heels of the true reports of the invention of the spyglass in the Netherlands, the stories were for some time closely associated.[30]

But the *Discoverie* was for the most part concerned with the behavior of the *patres* in Prague and Graz. Cambilhom claimed to have been a novice at the college in Graz until around 1605, and what he recounted about life there sounds like a hyperbolic version of Kepler's overwrought portrait of the academic debate, with a crucial difference: the conflict was in-house. Cambilhom alleged that under the pavement of the Jesuit church in Graz, there were prisons and storehouses, the latter being

> full of coards, swords, hatchets, pincers, fetters, boults, and ladders, which serve to mannacle, torture, and miserably to torment those which fall into [their] hands. . . . By these instruments they make captive the understandings of their poore Schollers, under their Jesuitical tyrannie: for if they finde any one that they suspect not to be constant in the resolution which he hath taken to be a good Iesuite, or if they feare that escaping, he will discover their secrets, they presently clap both on his heeles, and after they have made him to endure hunger and thirst, they put him to death with the most cruel torments.[31]

Cambilhom recounted other enormities in connection with the Clementinum: the institution at Prague, he wrote, had a vast collection of costumes for men and women of all rank and profession, and though the Jesuits maintained that these items were for their theatrical productions, their primary purpose was to allow members of the society to circulate undetected in public, to act as spies at court and in non-Catholic circles, and to beautify shabby girls destined for busy nights of "feasts and dansings" with them.[32] Both at Prague and at Graz, Cambilhom alleged, such costumes were also used for the education of novices: in the underground vaults of the Bohemian college, the priests wore diabolic masks and fearful attire in order to identify those students who were too timorous for sustained study of the occult,[33] while in the Styrian institution, a subterranean cache of sinister "garments fit for hangmen, hats, doublets, and hose" nestled alongside other instruments of torture.[34] In Prague, moreover, the Jesuits kept a supply of chains, iron scourges, cannonry, muskets, lances, battleaxes, and "great heapes of Bullets" in the space above the vault of their temples, Cambilhom wrote; he noted that the same sort of arsenal was maintained by the Jesuits of Cracow, adding, "I doubt not but the like is to be found in other coledges."[35]

The piquancy of some of Cambilhom's revelations did not detract from the credibility the treatise enjoyed in some quarters; a pamphlet describing the Prague Uprising of 1611, for instance, recapitulated the details about the Clementinum's lofty arsenal and storehouse of costumes by noting that "a shot rang from the top of the Jesuit College onto residents of the Old City, and the common man built it up in his head that a murder spree was about to take place whose purpose was to stamp out his religion," and that "some of [the Jesuits] escaped from the convent in clothes that disguised them."[36] Though the Clementinum was searched on this occasion for precisely the sort of weaponry Cambilhom depicted, none was found, and yet the *Mercurius Gallobelgicus* and several manuscript newsletters sent to Italy in the immediate aftermath of the invasion and uprising maintained, in fact, that such arms and ammunition existed in abundance.

> The Jesuits, in order to escape the mob, gave the keys [to their cloister] to the [Bohemian] Estates, and committed themselves and their college to its protection. When in addition to lances and other weapons, around six hundred firearms with a considerable supply of gunpowder and cannonballs were found there in the college, a great many musketeers were introduced into that very well fortified place, and the Jesuits slipped off to their protectors.[37]

The story was repeated in the *Nouvelles d'Allemagne*, with additional emphasis on the "great murmur" that rose from the populace of Prague when the news of the firearms became known.[38] Though Paolo Sarpi reported with evident surprise in a letter of June 1611 that he had learned from the Venetian ambassador in Prague that the rumor was not true, what is crucial here is not its credibility, but rather its durability, for some version of Cambilhom's story seems to have been repeated in each denial.[39] Despite the resonance between Kepler's presentation of an academic debate and Cambilhom's depiction of the education of novititiates in the Society of Jesus—a resonance muted by the differences in genre and the presumed purposes of these works—it seems to me very unlikely that Kepler believed the ridiculous particulars of the *Discoverie of the most secret and subtile practises of the Iesuites*. I am arguing, rather, that in elaborating upon Galileo's telling allusion to Bohemia in the course of his *Starry Messenger*, Kepler drew upon this popular impression of local news, the focal point of which would have been the Jesuits of Prague. Put differently, popular rumor provided him with a ventriloquized voice in matters where it was best to remain silent, and one so impossibly strident that it seemed to bear only the faintest resemblance to his own.

The Rasping of an Indistinct and Unclear Voice

It is worth noting in this connection that Kepler alluded several times in the course of his *Discussion* to his own reticence, a claim which is a rough analogue to his reliance on the raucous voice of public opinion. While he assumed this posture of timidity with peculiar facility and always in explicit contrast to what he read as Galileo's blunter manner and bolder approach to optical problems, the assessment appears to have very little to do with either Kepler's bold experimental style or the relatively unrestrained manner in which he described actual procedures and articulated more abstract conjectures. Though it is undeniable that at least some of what Kepler portrayed as his own early diffidence is a means of asserting his primacy over Galileo's belated and roughshod path through the same intellectual terrain, his emphasis on a certain indirection and restraint applies above all to his treatment of the confessional turmoil in Prague. When, for instance, Kepler noted that despite Rudolph II's enthusiasm he had failed to construct a telescope himself because he had both mistrusted the claims of Giambattista della Porta's *Natural Magick* and feared the distorting effects of the atmosphere, he congratulated Galileo for his energy and for his decision to rely on actual observation of, rather than on mere conjectures about, the instrument's capabilities. In proclaiming to Galileo that the "sun of truth having risen with your findings, you drove away all those specters of perplexity, along with their mother, Night, and you showed what could be done by the act itself,"[40] Kepler cobbled together a rather conventional figure of speech, albeit one not especially appropriate to a procedure mostly used, thus far, for nocturnal observation, and emphasized his own earlier work in optics. His depiction of those efforts put to flight by Galileo's findings as *omnes istas titubationum larvas*, "all those specters of perplexity," is somewhat ambiguous, for this awkward phrase might also be translated as "all those *masks* of perplexity," the slippage of the meaning of the word *larva* being at issue in at least two other early modern optical works.[41]

And yet nowhere in the course of his *Discussion with the Starry Messenger* did Kepler remove this "mask of perplexity"; he merely modified slightly the persona of one who—as the word *titubatio* suggests—wavers, stammers, or in some way falters when called upon to express himself, a man living "like a private individual on the world's stage." In returning to the portrait of Galileo as a man of almost comic boldness and directness—one perhaps quite at odds with the Italian's manner—Kepler presented himself as entirely different in affect, though similar in intellectual interests, and undeniably prior in certain optical studies.[42] In a detail that is disconcertingly out of place in the fiction of a "discussion" with Galileo, Kepler set aside both first- and second-person pronouns, and offered the following vignette: "Let Galileo then stand with Kepler, the former observing the moon, having directed his face toward the heavens, the latter the sun, turned toward a writing tablet so that the lens not burn his eyes. Each man has his own technique, and let their association produce one day the clearest theory of the distances [between the sun, moon, and earth]."[43]

The image is accurate in that Kepler's method of solar observation through projection onto paper was quite influential: most early students of the sunspots would adopt the technique described in the *Unusual Phenomenon, or Mercury's Solar Transit* of 1609, and some, such as the Jesuit Odo van Maelcote, explicitly acknowledged their debt to the imperial astronomer.[44] But the passage is also remarkable for its presentation of him as *aversus in tabellam*, "turned toward a writing tablet." Not surprisingly, the noun *tabella* lent itself to a variety of interpretations: sunspots and other phenomena were projected onto white paper, blank books, and canvaslike surfaces, and described as a type of printing or painting.[45] Of greater significance, however, is the suggestion that Kepler was turned toward a written composition or a public record, a category of documents into which tabloid news would fall. Such an interpretation would have been reinforced by the fact that while *tabella* was a rather imprecise noun, *tabellarius* was a good deal more specific, and referred to a courier, the bearer of written news. Put differently, Kepler's primary intention was to present his astronomical work as the complement of Galileo's, and the fruits of their imagined collaboration as an accurate depiction of the relationship of the sizes and distances of the sun, moon, and earth; his secondary aim was to allude to the attention he gave such news reports in his depiction of the lunar world.[46]

I have been arguing, then, that Kepler's reliance on rumor and news constituted a resolute turn away from methods involving direct observation, and toward whatever might be projected onto the paper or canvas of the dark room: a sort of ventriloquism. In the early modern period, the process of "speaking from the gut" appears to have been more strongly associated with either oracular inspiration or demonic possession, though occasionally, and often within a Protestant context, the voice of the ventriloquist or *engastrimyth* figures merely as a tool for fooling others, and as an instrument distinct from a particular being.[47] It is this sense, closer to our own understanding of ventriloquism, that is of interest here; something like it reappears in the notes Kepler wrote in the early 1620s to accompany his *Somnium*, a meditation on lunar astronomy composed in 1609.

In the *Somnium*, the long section on the physical characteristics of the moon is recounted to the narrator and his mother by a lunar daemon, one whose speech is introduced as "the rasping of an indistinct and unclear voice," and further qualified as "in the Icelandic tongue."[48] The notes used to explain and elaborate the *Somnium* itself alternate between collapsing all distance between the pronouncements of this being and Kepler's previous findings, and presenting the former utterances as an entirely separate narrative ungoverned by astronomical research programs. At times, in fact, Kepler guessed (perhaps rightly) that "the allegory [was] growing chilly," and he simply carried on with the physical description of the moon; elsewhere he expressed some confusion about what he had originally meant to convey, in allegorical terms, in the *Somnium* proper.[49] Of the odd Icelandic voice that both was and was not his own, he stated,

It is not impossible, I believe, with various instruments to reproduce individual vowels and consonants in imitation of human speech. Yet whatever this is going to be, it will resemble rumbling and screeching more than the living voice. And in this device, I think, there are built-in traps for the superstitious and gullible, so that they sometimes suppose demons are talking to them when instead the technique resembles magical tricks. And yet in my opinion it is more appropriate for others to assert on a sound basis that anything of

this sort actually happens than for me to deny it without support
from any experience of my own.[50]

While this note implies that Kepler viewed his articulation of the lunar
daemon's rasping and foreign message as a kind of ventriloquism—a position
not very different, then, from his sometime adoption of the raucous voice of
popular opinion in the *Discussion with the Starry Messenger*—the preceding
note concerns the obscure conditions from which such observations emerge.
As Kepler told it in the *Somnium* proper, the narrator and his mother covered
their heads with their clothing prior to the approach of the lunar daemon,
and in the corresponding note the astronomer stated that this gesture recalled
the equally makeshift observatory at the Imperial Castle of Prague that he
and several envoys of the Count Palatine of Neuburg had had to fashion of
their cloaks during a solar eclipse of October 1605.[51] This detail is crucial, for
as in his *Discussion with the Starry Messenger*, Kepler evoked a dark room, and
one in which observers turned away from the aperture and the phenomena
beyond it, and toward a tablet or canvas to obtain information about the
world outside.

Though Kepler was interested in using the eclipse to determine with
greater precision the eccentricity of the moon's orbit, it is clear that other
observers viewed the event in quite different terms.[52] The temporary scientific
collaborators Kepler found in the legates from Neuburg were in Prague in
order to press for the Count Palatine's claim to the strategic territory of
Jülich-Clèves,[53] and presumably shared the astronomer's interest in what
diverse observations would reveal about the lunar path. But there may not
have been, finally, great differences between the popular reaction to an event
whose chief feature was a zone of darkness across Southern Europe and particular obscurity at Rome, and the political calculus such envoys were then
making about the role of Spain, Italy, and especially the pope in the succession crisis.[54]

It is also the case that the role of the ventriloquist—or, to use the more
current term of early modernity, the *engastrimyth*—was familiar enough to
be adopted within the context of news in an English poem of 1625. Where I
have presented Kepler as the ventriloquist, arguing that in articulating rumors
previously circulated in tabloids that he assumed a voice so strange as not to
seem his own, this work makes the bodiless utterance something more like
the *vox populi* itself. At issue in William Crosse's *Belgiaes Troubles, and Triumphs* was a baseless tale about General Spìnola's approach:

This rumour first was grounded on the voyce
Of Fames *Engastromists*, the vulgar noyse,
Who trumpet out for loud resounding Fame,
Things not done for things done, and make the same
Which but appearing was, apparent true
To the deceived worlds deceiving view:
Who taking it, this Marchandise doth sell
To those Retaylors, whose broad eares doe dwell
In Tavernes, Barbers shops, and publike Marts,
Where lyes are sold to hollow Spungeous hearts,
For Beere, for Wine, and that curst Indian weed,
Whereupon these puffes of noveltie still feed.[55]

Crosse located the ventriloquist at the origin of any news cycle, for these authorless rumors were simply the raw material on which the nascent industry relied. Because tales such as these lacked an apparent source they were at once difficult to control and easy to appropriate, and they flowed freely and without cost into taverns, barber shops, and markets, before being converted to print and then to coin by the newsvendor. For Kepler, by contrast, the act of ventriloquism—speaking in an unrecognizable voice—occurred not when the rumor emerged, but when it made its raucous way into his work.

The Particular Character of Their Provinces

To summarize, Kepler appears to have adopted the motifs of both ventriloquism and the oblique "turn toward the writing tablet" in the presentation of unpalatable notions in his *Somnium* and his *Discussion with the Starry Messenger*, but it is clear that the fictional nature, variegated tone, and quite uneconomical textual apparatus of the former text allowed for much greater departures from a single and chiefly scientific perspective. Thus while most of the *Discussion* concerns the telescopic appearance and true nature of celestial objects, Kepler offered in passing a crucial allusion to the particular terrestrial setting from which his response emerged. In noting, for instance, that he first learned of the *Starry Messenger* when imperial councilor Johannes Matthias Wacker von Wackenfels "announced it to me from his carriage in front of my house," Kepler incorporated the Clementinum into his story, the growing Jesuit compound being directly across the street from him.[56]

Consider in this context Kepler's remarkable meditation upon the "Bohemian" crater described by Galileo. In speculating upon what forms of life might be supported on the moon, and concluding that the craters were in fact of artificial rather than natural production, Kepler reasoned that the moon dwellers had constructed circular structures above the ground in order to avoid the extremes of heat and cold characteristic of that globe.

It is perfectly reasonable to believe that if there are living creatures on the moon . . . they conform to the particular character of their provinces, which have mountains and valleys of much greater size than ours, and that they are equipped with much larger bodies, and therefore accomplish immense works. And as their day is the length of fifteen of ours, they feel an intolerable heat; lacking, perhaps, rocks that would serve as ramparts against the sun, they might rely instead on a sticky sort of clay. Here is the sort of construction they would favor: perhaps in order to find water in the depths, they dig up vast areas, the displaced earth then being piled up in a circle, such that they might retreat into the deepest shadows, behind these excavated mounds. The lunar creatures move around following the motion of the sun in order to stay within the shadow. And this arrangement is for them a kind of underground city: their houses are numerous caverns dug out of that circular foundation, and their fields and flocks are in the middle, such that when fleeing the sun they need not go too far from their booty.[57]

This hypothesis, while consistent with that presented in the *Somnium*, offers peculiar overtones. The moon dwellers appear, in Kepler's description, to have an odd air of militarism: they need *munitiones* or ramparts against the sun, and they evidently hesitate to distance themselves from the spoils or *prædæ* previously obtained in some undisclosed battle. And though they roam about above the moon's surface, their dwellings are rather confusingly said to be "for them a kind of underground city." In May 1610, however, Kepler's lunarians would have been familiar enough to the newsreading public, for that audience would have recognized in these details those contemporaneous reports of the vast subterranean crypts, vaults, and cellars of weaponry and booty allegedly maintained by Jesuit priests throughout Bohemia and Austria, but concentrated especially in Prague and Graz. Thus while

Galileo had transformed the crater Albategnius into Bohemia, Kepler responded by accentuating the Italianate face the Jesuits in particular had given that region, for monumental structures, often ornamented with the "sticky sort of clay" of stucco or terra cotta, or built with brick rather than constructed with stone, were variously associated with an Italian and Catholic presence and an aggressive Counter-Reformation movement.[58] This is not to say that Kepler's celebrated discussion of the moon's features was without physical merit—for he intended it as a coherent meditation on the atmosphere and conditions of the lunar globe—but rather to emphasize his simultaneous reliance on the backdrop provided by journalism, that oblique position he characterized as a "turn towards a writing tablet."

Like the tabloid image of the Jesuits, therefore, Kepler's lunarians divide their territory into *provinciae* or provinces. Central to their existence, literally, on the one hand, and figuratively on the other, are pastoral cares, *pascua in medio*; the focus, both of Kepler's remarks and of public scrutiny, is on the *domus* or centralized houses from which their authority emanated.[59] And because one of the wilder and most durable details of Cambilhom's *Discoverie* was the allegation that the underground vaults served as burial grounds for unpromising or uncompromising novices and other victims, Kepler's allusion to the lunarians' *tumuli egesti*, a phrase which he intended to mean "excavated mounds," could also have been translated, by those familiar with and convinced by these news reports, as "dug-out tombs."

The manner in which Kepler describes the industry of his lunarians is tellingly reminiscent both of the grandiose monumentality of the Italianate face of Catholic Prague, and more specifically, of the Jesuit Order with which such architectural and cultural efforts were associated: *Nam profecto consentaneum est . . . immania etiam opera patrare*, "It is perfectly reasonable to believe . . . that they accomplish immense works." *Opera* was a favorite Jesuit term for designating tasks in the society; the comparatively rare verb *patrare*, then thought to derive from the word *pater*, "father," and associated with the noun *patratus*, a type of ancient priest charged with the declaration of war and peace, would have valuable connotations for an order whose fathers were considered, especially in Bohemia and Austria, much too closely involved in the hostilities of the period. It was, in fact, a commonly reported rumor that, when soldiers hostile to the Protestant Evangelicals entered the Old City of Prague in 1609 and again in 1611, they had been let in or summoned by its Catholic and predominantly Italian minority, and lodged in the Jesuit college.[60] And even those who discounted the stories about the society's militarism in Bohemia would have agreed that these fathers accomplished immense architectural works because they worked closely with the Italian stonemasons of Prague—a group that represented more than 25 percent of all such artisans in the city—to encourage the importation of foreign structural and decorative elements and to forbid, from the late sixteenth century onward, the employment of any non-Catholics in their guild.[61]

There is, finally, some hint in the *Somnium* proper, written around 1609, that Kepler had originally imagined the targeted population of the moon as a distinct social group like the Jesuits, and that by the time he appended the notes, in 1620, his intentions had shifted slightly. As the text and his personal letters make clear, in the early 1620s Kepler's preoccupations were with his mother, who died in 1622, but who had risked burning for witchcraft several years earlier.[62] In the *Somnium*, among those humans recruited for a trip to the moon there are indeed "dried-up old women, experienced from an early age in riding he-goats at night or forked sticks or threadbare cloaks, and in traversing immense expanses of the earth."[63] But despite their importance to Kepler from around 1616 to 1622, the witches constitute a minority population; most of those judged suitable in 1609 for the lunar voyage embodied a kind of vigorous, militant, seagoing masculinity consistent with Ignatius of Loyola's sinewy soldiers and sailors of Christ, or some satiric treatment of that group. As the daemon explains, "We admit to this company nobody who is lethargic, fat, or tender. On the contrary, we choose those who spend their time in the constant practice of horsemanship or often sail to the Indies, inured to subsisting on hardtack, garlic, dried fish, and unappetizing victuals. . . . No men from Germany are acceptable; we do not spurn the firm bodies of the Spaniards."[64]

Even after his mother had died, it would have naturally been impolitic for Kepler to insist on the Jesuitical identity of the lunar colonists, and in his notes, he merely pointed to the usual distribution of narrative attention between a straightforward focus upon the actual physical facts of the moon, and that oblique turn toward the writing tablet. Thus in a note to this section Kepler stated that there were perfectly good reasons for restricting the lunar voyage to those without much bulk, inured to hardship, and used to travel—they were, very unenviably, to "be hurled twelve thousand miles upward in the course of an hour"—but he also warned that his remarks were not limited to those considerations.[65] "Looking straight ahead, I concentrate on physical reasoning; out to the side, I shoot satirical arrows here and there at spectators who feel sure of themselves."[66] The notes of the 1620s go on, confusingly, to present Kepler's teacher, Michael Maestlin, as the ideal body type for lunar voyages, as if his slight build, the index of a refined mind, eased the original restriction on German participants.[67]

How wounded any smug spectators felt by Kepler's "satirical arrows" when the *Somnium* was at last published is not obvious, but many would have recognized the implicit debate over the alignment of certain intellectual styles with specific body types and nationalities as one that had involved Jesuits. Quite clearly, the recent appearance of the German Jesuit—rather than the better established stereotype of the crafty Spaniard or subtle Italian—had complicated matters; during the long debate over Christoph Clavius's reformation of the calendar, for instance, Joseph Scaliger had described the Jesuit from Bamberg as the embodiment of a plodding zeal, "he is a German, a ponderous, patient mind, and that is how astronomers should be. No outstanding mind can ever be a great astronomer."[68] What Scaliger neither said nor needed to say was that the correlative of Clavius's heavy intellectual tread was the apparent weight the arguments of the society's most prominent mathematical astronomer had with at least some of his readers. More troubling from the standpoint of those who wished to maintain useful stereotypes was the emergence of clever, rather than merely massively learned, German Jesuits: such, it appeared, was the society's most vigorous polemicist, Father Jacob Gretser of Constance. Cardinal Jacques-Davy Du Perron was said to have remarked in Gretser's regard, "he is fairly witty, for a German."[69]

While the notes Kepler wrote in the 1620s clearly have to do with the debate over the relationship of body, mind, and national identity, and while he appears then to have had reason to obscure the identity of the original lunar colonists of the *Somnium*, in 1609 the connection between the Ignatian type and "those experienced from an early age in riding he-goats at night or forked sticks or threadbare cloaks," would have been reasonably clear to local readers. Apart from the charges made by Cambilhom about the society's familiarity with the occult, in 1607 and 1609 two separate but similar news stories circulated about the efforts made by Jesuits in Prague to dissuade young girls and matrons from dabbling in practices involving a magical mirror and a crystal ball, and contemporary accounts likewise relate that they also worked in 1609 to persuade a woman in childbirth to exchange an ineffective amulet and a *grimoire* for the superior protection of a prayer to "Blessed *Pater* Ignatius."[70] Though these stories reflect a kind of benign interest in the spiritual and physical welfare of actual and potential sorceresses,

Kepler knew even then that the authorities could take a less indulgent view of practitioners, for he had also been reading the *Six Books of Magical Investigations* of Martin Del Rio, a Jesuit from the Spanish Netherlands, between 1606 and 1609.[71] Significantly, Del Rio had been at the Jesuit house in Graz, and was specifically mentioned in this connection by Cambilhom.[72]

The Lunatic Nature of This News

Kepler had lived directly across from the Clementinum since 1607, and maintained amicable relations with several members of the society in Prague. Apart from what he would necessarily have observed and heard of his neighbors, there is perhaps some evidence that he also read the various charges brought against the order in the refutations published by the "fairly witty" Father Jacob Gretser. Gretser, writing from Ingolstadt, had several occasions to address Cambilhom's work, for the *Discoverie of the most secret and subtile practises of the Iesuites* had a long afterlife not just in the piecemeal versions offered by the press, but also in several other anti-Jesuit tracts published throughout Europe in the same period, and in the polemical sermons and works of Lutheran preachers in nearby Augsburg in 1609 and 1610.[73] Never hesitating to review in great detail each of the rumors about his order, Gretser effectively gave these stories even greater currency; a remark by Paolo Sarpi implies that those who were unable to obtain or to read the ephemeral vernacular pamphlets which occasioned the Jesuit author's lengthier refutations were able to piece together their content from references in his Latin works.[74]

Gretser portrayed Cambilhom, the authors of similar pamphlets, and the Lutheran preachers of Augsburg as "new Athenians," arguing that like their counterparts in Acts 17: 21, these men spent all their time either relating or absorbing the latest news. He offered the shrewd suggestion, moreover, that the similarity of the various charges made across Europe against the Jesuits did not function as evidence against his order, but rather as an index of a widespread social dependence on news, innuendo, gossip, and rumor. Gretser addressed all these rivals as a well-connected collective of "newsmongers," "smoke sellers," *Neue Zeitung Schreiber*, and composers of *lettere d'avvisi*, and reminded them of the penalties awaiting libelers publishing within the Holy Roman Empire under the *Constitutio Criminalis Carolina*,[75] and of the late injunctions of Popes Pius V and Gregory XIII.[76]

Within a tract published in October 1609 Gretser devoted himself specifically to the charges brought by Cambilhom, suggesting that the author must have deployed a starry mirror himself to spy out Father Coton's doings, and sarcastically asking if the French Jesuit, whom he knew to have poor eyesight, dragged the instrument about with him everywhere.[77] What is worth noting here is that Gretser's mocking question—how exactly did Coton use this optical device?—is precisely that implied in the more serious exchanges earlier in this same year between Paolo Sarpi, Jacques Badovere, and Galileo, and more generally in any discussion of the emergence of the telescope, as both the *Starry Messenger* and Kepler's response to that work acknowledge.[78] Gretser's interest in the question, moreover, would not have been idle: a letter written to him from Augsburg by his close friend Marcus Welser in fall 1610 about the nature of the moon in the wake of the *Starry Messenger*, for instance, makes clear that the Jesuit writer was frequently consulted on questions of this sort. It was in fact Gretser who facilitated Welser's contacts with the man who would become Galileo's rival in the debate over the sunspots, Father Christoph Scheiner.[79] And because Cambilhom claimed to have seen the Jesuit Martin Del Rio in Graz in a period when that priest had been elsewhere, Gretser was not long in concluding that the writer likewise had eyes "like a lynx," a kind of visual acuity that allowed him to discern weapons where there were none, and people long after their departures.[80]

But apart from its general relevance to optical and astronomical contexts, Gretser's diatribe against Cambilhom also has specific consequences for Kepler's remarks about the inhabitants of the moon. Gretser, too, had used the word *patrare* in relation to Jesuit behavior: as for the priestly orgies with poor girls which Cambilhom had described, he wrote *Nil tale ullo in collegio patratur aut patratum est*, "Nothing of the sort is brought about or has been brought about in any of our colleges." *Et qui patrari credit, ille de suo ingenio Jesuitas aestimat*, "And he who thinks it has been brought about, ascribes to the Jesuits his own natural inclinations."[81] And whereas Kepler appeared to assert that the newly developed telescope revealed to an unsuspecting earth the *opera immania* or giant works the lunar inhabitants were bringing about in their peculiar dwelling places, Gretser argued that such enormities—*haec tam immania facta*—as Cambilhom described in his *Discoverie* could not possibly have been concealed from the world for so long, and were therefore untrue.[82]

Thus while Kepler's treatise embodies both a serious engagement with optical and astronomical novelties, and a sometime conflation of that information with much less reliable news concerning the Jesuits of Prague, Gretser's effort was to show that Cambilhom's revelations were, like those of other anti-Jesuit tracts, neither new nor accurate, but hyperbolic versions of older fictions. Whereas greater and greater magnification was in general among the first users of the telescope the index of reliability—the sudden addition of visual particulars being what made hypotheses about distant bodies persuasive to many—Gretser sought to uncouple the analogous level of detail in Cambilhom's narrative, its picayune features and correlative claim to an insider's knowledge, from its bid for novelty by insisting that the treatise was part of a well-established genre. And while Kepler's conclusions about the lunarians were highly specific to their environment, Gretser sarcastically demanded why the arsenals observed by Cambilhom in Cracow and Prague were not part and parcel of every Jesuit institution: "I am almost angry with the reporter because he has only named two colleges among so many. Why did he not add a third name? Why not a fourth? Why not one in India? One in Japan? One in Brazil? One in the Peruvian Province? One in Mexico?"[83]

In August 1610, three months after the publication of his *Discussion with the Starry Messenger*, Kepler mentioned in a letter to a mentor and friend from Tübingen that he had come upon one of Gretser's theological tracts.[84] This work was published in 1609 and again in 1610, and though its target was a Lutheran preacher, it had nothing to do with the rumors about the misdeeds of the Jesuits. In January 1609 and twice in the latter part of 1610, however, it was bound and sold with two of those treatises that did, the first being Gretser's attack on a satirical pamphlet circulating in 1609 in Latin and Italian under the wholly misleading title of *A copy of a letter written in Bologna, in which the excellency and perfection of the Jesuit Fathers are exhibited*, and the second his polemic against "a certain anonymous storyteller, or, as people say, newswriter, who explicitly and implicitly attributes the murder of the most Christian King of France and Navarre Henri IV to the Jesuits."[85]

The tenor and means of *A copy of a letter written in Bologna* differ somewhat from Cambilhom's work, though both offer portraits of the Jesuits as rapacious and seditious meddlers. Through a series of increasingly extravagant promises and wheedling excuses, the fictive letter writer in Bologna told his young addressee that

> the houses of private individuals will be open to you, and the most
> important civic issues will be referred to your judgment, and you
> will be party to anyone's secrets, and when you devote yourself to
> this, as is fitting, having attracted general esteem through that out-
> ward show of simple piety, through your judgment as teacher or

preacher you should be not just in appearance but in actual fact like an Emperor and Monarch.[86]

The fictive recruiter went on to envisage a sort of global domination enjoyed by the society—"just like the circuit of the Sun, it will be glorious with its sons in every part of every country, and blessed in heaven"—and the letter asserted that the Jesuits had gained control of "Poland, Transylvania, Austria, Bavaria, Spain, and even the Indies" before culminating in the vision of "one flock, and one Shepherd."[87]

We cannot know if Kepler's reading of Gretser's indignant defenses of the Bolognese Jesuits predated his composition of the *Discussion with the Starry Messenger*, but we should note that in January 1609 the polemicist developed a number of motifs central to our discussion in his response to *A copy of a letter written in Bologna*. The energetic Gretser appears, for one thing, to have insisted in the latter pages of this tract on what we might call the "lunatic" nature of the various accusations against his order. Drawing on an author crucial to both Kepler's *Discussion* and his *Somnium*, for instance, Gretser compared reports of Spanish ships bearing untold wealth for Jesuit coffers to Lucian of Samosata's fantastic and overly precise accounts of what could actually be observed of the earth from the moon.[88] He further stated that the accusation that the society was attempting to submit all states to ecclesiastical rule was like the suggestion that these priests endeavored to draw our satellite down from the heavens in order that the sun alone might shine in the firmament.[89] And though *A copy of a letter written in Bologna* asserted that Father Pierre Coton was at present ruling France and its king, it did not specify his means of doing so; Gretser, however, alleged that those men so jealous of his colleague as to spread rumors of his magical practices and starry mirror were like dogs howling at the moon.[90]

Gretser's maneuver is especially remarkable in the way it combined the particulars of *A copy of a letter written in Bologna* with details from other anti-Jesuit publications of the period, as if they were not independent news accounts but either had all emerged from a single and mendacious source, or were reiterations of and variations on one original lie. Thus while the purported letter from Bologna fixated on the worldliness of the Jesuits, and noted that the order preferred to establish itself in populous and wealthy cities, that its members wore habits of the finest wool and linen, that those in its higher ranks ate extremely well, that they sought out connections and benefices from the rich and powerful, and that they meddled in the affairs of state,[91] Gretser connected such statements with the more vivid rumors about gold-bearing ships, magical practices, and starry mirrors rehearsed in other such texts.[92] Though *A copy of a letter written in Bologna* lacked some, though not all, of the lurid foolishness of Cambilhom's tract, it acquired an air of family resemblance in Gretser's response to it. Given that in 1608 and 1609 Galileo's Venetian friend Gianfrancesco Sagredo had composed fake letters designed to expose the Society of Jesus as rapacious both in nearby Ferrara and in their missions in the Far East, and that his close associate Paolo Sarpi was likely the force behind a spate of anti-Jesuit tracts, Gretser's gesture to a web of related attacks might have seemed plausible even to insiders in the Veneto who were otherwise unsympathetic to the order.[93]

If by chance Kepler did encounter Gretser's reply, he would have later associated the targeted institution in Bologna, the Jesuit *Collegio dei nobili*,[94] with the very place where his former student, the disgraced Bohemian Martin Horky, hid in June 1610 after he had sought to discredit Galileo's findings in an embarrassing little pamphlet printed in Modena and circulated in Prague by Rudolf II's treasurer, Matthäus Welser. Though Kepler's dismay over Horky's pamphlet alone is enough to explain his alarmed tone in a letter written to his former student in August 1610, the additional knowledge hypothetically supplied by Gretser's response to *A copy of a letter written in Bologna* could not have enhanced his views of that institution. As the astronomer saw it, the Jesuit college was no place for the son of a Lutheran preacher: "If you want to follow his paternal advice"—rather than that of the *patres*—"get out of those lodgings as soon as you can, whatever it takes."[95]

Horky's entire strategy—his brief tenure as secretary in the house of the prominent pro-Jesuit astronomer Giovanni Antonio Magini, his alleged survey of his employer's private correspondence, his nocturnal attempt to make a wax impression of Galileo's telescope during the latter's visit to Magini, his bumbling pronouncements about the death of Henri IV, his pompous and pious address to the "Learned Men of Bologna" in his *Brief Excursion Against the Starry Messenger*, his ejection from Magini's house and circle, his theft of books of secrets from Magini's library, and his eventual refuge in the *Collegio dei nobili*—sounds like the plot of an early modern novel. An obvious analogue would be the picaresque passages of Tristan l'Hermite's *The Disgraced Page* of 1643, where the hero's attempt to duplicate certain tricks in della Porta's *Natural Magick* earns him a reputation for sorcery, the rage of his preceptor, and a beating.[96] But it reads even more like a series of tactics attempted by someone who took literally the promises in *A copy of a letter written in Bologna* about access to private houses, to civic matters, and to secrets, about the show of simple piety and the eventual life as "Emperor and Monarch," and who sought somehow to acquire the baseline requirement of nobility after the fact when his shenanigans failed to go as planned.

It is perhaps significant that when Kepler wrote to Galileo about Horky's pamphlet in August 1610, he focused on this lack of status as a corollary to the young man's temerity, but he also strongly suggested the complicity of the Italian astronomer's other enemies. Given that Kepler had already been led to believe that Giovanni Antonio Magini, directed by the Jesuits of Bologna, had been behind Horky's clumsy maneuvers, and that he wrote this letter on the same day and just before he warned his former student "to get out of those lodgings," it is credible that he conceived of these hidden enemies as either the *patres* themselves or allies like Magini, and that he imagined both in terms absolutely consonant with the tabloid image of the order.[97]

> I marvel at the brazenness of this young man, who, while local schol-
> ars just muttered [about the *Starry Messenger*], attacked it aloud and
> all alone, though he was a foreigner and inexperienced. I believe that
> being a newcomer and bearing an obscure name added to his audac-
> ity, just as masks embolden actors. Might you not have some Italian
> rivals who oversaw the work of this outsider?[98]

An Established Outpost

The latent and fantastic suggestion of Kepler's *Discussion with the Starry Messenger*, that the Society of Jesus might take over the moon as well as the earth, was already perfectly well known to those European readers who consumed anti-Jesuit tracts. It dated at least to the popular French satire *News from the Regions of the Moon*, published in the 1590s as a confidential newsletter about Jesuit prospects on the lunar body. As the pseudo-Jesuit author reminded the Spanish monarch, "It is by our works and tricks that you are now Lord of the Indies," and he believed that missionary efforts would be even more fruitful on the moon. His Majesty would do well, he noted, to raise an army for the immediate subjugation of the earth's satellite, for

> the conquest will hardly be more difficult than that of the Indies,
> and you will gain more there than in invading France. And on the

moon gold is not found underground, as it is in the Indies, but rather
in every man's breeches. And when you have overrun some corner of
this [lunar] land—which will be very soon, God willing—we affec-
tionately beg you to install us in the first little settlement, and to
found a boys' school, and to leave the rest to us. We will preach so
well, and wield the rod so skillfully, that soon you will be king; if it
is only a matter of seducing the people, getting them to rebel against
their own prince, and deposing him or bringing in assassins, we will
make quick work of it. The people of the moon will be yours long
before they catch on to any of our tricks.[99]

Though not all associations of the Jesuits with the moon adopted the
ribald tone of this early French satire, such conceits were common, and
invariably associated with tabloid news. Let me conclude, then, by pointing
out that within ten months of the publication of Galileo's *Starry Messenger*
and eight months of Kepler's response to it, John Donne proposed a lunar
colony for the society in *Ignatius His Conclave*.[100] This satire, issued in Latin
in January 1611 and in English in May that year, began with a sort of vision
of the heavens, one proposed as a newer but less audible message than those
of Galileo and Kepler, for here the narrator preferred

> to be silent, then to doe *Galilæo* wrong by speaking of [the firma-
> ment], who of late hath summoned the other worlds, the Stars to
> come nearer to him, and give him an account of themselves. Or to
> *Keppler*, who (as himselfe testifies of himselfe) *ever since* Tycho
> Brahes *death, hath received it into his care, that no new thing should
> be done in heaven without his knowledge*.[101]

While the speaker could not or would not relate what he observed of the
heavens, he did offer a detailed account of a series of exchanges between
Lucifer and the no less demonic founder of the Jesuit Order, Ignatius of
Loyola, who was fit to enter a special zone in Hell reserved for innovators.
Various applicants for this highest of infernal honors included Copernicus,
Paracelsus, Machiavelli, and Aretino, but all were found wanting in novelty,
and Loyola proposed instead first Father Christoph Clavius of Rome for his
resistance to astronomical truth and his meddling with the calendar, then
some randomly selected and untrained Jesuits for their attentiveness to the
dying, next Fathers Pierre Coton of Paris and Roberto Bellarmine of Rome

for their political meddling, and finally Father Matthäus Rader of Augsburg for demonstrating an encyclopedic and methodical smuttiness in his expurgation of classical texts. At length Lucifer, recognizing that no region of hell was suitably large for the diabolical Loyola's followers, proposed to

> write to the Bishop of *Rome*: he shall call *Galilæo* the *Florentine* to him; who by this time hath thoroughly instructed himselfe of all the hills, woods, and Cities in the new world, the *Moone*. And since he effected so much with his first *Glasses*, that he saw the *Moone*, in so neere a distance, that he gave himselfe satisfaction of all, and the least parts in her, when now being growne to more perfection in his Art, he shall have made new *Glasses*, and they received a hallowing from the *Pope*, he may draw the *Moone*, like a boate floating upon the water, as neere the earth as he will. And thither (because they ever claime that those imployments of discovery belong to them) shall all the *Iesuites* bee transferred, and easily unite and reconcile the *Lunatique Church* to the *Romane Church*.[102]

It is perhaps chance that made Donne's meditation on an innovative vision of the heavens in the post-telescopic era extraordinarily up-to-date. In the spring of 1611, as *Ignatius His Conclave* emerged from the press, Galileo was enjoying a very warm reception in Rome, particularly at the Roman College, where the findings of his *Starry Messenger* were for the most part ratified by the Jesuit astronomers; Father Odo van Maelcote, their spokesman, described himself as a *claudus tabellarius* or "limping messenger," one whose slower pace guaranteed the accuracy of the tidings he bore.[103] As Maelcote noted, Father Clavius had expressed some doubts about the substance of the moon, and it is fitting that in Donne's satire Ignatius proposed not to waste this scholar, but rather some lesser Jesuit mathematician on the fledgling lunar colony.[104] *Ignatius His Conclave* gestured briefly at this point towards Cambilhom and Gretser, whose work I have been proposing as part of the general background to Kepler's depiction of the moon. The society's founder announced a plan to publish a lunar newsletter once the colony has been established, and to fill it with magnificent lies, as had been those from the Jesuit missions in the Indies, "until one of our *Order*, in a simplicity, and ingenuity fitter for a *Christian*, then a *Iesuite*, acknowledged and lamented that there were no *miracles* done there. . . . It is of such men as these in our *Order*, that our *Gretzer* saies [in his attack on Cambilhom]: *There is no body without his Excrements*, because they speake the truth, yet they speake it too rawly."[105]

The lunar business ends there, but it is worth noting that Donne's narrator and reporter, otherwise rather colorless, has assumed the role of Cambilhom, offering a hyperbolic version of an insider's knowledge of the society's doings to a largely hostile public. Donne himself would have a position more like Kepler's, who in "turning toward the tablet," or tabloids, let another speak for him, and precisely in the overlap between Jesuit activity and lunar observation. For this reason it is above all curious that Kepler, unaware of Donne's identity, and somewhat confused in his chronology, alleged in his *Somnium* that an early manuscript copy of his work "fell into the hands of the author of the bold satire *Ignatius His Conclave*, for he stings me by name at the very outset." What Kepler proposed as the timing of the incident—that his pilfered manuscript traveled in 1611 from Prague to Leipzig to Tübingen and then somehow to London—before Donne published *Ignatius His Conclave* is clearly impossible, and the substance of the alleged theft, the notion that an Icelandic volcano served as a portal to Hell, very meager indeed.

But Kepler's sense that Donne had somehow gained something from him in late 1610, and that it involved both his astronomer's vow "that no new thing should be done in heaven without his knowledge," and "the rasping of an indistinct and unclear voice" like Icelandic, but even more like Cambilhom's raucous fictions, is perfectly sound. Such a debt—Kepler's to the humble Cambilhom, and Donne's to Kepler—also accounts for the change in genre from tabloid news to scientific treatise to diabolical dialogue. After all, as Kepler had it in the *Somnium*, those who encounter something spoken "from the gut"—and this would include all those who read some transmuted version of Cambilhom's work, for "no body is without his excrements"—were apt to believe that "demons are talking to them."

Donne's influential presentation of the moon is echoed in Phineas Fletcher's *The Locusts or Apollyonists*, an epic poem devoted to the Gunpowder Plot, written in separate Latin and English versions, and begun in late 1610 but not published until 1627. The Jesuit colony is not, in this case, a project for the near future, as in *Ignatius His Conclave*, but rather a kind of old safe haven that the society would have used had its Counter-Reformation efforts not been so successful.

Had not these troopes with their new forged armes
Strook in, even ayre, earth too, and all were lost:

Their fresh assaultes, and importune alarmes
Have truth repell'd, and her full conquest crost:
Or these, or none must recompence our harmes.
If they had fail'd wee must have sought a coast
I'th' Moone (the Florentines new world) to dwell,
And, as from heaven, from earth should now have fell
To hell confin'd, nor could we safe abide in hell.[106]

Fletcher's suggestion that there was something slightly obsolete about the prospect of Jesuits on the lunar body would be a way of either minimizing his debt to Donne, or of showing Donne's own debt to works such as the French satire *News from the Regions of the Moon*. Despite the enormous interest which Cambilhom's revelations about gunpowder excited in England, there seems little direct acknowledgment in *The Locusts* of that unsung source, and certainly no particular fascination with the issue of appropriating news.[107]

Matters are otherwise, of course, in Ben Jonson's *Staple of Newes*, a comedy of 1626.[108] Here, what had originally been quite condensed figures of speech, typically some sort of marginal epithet or metaphorical phrase, generates a fresh batch of lies at once outrageous and faintly familiar. Thus the renowned commander of the Spanish forces, Ambrogio Spìnola, has just become general of the most militant order, the Society of Jesus, while machinating Jesuits now form a corps of engineers.[109] Galileo is duly reported to have a study equipped with a "burning glasse" because he, like several contemporaries, was known as "the Archimedes of his age." The actual superior general of the Society of Jesus, the haughty Muzio Vitelleschi, has taken up the position of "cook"—a term adopted then as now to describe one who offers a biased version of events—and in this new role assaults the enemy with powdered eggs whose yolks burn like wildfire, the requisite explosives of most anti-Jesuitical accounts.[110] As Jonson's newsagents tell it, Spìnola is now on the moon, and he uses it as a base from which to attack Protestant Europe; his efforts are supported by the machinators below, their cook, and by Galileo, whose glass will ignite any ship out at sea.

The irony here is that at this moment Galileo's relations with the Jesuits were extraordinarily vexed, and their views of the substance and nature of the moon, in particular, increasingly unrelated.[111] But more to the point for us, and for those who first encountered Jonson's play, is the way in which what was just a crater "something like Bohemia" in Galileo's account, and what was "indistinct and unclear" in Kepler's *Discussion with the Starry Messenger*, have been managed. Such passages had seemed to enjoy an uneasy and improbable coexistence in both an astronomical discussion and popular rumors about the Society of Jesus, but have here been brazenly fused into a single literary entity and a brand new thing, a story, briefly told and quickly consumed, about Jesuits on the moon. It's news.

CHAPTER 2

Medici Stars and the Medici Regency

The French stories speake of divers auguries and predictions of
[Henri IV's] death. . . . To this may be added the speech of *Francisco
Corvini* a Toscan astrologer, who the night before *Henry* the fourth
was slaine, leaning upon a Balcon in Florence, which is neer upon
600 miles from Paris, and prying into the motions of the starres, he
suddenly broke out of his speculation into these words; *Tomorrow
one of the greatest Monarchs of Christendome will be slaine*; and the
very next day the mortall stab was given by *Ravaillac*, who had been
seen often a little before at *Brussels*, with Marquis [Ambrogio]
Spìnola, which many wondred at, [the assassin] being so plaine a
man.

—Howell, *Lustra Ludovici*, 4

So James Howell recalled, in the first pages of his *Lustra Ludovici* of 1646,
the assassination of French king Henri IV by François Ravaillac in May 1610.
Howell's relative indifference to accurate details, his preference for new material, and the lack of evidence for the very existence of this particular Francisco
Corvini suggest that no such "Toscan astrologer" ever leaned upon a Florentine balcony and divined the imminent death of the most powerful European
monarch. What Howell's vignette does capture, however, is the peculiar meshing of the political and the scientific in the spring of 1610: news of the king's
death reached most Europeans just as a Tuscan astronomer made public his
celestial discoveries in the *Starry Messenger*. Howell's spectral astrologer—
condemned to obscurity by his casual posture and his ineffectual, one-time
utterance on the very eve of the assassination of a ruler—may be understood

as the antitype of the rational and orderly astronomer whose news consisted, in large part, in his painstaking records of the appearances of the satellites named for the Medici every night from January 2 to March 7, 1610.[1] And yet it is perhaps unwise entirely to dissociate the two figures: Galileo's treatise was originally directed to a reading public vaguely designated as "astrologers," subsequent to its publication he was called not *corvinus* ("raven-like") but *corvus* ("raven") for his alleged habit of adorning himself with others' feathers,[2] and even those most favorable to him, such as his friend Angelo Grillo, described the instrument as fit for "prying into the motions of the starres."[3]

Howell's version of the assassination is only one of many efforts to make sense of the simultaneity of astronomical developments and political events of the spring of 1610, the most common of which is the construction of a macronarrative designed to account at once for the sudden emergence of the regency of Marie de Médicis in Paris and that more distant satellite state about Jupiter. The conflation of the stars and regency finds its most succinct expression in an allegorical frontispiece designed in 1611 for Nicolas-Claude Fabri de Peiresc's tabulation of the positions of the four satellites, where Marie is enthroned and surrounded by stars named for the Tuscan grand dukes Cosimo I, Ferdinand, Francesco, and Cosimo II de' Medici. The image is easily explained in terms of patronage: the regent would be the dedicatee of Peiresc's proposal to use the satellites' relative positions as an aid to finding longitude. It leaves tactfully unaddressed, however, any causal connectives between the astronomical and political events.[4]

This chapter considers the various ways in which French writers refashioned the separate narratives brought together by the news in the spring of 1610.[5] Their accounts do not focus merely upon the extraordinarily malleable "facts" of the assassination, but also upon the manner in which certain details about the king's death allegedly became known and were circulated prior to the incident, and at a great distance from it. The attention of those looking back to May 1610, in other words, is as much devoted to the mechanisms of newsgathering, dissemination, and consumption as it is to that cardinal event in early modern European history.

Because the emblem of particular types of early modern reportage was frequently an optical device, the conflation of the assassination and the publication of the *Starry Messenger* is not a surprising one. To some extent, the astronomical treatise announcing the presence of Medici stars is an updated supplement to the astrological prognostications of the king's death, since the latter would by implication point to the emergence of the Medici regency in France. But there is more at stake in the connection of Galileo's treatise with the regicide, for the insistence upon the widespread foreknowledge of the assassination among certain sectors of the population, and upon the rapid transmission and diffusion of such news from a popular milieu back to a noble elite betrays a pronounced anxiety about the role of media and technology in the formation of a public sphere, and about the novel sort of access to information that would, in theory, accompany technical and artisanal expertise.[6]

This emphasis on the mechanisms surrounding advance news of the event differs significantly from the reactions to new military instruments, for the latter typically warned that recent inventions thwarted conventional restrictions of popular power, and gave individuals of the lowest social standing sudden access to force and a brief moment of agency, usually in the form of an assassination. Henri's killer, however, used a primitive instrument—a knife—and public anxieties accordingly shifted to the more sophisticated means by which news of the event allegedly traveled far from Paris weeks before François Ravaillac struck, as if this foreknowledge were the real index of an unsettling technological advance not controlled by the state but rather by its enemies. Put differently, the relatively simple method of the assassin meant that he, like the knife itself, was merely an "instrument" of more complex forces. Though conjectures of this sort occasionally produced unlikely sightings such as the furtive encounter of Ravaillac and Marquis Ambrogio Spìnola, military commander of the Spanish Netherlands, the phenomenon of advance reports of the deed received more attention in early modern accounts.

In ascribing foreknowledge of Henri's death to political elites and to "the people" alike, and associating such advance notice with access to optical instruments, as if spatial distances had been converted to temporal ones, these writers articulated crucial developments in the emergent news industry. The recent access of a much broader public to news is at stake here: from the early seventeenth century news concerned events well beyond one's own local and national horizon, and was available in a variety of formats to almost all strata of society. While civic eyes and sometimes the long arm of local authorities were on printers and middlemen engaged in the physical redaction and commercial distribution of news, my focus will be on more elusive figures, the presumed sources of oral reports circulating days, weeks, and even a year before the royal assassination.

Astrology and Assassinations

In June 1610, the "Pope of the Huguenots," Philippe Duplessis-Mornay, weary of the rumors concerning France's political and religious future in the wake of Henri's death, wrote from his home in the provinces to his nephew, the French ambassador to Spain, "You do not lack people who can write you from the court, but I see it only as if through a Dutch telescope; rather than drawing myself close to these matters, I draw them close to me."[7] In another letter written the same day, Duplessis-Mornay offered an implicit comparison between his own remote vantage point and a telescopic view of matters in Paris: "we have this advantage as well, of being above the vapors and the thunderbolts that sometimes envelop and deafen those who are closer."[8] It is the simultaneity of the assassination and the publication of astronomical observations that had given no hint of that event that explains the Protestant leader's tendency to present the telescope in terms of what it could *not* offer, a precise image of the current state of French affairs.

But some writers in that same period saw little difference between the type of information the telescope might convey to astronomers and the validity of astrological prognostications; in their view, the new instrument could do nothing but enhance such predictions. In a letter to his friend William Camden written in July 1610, the English astrologer Christopher Heydon defended one of the most famous forecasts of Henri's death, that made by a doctor and astrologer named La Brosse,[9] as something more than a lucky guess. What is more surprising than Heydon's desire to justify his own practice of judicial astrology is his reliance upon Galileo's recent treatise and telescope as a means to persuade his correspondent; he concluded his argument about La Brosse's prediction, evidently with some success, with the observation that "certainly Mathematicians never err, but when they are abused by false information. I have read *Galilaeus*, and, to be short, do concur with him in opinion."[10]

While he emphasized another aspect of the *Starry Messenger*—the wholly unexpected discovery of the Medici Stars—René Descartes, then about sixteen or seventeen years of age, likewise associated the assassination with the astronomical treatise in his "Sonnet on the death of the King Henry the Great, and on the discovery of certain new Planets or Stars wandering about Jupiter, made that very same Year by Galileo Galilei, the celebrated Mathematician of the Grand Duke of Florence." In an academic exercise written for the first anniversary of the king's death, Descartes pointed to no causal connection between the emergence of the Medici Stars and of the Medici regency in France: the title of the work avoids the name entirely, and the poem itself ends with the peculiar suggestion that the late Henri has taken his place in the orbit about Jupiter, presumably among the four satellites, or perhaps as a companion to them.[11]

Descartes's contemporary, the Bohemian Martin Horky, provides us with a decidedly idiosyncratic association of the royal assassination and the reception of the *Starry Messenger*. Then a student in Bologna, Horky was particularly involved in the first controversies surrounding the telescope because he lived in the home of Giovanni Antonio Magini, with whom Galileo stayed during a visit to the city in late April 1610. In a letter to his sometime mentor, Johannes Kepler, Horky began with a colorful lamentation for the late king, and proceeded directly to the question of accurate observational data:

> O times! O men! O evil act! O customs! O crime! Henri IV, the
> light and leader of France and the King of Navarre, has been killed,
> as we learn from reliable reports. The gemstone, the crowning glory,
> the flower, the golden guide, the support of studies and of students
> perished the other day after he passed the crown to his young son,
> whom the French call the "dauphin." He did indeed die on the
> fourteenth day of this month: see the truth on page 93 of Johannes
> Rudolph Camerarius's *One Hundred Nativities*, most learned man.
> O unlucky, and to me unknown hour [of the king's birth]! If only
> the fates would allow me to learn it quickly! If only I could see the
> position of the seven celestial planets at that very moment in time! I
> fear the same for myself: I have written most harshly against the
> *Starry Messenger*, but the sire of this *Messenger* took all that without
> my knowledge when he spent the night at our house here in
> Bologna.[12]

Horky's train of thought, as always, is difficult to follow, but his subsequent letters to Kepler and his eventual treatise seem to suggest that he either saw or simply wished others to see Galileo as the antitype of the martyred French king. Thus while the reports of Henri's death confirmed, in a sense, at least some of the prognostications of the astrologers, Galileo's news of Jupiter's moons, entirely unprecedented, was in Horky's view without foundation. "You were correct to call [your treatise] a 'Messenger,'" he would write in his *Brief Foray Against the Starry Messenger*, "for most messengers sell lies."[13] Horky's several references to Galileo as one who, like the raven, draped himself in the plumage of others[14] makes the *corvus* the unfortunate counterpart to the noble *Gallus* or rooster, and his portrait of Henri as the "support of studies and of students" stands in contrast to his contemporaneous description of the astronomer as a sham professor whose fingers were gout ridden "because he secretly bore away philosophical and mathematical money."[15] Before dying the good French king had providentially transferred the crown to his son, Horky maintained, while Galileo had sired an illegitimate little messenger to continue his nefarious work.

The manner in which Horky described what he expected of the telescope prior to Galileo's visit to Bologna, and how the device failed him, is also designed to emphasize the bizarre contrast between the French king and the Italian astronomer. As he related in the *Brief Foray*,

> Often I wished I could see that novel and unheard of spectacle, proposed to all mortal men, of the four new planets, and often I prayed to the thrice-great God, that He would provide through His divine wand an ecstatic hour for seeing through Galileo's perspective glass, the glass with which he said he had discovered four new planets. And behold, here, an unvanquished king, an immortal mathematician, a teacher without limits, had been made, just as I had wished.[16]

Horky's first exposure to the telescope came on April 24, 1610, just as he learned of the assassination of Henri IV, and he suggests that he initially conceived of the "ecstatic hour" and "unvanquished king" as correctives to the "unlucky hour" of the monarch's birth and the very fact of his death. But it is perhaps Horky's nastiest depiction of Galileo, made in the course of a private letter to Kepler on April 27, 1610, that is the most telling index of his peculiar portrait of the astronomer as an inadequate substitute for the martyred French ruler. Horky reported that "all of [Galileo's] flesh, his whole hide, was covered with the French disease [*flore Gallico*]," and that "his optical nerves, because he had observed the [degrees and minutes] around Jupiter too curiously and too pompously, were all ruptured."[17] The secondary symptoms attributed to Galileo—hair loss, melancholy, and cardiac troubles—were commonly, though not exclusively, associated with venereal disease, as if the contagious illness from which Henri IV had long suffered had also infected him.

While his insistence that Galileo's "optical nerves" were ruptured—by which he surely meant to signal the astronomer's bloodshot eyes—appears plausible under those circumstances of intensive observation, Horky's notion that there was something excessively "pompous" about this work is puzzling. But if we view Galileo as a pretender to the throne, as an audacious successor to Henri IV, which Horky evidently did, and his *Starry Messenger* as profoundly flawed in its failure to forewarn readers of the royal assassination, then his observational efforts might be described as a sort of prideful overreaching, and as a hubristic attempt to become an unconquerable king.

Horky's obscure sense of the relatedness of the French assassination and the Italian astronomical treatise is an important cultural issue, and the implicit contrast between Galileo's excessive "curiosity and pomposity" about optical instruments, and Henri's regal disregard for such devices, was in fact a trope in this period. The popular sense that the telescope was somehow connected with foresight, as if the spatial component could be exchanged for a temporal one, such that he who saw distant things was somehow viewing the future, was a persistent early feature of the device. Thus when the recent invention of the Dutch telescope was reported in the *Mercure François* of 1609, the succeeding story, devoted to financial information, began "It would have taken a good many of those glasses to see all the bankruptcy frauds that year."[18]

The Vision Thing

The motif of individual foresight is invariably paired with the collective or contagious blindness of the central protagonists of the political drama. In all probability, Henri IV had first learned of the invention of the telescope in early January 1609, when he and the duc de Sully received notice of the device in letters from prominent statesman Pierre Jeannin, then negotiating the truce between Spain and the United Provinces in The Hague. Jeannin appears to have conceived of the invention, at least initially, as an instrument for reading "minute letters" at a distance; in this particular, he was influenced by discussions of optical aids from Roger Bacon in the mid-thirteenth century to Leonard and Thomas Digges and William Bourne in the 1570s, and Giambattista della Porta in 1589. His letters to the French king and to his chief minister, written on December 28, 1608, likewise draw on the medieval pedigree of the perspective glass, but present the telescope not only as the emblem of some diplomatic snooping that had just taken place in The Hague, but also as a valuable instrument of military reconnaissance.[19]

It was reasonable for Jeannin to assume that either Henri or those in his courtly orbit would find the new invention of interest. The king had been trying since June 1607 to extend *brevets de logement* to a select number of his best artists, engineers, and craftsmen, such that they could establish, or in some cases, remain in the workshops provided on the ground floor of the Grande Galerie connecting the Louvre and the Palace of the Tuileries. Parliament's resistance to the concessions accorded this group had been overcome only in late December 1608, a few days before Jeannin wrote to the king and Sully, and the privileges were formally awarded in early January 1609. In addition to goldsmiths, painters and sculptors, clockmakers and enamellers, this group of twenty individuals included three makers of mathematical instruments and a prominent professor of mathematics and optics.[20] The king's proclamation of December 1608 had much to do with the Twelve Years' Truce—"among the infinite blessings that come with peace, that which arises from the cultivation of the arts is hardly the least," it began—and seemed tacitly to recognize the importance of Jeannin's recent diplomatic efforts in The Hague.[21]

Regrettably, neither Henri IV nor the duc de Sully seemed interested in Jeannin's news. In early January 1609, the French king, preoccupied by the flooding caused by the harsh winter, replied simply that he would take pleasure in seeing the glasses Jeannin mentioned, but added that at present, he had greater need of something to help him make out what was close at hand, rather than far away.[22] It is possible that the king, then fifty-seven and unable to read without eyeglasses, had simply misconstrued the Dutch device as an aid for myopes, and stipulated instead that he suffered more from presbyopia. It seems to me more likely, given both the context of the allusion—the pressing problems in France—and Henri's fondness for figurative language, that he meant to suggest that his current worries did not allow him much time to examine amusing optical trifles sent from the United Provinces. Of Sully's reaction there is no record; though he noted in his memoirs that he and the late king had planned, as was then common, to design a sort of museum of maps and scientific and military instruments, no particular mention was made of the role optical devices would play in battles and sieges.[23]

The Queen's Glass

Henri's indifference notwithstanding, by the spring of 1609, one could buy some version of the Dutch telescope in Paris; the *Mercure François* commented on the origin and design of the device, and the chronicler Pierre de l'Estoile examined it in a lens maker's shop on the pont Marchand.[24] The *Mercure* suggested that there were significant differences in quality in the telescopes—"there are some craftsmen who make better ones than others"— and thus implied that a certain number of the instruments were already available to the Parisian public. It is probable that these, like that described by l'Estoile, were sold on or near the short-lived pont Marchand, a brand new bridge replete with fifty-one different shops, designed to replace the low, dirty, and ill-famed pont aux Meuniers. The latter, "being covered and very dark," was said by a 1585 visitor to be "the refuge of merchants who sell false jewels made of glass, and earrings, necklaces, and chains made of gilded copper," as well as the venue for legitimate jewelers, the most prosperous of whom was Huguenot Thierry Badovere, father of Galileo's student and friend Jacques Badovere, the crucial source of the astronomer's information about the Dutch telescope.[25]

The pont aux Meuniers had collapsed in 1596, allegedly under the weight of ill-gotten gains taken from its merchants—chief among them Thierry Badovere—during the Saint Bartholomew's Day Massacre in 1572.[26] "A singular and exquisite structure," l'Estoile called the new pont Marchand, and this elegant commercial venue, the result of Henri IV's and Sully's efforts to restore the French capital, would have been well situated for lens makers.[27] Beginning at the Place du Châtelet, it was very close to the rue de la Verrerie, where glassworkers had congregated for several centuries. It formed an acute angle on the Île de la Cité side with the old pont au Change, whose proximity meant that both burned down in 1621, and led to the quai de l'Horloge, where mathematical instruments, maps, and optical devices of various kinds would soon be sold, the importance of this last commodity resulting in a brief name change to the "quai des Lunettes" in the early nineteenth century.[28]

What l'Estoile and the *Mercure* portrayed as popular and commercial interest in the telescope in 1609 is matched by several contemporaneous sources concerning the reception of the new device at the French court. These relate exclusively to Marie de Médicis, rather than to her husband or to Minister Sully. Thus in October 1609, François Malherbe noted that while

French telescopes were in general inferior to those manufactured in the Netherlands, he had nonetheless "heard much made of the one owned by the Queen, but what they fashion for the gods is done with a good deal more care than what they make for men."[29] Whether the queen had then much time or energy to use the instrument is questionable, for Malherbe's next letter, written nine days later, is almost wholly concerned, "for lack of better news," with the advanced stage of her pregnancy.[30] But in July 1610, the Tuscan ambassador to France wrote home to procure a Galilean telescope for the newly widowed Marie. Matteo Botti stated that she had been trying for more than a year, but without success, to have a good telescope made in Paris, and he urged the grand duke's advisor to seize the occasion to glorify the Tuscan state with this gift. Despite the queen's particular dissatisfaction with her French telescope, Botti confided, "here they already make much of ordinary spyglasses, and the shops are full of them."[31] By mid-September 1610 the ambassador reported that "Galileo's glass" had arrived in Paris, but that Marie found it did not reveal much more than the others at her disposal.[32] Some eleven months later, however, when yet another telescope had been sent from Tuscany to France, Botti wrote to both Galileo and Cosimo II de' Medici that Marie was so pleased with the instrument that she even knelt upon the pavement of her chambers in his presence in order to better observe the moon.[33]

There are several reasons that the instrument was presented exclusively in terms of Marie de Médicis' interest. First of all, when Jeannin's letter reached him in January 1609, Henri IV was only eighteen months away from his death, and would have relatively little opportunity to examine the new invention. Second, Marie appears to have had a longstanding interest in optical devices: a letter of mid-January 1609 relates that a secretary, Jean Baptiste Duval, had composed at her request an Italian translation of a Latin treatise published by Giovanni Antonio Magini in 1602, Ettore Ausonio's *Theory of Concave Mirrors*, and that she had sought in vain to have a large glass of this sort made by an artisan in Paris.[34] Finally, after the emergence of the *Starry Messenger* in spring 1610, the public would have associated the Dutch telescope more closely with the Medici, and would have found the Italian-born queen's interest in acquiring superior versions of the instrument entirely natural.

Any of these factors alone would be sufficient to explain why it is Marie's interest in the telescope, rather than her consort's, that finds representation in this period. Cumulatively, however, these isolated explanations worked to suggest, in the immediate aftermath of the assassination, that Henri IV's unforeseen death was the consequence of his inattention to the new invention, and his wife's survival the correlate of her optical enthusiasms. From this perspective, it is surely significant that the wife of King Henri II, the Florentine Catherine de Médicis, was popularly believed to have used a sort of magic mirror to learn the political destinies of her children and their successors, among them the future Henri IV, who would marry Catherine's daughter Marguerite in 1572, and secure a royal divorce from her in 1599, in order to wed Marie the following year.[35] The singular manner of Henri II's death in 1559—of a wound to the eye in the course of jousting festivities[36]—and the longevity and political importance of Catherine form, moreover, a rough template for some impressions of the events of May 14, 1610, where Henri IV's apparent blindness to threats on his life left France in the hands of his still youthful Florentine widow, who had had the foresight to have herself crowned regent the day before.

A final and curious touch served to associate Catherine de Médicis with the satellites of Jupiter. In the winter of 1564–1565 the great French poet Pierre de Ronsard, following in the tradition of Matteo Franzesi's light-hearted verses "On the News" of 1555, had written *Les Nues ou Nouvelles*, an extended comparison of the highly mutable images and menacing noises of thunderclouds with the confused impressions and startling rumors circulating among Huguenot and ultra-Catholic factions throughout France in that period. Ronsard's notion that clouds and news had much in common—premised on the fact that the French word *bruit* referred at once to noise and to rumors, and that both temporary aerial phenomena and printed materials were called *impressions*—would have gained some traction in the wake of the debate over the sunspots in 1611–1613, where the clouds said to be covering the sun constituted the most surprising piece of news Galileo and his confederates had to offer, but the real interest of the poem is in its final lines. Closing with a comparison of the gradual shift in power from Catherine to her young son Charles IX to Jupiter's eventual acquisition of thunderbolts, Ronsard predicted that "After the many long years / Ordained by Destiny / You will both go to the heavens; / Like two flares you will shine, / Two new stars, favoring with fortune / Your heirs, the Kings and your France."[37]

The suggestion that a Medici ruler and her son had long since emerged in the heavens as two new stars near Jupiter would have made Galileo's novel discovery of four satellites in January 1610 seem a somewhat belated gesture. And *Les Nues ou Nouvelles* would have figured as a kind of news itself, for though it had appeared as a separate pamphlet in 1565, Ronsard's poem came to the attention of the greater part of the public only in editions of the turbulent first decades of the seventeenth century.

Miniaturized Landscapes

Among the most complete descriptions of the events leading to the king's assassination are Pierre Matthieu's *Histoire de la mort déplorable de Henri IV, Roy de France et de Navarre*, and Nicolas Pasquier's *Lettre à Monsieur d'Ambleville*. As he was the royal historiographer for both Henri IV and the dauphin Louis XIII, Matthieu's account, published in French in 1611, 1612, 1613, in English in 1612 and 1621, in Italian in 1615, 1625, 1628, and 1638, in Spanish in 1628, and in Latin in 1613, would have been the official version of the king's final days. For some readers, it was familiar enough to be recognized without the author's name: soon after its emergence in the Italian translation, Alessandro Tassoni asked a correspondent for "one of those booklets on the death of Henry IV, king of France," and it was minimally identified in Galileo's library as well.[38] Matthieu had completed a copious chronicle of the king's life, in fact, in February 1609, and a minister who saw the draft dryly observed, "it is to be hoped that His Majesty will furnish matter in the remaining days of his life so as to fill up that blank folio."[39]

Nicolas Pasquier's letter, purportedly written within days of François Ravaillac's execution, was first published as a part of an epistolary collection in 1623. His role as *maître des requêtes de l'hôtel du roi* would have allowed him access to a number of details bearing on the king's death; his literary aspirations encouraged him to develop the narrative into a rather lengthy letter. My discussion of Matthieu's and Pasquier's accounts is complemented by two shorter versions published several decades later in the seventeenth century, the abbé Michel de Marolles' brief allusions to the cardinal event of his childhood, and Pierre Gassendi's analysis of the royal assassination in his *Life of Nicolas-Claude Fabri de Peiresc.*

These four writers faced certain common constraints. No explanation of Henri's proposed military involvement in the Jülich-Clèves succession crisis, the preparations for which occupied him even at the moment of his death, could acknowledge that this campaign in the lower Rhineland, very likely to provoke Archduke Albert, ruler of the neighboring Spanish Netherlands, also had something to do with the king's infatuation for a sixteen-year-old girl then sequestered by her alarmed husband in Brussels.[40] Nor could excessive emphasis be placed on the uncanny coincidence of the queen's coronation on the eve of the assassination, a circumstance that arose from her fear of a royal divorce. And while these writers offered long discussions of the many prognostications of the king's death, none could resort to simple astral fatalism to explain the turn of events in May 1610. Not much could be made, finally, of the king's apparent disregard, even after nearly two dozen other attempts on his life, for his own security and for the stability of a deeply divided nation and continent.

Pierre Matthieu's narrative of the assassination begins with the story of the queen's coronation on May 13, where Marie's need to straighten the wobbling new crown before it slipped from her head was later viewed as a most provident gesture.[41] Matthieu proceeded from that ceremony to the many warnings of the king's impending death, the most celebrated of which was that offered by the astrologer La Brosse to the duc de Vendome and to the irritated monarch that evening, that "if he could avoid an accident that menaced him, he would live for another thirty years."[42] A variation on this story appeared in a letter written on May 19 in Paris and addressed to William Trumbull, the English agent in Brussels; there, La Brosse's warning would have allowed the king another twenty-five years of life, had he been able to escape the threat of May 14.[43] Though François Malherbe would follow his initial report to Nicolas Peiresc of La Brosse's prognostication with the observation that the duc de Vendome later admitted inventing the story, the tale appears in many accounts of the assassination.[44]

Matthieu also recorded other prognostications by Cyprianus Leovitius and Paracelsus, the "mountains of ice" in France's rivers in the winter of 1608, the revelations of Catherine de Médicis' magic mirror, ominous celestial signs, troubling dreams, a weeping statue, the wilting hawthorn tree in the courtyard of the Louvre, and rumors of certain verses, written in the style of Nostradamus and posted on the pont Neuf, "that spoke clearly of the death of the king."[45] Matthieu presented a king who trusted more in divine will than in dubious omens, and yet who was strangely restive, and singularly given to prophetic speech, in the last days of his life.[46]

The chronicler's description of the king's activities and intentions on May 14 is provocative. Though racked with indecision about whether or not to venture out into the city that day, Henri wished to consult with an adviser, Pierre Fougeu, seigneur d'Escures, about leading a vast army toward the lower Rhineland that spring in order to ensure that the disputed territories not fall into Habsburg hands. Matthieu was at pains to avoid the issue of Henri's designs upon the youthful Charlotte de Montmorency, princesse de Condé and daughter of the king's constable, then beyond royal reach in Brussels, and he focused on the logistics of the military expedition to Jülich-Clèves, but with the understanding that this conflict could easily lead to a full-scale campaign against the Spanish Netherlands. The eventual confrontation was to be with General Ambrogio Spìnola, commander in chief of the forces in the Spanish Netherlands, who, according to a diplomatic report of April 29, 1610, "doth threaten to at least look the French army in the face."[47]

D'Escures, one of Henri's most trusted ministers, and brother of the cartographer and topographer Jacques Fougeu, had come to the Louvre earlier in the day to advise the king on the route to take along the River Semois, and had seemingly assured him about the state of his army, his munitions, and the lack of resistance he would find in the sovereign territory of Château-Regnault in the Ardennes.[48] Though others had warned Henri that the border territories were in control of General Spìnola, and that they were "a Country inaccessible for Mountains, Rocks, Woods, Streights of Passage," Matthieu reported that d'Escures had nothing but encouraging news for the king, and had even insisted on "the general amazement of Luxembourg and of other provinces in the Netherlands, who imagined an invincible prince."[49] The king was further pleased to learn of a report that Spìnola would be seeking to meet him en route and to block his passage through the bishopric of Liège, but that the inhabitants of that nominally independent region were already proclaiming, "Vive la France!"[50] Clearly intending to convey both Henri's military and amatory ardor, the royal historiographer stated that it was d'Escures's optimism about the eventual invasion that convinced the hesitant king to leave the Louvre and hurry to the Arsenal to consult with Sully.

> His wish to see it [*Le désir de la veoir*] grew, when he saw that his plans were outstripping, as it were, his desires, and that everything was conforming to his will, and that it seemed as if Fortune were presenting towns and provinces to him in nets. He had been so faithfully and suitably served that those who commanded the chief strongholds he had had reconnoitered had neither news nor shelter, and he was more assured of taking them than resolved to attack them.

All the officers of the crown, the principal lords of the kingdom, and all the bravest and most gallant men in the provinces were around him, and though they did not know, in truth, where the wind would take them, they were content to embark on a ship where the captain was not only capable of managing the rudder, but had also always been favored and cherished by Neptune.[51]

It is Matthieu's report of the king's meditation, with its fanciful conclusion in a voyage, which merits attention: here, the distant towns and provinces under Spìnola's protection seem suddenly reduced to submission to the helmsman Henri IV. Given the absence of any feminine antecedent in the preceding paragraph, the phrase *le désir de la veoir*, and its inevitable homonymic double *le désir de l'avoir*, refer rather transparently to the king's designs upon Charlotte de Montmorency.[52] More crucial than these allusions is Matthieu's emphasis on the visual nature of the royal reverie.[53] The miniaturized strongholds and manageable provinces are doubtless the effect of the manuscript map provided by d'Escures; both the king and Sully were devoted to cartography and topography, and the development of those disciplines, and the production of maps of unprecedented accuracy under the aegis of d'Escures's brother are a hallmark of Henri's reign.[54]

The king's imagined landscape, his improbable voyage, and the encounter with the military leader of the Spanish Netherlands also recall the first descriptions of the telescope made available to the French public in news reports of 1609, 1610, and 1611, descriptions beginning and ending with notice of Spìnola's embarkation from The Hague. Significantly, these journalistic accounts had included that general's celebrated remark to his longtime Dutch enemies, "I will no longer know how to be in safety, for [with the telescope] you will see me from afar."[55] Henri's "vision" of Spinola, by contrast with that of the Dutch, would pose no real threat, for his was nothing but a cartographic image, and a fantastical one at that, rather than an actual telescopic observation of his adversary.

Matthieu's description of the king's fatal ride through the streets of Paris to the Arsenal insists less upon a lapse in judgment on the part of the heedless monarch than on a kind of mistaken focus. The chronicler once again made d'Escures's military map the subject of Henri's attention. As the royal carriage tried to make its way through the crowded rue de la Ferronnerie, the king dismissed his guardsmen, and

This tiger [François Ravaillac] was coming along the same road, his cloak hanging on his left shoulder, a knife in his hand, his hat covering it. The posture of the King made him bold; if [Henri] had not been turned the other way, I believe that the reverence and the majesty that God's hand imprints on the foreheads of kings would have held him back. [Henri] had his right arm around the neck of the Duke d'Espernon, to whom he had given a paper to read; his left arm was on the shoulder of the Duke de Mombason, who had turned his head away so as not to seem curious about what the King was saying aloud. The king had leaned forward to say to the Duke d'Espernon and to the Marshall of Lavardin these words: "When we return from the Arsenal, I will let you see the map that d'Escures has made for the passage of my army; you will be very pleased with it, as I was."[56]

"At these words," Matthieu relates, Ravaillac sprang forward and struck his victim with two blows, the second of which proved fatal.[57] The entire episode suggests some sort of visual inattention, an impairment that fell just short of blindness. Henri's royal gaze, strong enough to "reduce" the stronghold of Sedan to submission in 1606,[58] was misdirected here; his eyes were too weak to read the paper held by the duc d'Espernon; his mind was, in any case, not on reading, but on a kind of void, the absent map of a remote landscape associated with a dubious military campaign. His three companions appear marked by a similar kind of handicap. The duc d'Espernon had assumed a posture of courteous incuriosity that blinded him to the approaching danger; the other two men, too interested in the discussion of the map, were unable to detect Ravaillac. It is as if d'Escures's topographical drawing, or even the mere mention of it, had somehow engendered blindness in all concerned.

Matthieu's vignette is, in psychological terms, a plausible one; after the assassination, the king and all involved in the proposed campaign may well have seemed to him truly benighted. And as Matthieu owed his early advancement as historiographer to Jeannin's support, he offered a version of events emphasizing particular types of impairment and insight.[59] The emblem of that blindness, the unseen map drawn by d'Escures, conforms to an already familiar image of the king: Henri's cartographic conquest constituted an implicit rejection of the emergent technology of the telescope. Thus the quite fantastical nature of Henri's "reduction" of neutral and enemy territory, a gratuitous reference to Jeannin as a valuable but neglected adviser,[60] the insistence on a kind of optical misprision at the moment of the ruler's death, all contribute to this impression of the king. There is, of course, no mention of the hawkish Sully in this account, but not surprisingly, certain passages in his letters evoke a similar faith in the evidence of maps. To choose an instance from October 1608, during the negotiations in The Hague over the Twelve Years' Truce, when Sully expressed indignation to the king over the tendency of the Dutch to manoeuver between the English and the French, he described their perceived ingratitude to Henri in cartographical terms: "they are threatening that the province of Zeeland, in its misery, will leap into the arms of [James I], as if the greatness of France were based upon the whims of a handful of men who are almost not depicted on maps."[61]

Advance Reports

It is worth considering in this connection Matthieu's discussion of the uncanny advance reports of the assassination. Stories of such rumors, ones that had anticipated the king's death by a few days, a week, or even a fortnight, abounded directly after the event, and were sometimes construed as evidence of a murderous conspiracy in which Ravaillac was only the most obvious participant. Thus when Protestant leader Duplessis-Mornay asserted around June 1610 that "they knew of our misfortune in Prague, in Madrid, and in Brussels, before it happened, and it was soon verified by ambassadors and by this damnable correspondence [of the Society of Jesus]," he was alluding to the very common belief that Ravaillac had been prompted to act by Jesuit agents.[62] The *Anti-Coton* of 1611, directed specifically against the late king's Jesuit confessor, Father Pierre Coton, likewise asked

> But whence was it, that at Bruxelles and at Prage where the Iesuites dominiere, the kings death was spoken of (some) twelve, or fifteen dayes before it came to passe? At Roan [Rouen] divers received letters from their friends at Bruxelles desiring to be informed whether the report that went of the Kings death were true, albeit at that time it was not so. Monsieur *Argentier* at Troyes received letters from the Tutor to his children at Prage, wherin he was aduertised that a Iesuite had given it forth, that the King was dead before it fell out to be so.[63]

It was a similar suspicion of conspiracy that led the Huguenot Jean Hot-man to observe "we have letters from Antwerp and Brussels of the 12th [of May] saying that the King had been killed, and the murder was only the 14th, and there are many other signs that something was known there."[64]

Matthieu, too, was especially interested in the rumors originating and circulating in the Spanish Netherlands, the territory of the king's projected campaign.

> [François d'] Aerssen, ambassador of the States-General [to France], told me that he could produce for the Queen people by the hun-dreds who swore that the rumor had been made public in Flanders [i.e., the Spanish Netherlands] before the couriers arrived. I heard it said to the Queen that her goldsmith had received letters from there, written around the same time, in which he was asked to verify if the King had been killed. Often the rumors of these great accidents are spread and planted among the common people without a certain author, and when one looks for the source of them, they are lost in the crowd like waves and ripples in a vast ocean.[65]

And in his *Relation of the Prince of Condé's Flight from France*, even Guido Bentivoglio, papal nuncio in the Spanish Netherlands, offered an ambiguous suggestion, or rather a succinct version of Matthieu's account, that something of the king's death may have been known in Brussels before the event itself: "But here, in the midst of so much trouble, and on the tossing waves of these affairs, all of the sudden a rumor broke out in Brussels, at first timid and variable, and then made strong and vigorous from the magnitude of the event, that the king of France had been murdered. And this was then verified."[66]

Advance rumors were clearly an analogue to the astrological prognostica-tions reported by Matthieu at the outset of his chronicle.[67] The distinction between the two sorts of news as learned and popular variants of a single phenomenon is merely apparent: though prognostications generally predated the event by many years, were presumably the product of learning, skill, and a print culture, and were typically associated with a single practitioner, the examples chosen by Matthieu were not without inaccuracy and ambiguity. By contrast, the advance tidings that emerged just a week or two before the assassination, circulating most often by word of mouth among the lower classes before being committed to paper as queries by tradesmen, clearly depended upon the occult knowledge of the unknown original source.

For his part, Matthieu found something unsettling about the whole business, and he sought to explain such doings not as Jesuitical, but as necromantic. In drawing this parallel, he seems to have insisted more on the distance of the reports from the event rather than on the temporal interval that allegedly preceded it. "Demons and sorcerers are involved in the prompt relay of news from very far away," he observed,

> The late king told me a story about the familiar spirit of a servant of a Count of Foix who, upon seeing his master sleepless and troubled over the outcome of some event, a battle or a siege, brought him right away all the details of it, such that this prince, without leaving his house in Corasse, where this spirit is still heard, knew everything that was happening throughout Europe. In just the same way Apollonius of Thyana saw from Ephesus that they were killing Domitian in Rome, and Caius Cornelius, without leaving Padua, was at the battle of Pharsalia.[68]

Matthieu's first instance of long-distance reportage, offered as an anecdote shared with him by the late king, derives, in fact, from the great royal historiographer Jean Froissart, who claimed to have heard it at the court of Count Gaston Phébus of Foix around 1389.[69] Familiar enough to early modern audiences to have been mocked in passing in 1580 by Michel de Montaigne, and later called a *grand nouvelliste* or "great newsmonger" by the Romantic François Chateaubriand,[70] the protagonist was a spirit who, traveling "with the wind or faster," lived in Southern France and routinely recounted events that had taken place just a day earlier in settings as remote as England, Germany, Hungary, Scotland, Flanders, Brabant, and Bohemia. Froissart's suggestion that the useful spirit was employed first by a cleric, then by the sire of Corasse, and then by Count Gaston Phébus of Foix would have had a certain irony after the assassination, for the late king was, as of 1607, the last count of Foix.

Though it was commonly held that Ravaillac was in league with the devil—evidence, of a sort, was presented at his trial, and a diabolical alliance was signaled in certain diplomatic reports[71]—there was little reason to suppose that Matthieu believed the many anonymous citizens who allegedly heard advance rumors of Henri's death had such powers or were confederates of the assassin. The unknown authors of these reports, by contrast, were in possession of some form of occult knowledge that kept them informed of the killer's doings in Paris. That Matthieu named Dutch diplomat François d'Aerssen as his source is not surprising; the Flemish-born Aerssen, then ambassador to France for the United Provinces, could reasonably have been expected to have some notion of affairs in Flanders, and the popular association of such long-distance reportage with a Jesuit plot would have been in no way alien to his views.[72] Aerssen's roles as a sometime protégé of the "Pope of the Huguenots," Philippe Duplessis-Mornay, as a favorite of the late king, and as a great trafficker in news would have only enhanced Matthieu's story: the famous private letter celebrating Henri's optical "reduction" of the fortress at Sedan in 1606 with *Veni, vidi, vici* had been entrusted to his hands. In a letter of mid-December 1608, Jeannin had explicitly associated Aerssen with retailing rumors of warmongers and of the masses—rather than listening to diplomatic channels—when, during negotiations for the Twelve Years' Truce, the latter had predicted from Paris the failure of the efforts overseen by Jeannin in The Hague.[73] And the fact that François d'Aerssen—along with his father, Cornelis d'Aerssen,[74] then the *greffier* or secretary of the States-General—had been accused by the French of treason and espionage in the summer of 1608 during these negotiations, and had yet managed to stay in the good graces of Henri suggests at once a certain enviable *savoir faire* and passing contact with France's enemies.[75]

In late 1608 and early 1609 François d'Aerssen also maintained close contact with Jacques Badovere, who was both Henri's envoy in the early stages of the crisis in Jülich-Clèves, and the man to whom Galileo then turned for news of the Dutch telescope, a detail which may suggest access to technological as well as political information. The interest and availability of the former sort of information is plausible, for in his capacity as *greffier* of the States-General Cornelis d'Aerssen had immediate knowledge of various patent applications for the new optical invention, among other items.[76] François d'Aerssen's bitter political rival, the Dutchman Cornelis van der Mijle, had been ambassador to the Venetian Republic when Galileo made known his development of the telescope, and would be eager to claim public attention in 1612 as the motivator and dedicatee of one of the earliest treatises on sunspots.[77] And though he referred often to the dangers of public discussion of political matters in the United Provinces, and deplored the popular recourse to anonymous publications, the younger Aerssen became widely associated in the coming decade with the publication of anonymous pamphlets for the instruction of "the simple."[78]

In Matthieu's account, the advance reports of the king's death available to "people by the hundreds" in Flanders and to the contacts the queen's goldsmith had in the region figure as the accurate counterpart to Henri's unrealistic view of the strongholds and towns near that area. Put differently, the news the Flemish folk had of the assassination in Paris was perfectly correct; the cartographic image the French king had of the territories bordering or within the Spanish Netherlands was a mere mirage. Furthermore, the very emblems of the tension between the unwitting perspicacity of the Flemish masses and the near blindness of Henri and his advisers are, of course, the telescope and the map.

Matthieu's intriguing reference to Marie's goldsmith, the alleged recipient of letters from Flanders asking for confirmation of the king's death before the event, is worth elaboration. Several of the goldsmiths of the French royal family, Josse de Langerac, David de Vimont, Corneille and Nicolas Roger, were from Flanders, and would presumably have had close contacts with the region. Relying on both Matthieu and Jean Hotman, Cardinal Richelieu stated in his *Mémoires* that "From Flanders someone wrote, on the 12th of May, to [Nicolas] Roger, a goldsmith and valet of the queen's chamber, a letter in which the king's death was deplored, though this did not take place until the 14th."[79]

The professional boundaries between glassworkers, gold- and silversmiths, and (to a lesser degree) jewelers were relatively porous, for these individuals shared certain tasks and often the same elite clientele. Thus early eyeglasses and telescopes were both frequently made by those who worked with precious metals, while a jeweler, as a purveyor of genuine rock crystal and of various types of glass, would be able to comment upon the optical quality of the lenses.[80] The various materials and techniques used to polish gems, Venetian *cristallo*, and mirrors were immediately adopted by early telescope makers.[81] It was a Flemish goldsmith who made the first telescope for the rulers of the Spanish Netherlands in the spring of 1609,[82] and a silversmith in Spanish-dominated Milan who wrought a tube for the brand-new device in this same period and showed it to an interested onlooker who set about duplicating it.[83] In a period when many marveled over the telescope, his early training as a goldsmith seems to have made the Jesuit Paul Guldin especially vigilant and even rather critical of the lens-grinding techniques of the artisans employed by the astronomers of the Collegio Romano in 1611.[84] And a decade after the invention, Galileo's best telescopes were made by an artisan originally trained in the production of *pietre dure*, Florentine mosaics of polished gemstones.[85]

It is worth noting in this connection that the only individual identified among the many who allegedly received advance warning of the assassination in the *Anti-Coton* of 1611 was a "Monsieur Argentier" of Troyes, who was said to have heard the news from his son in Prague. Though the story pertains to an extraordinarily rich and notorious merchant of Troyes, Nicolas Largentier, to readers outside of this orbit, the name would have suggested instead a silversmith with significant connections abroad.[86] The portrait of silver- and goldsmiths as artisans with remarkable foresight about royal personages, acquaintance with would-be assassins, and a network of professional contacts lived on: as the first anniversary of Henri's assassination approached, a story circulated about how a goldsmith whose shop was on the pont au Change prevented an attempt on Louis' life by gaining the confidence of a soldier hired for the killing and enchanted by a ring given to him by a priest. As the English agent in Paris, John Beaulieu, recounted it to William Trumbull in Brussels, the goldsmith's heroism did not consist so much in technical expertise—the sinister markings on the ring were discovered by others—but rather in his ability to coax a confession out of one who had not yet committed the crime, and in his decision to strike his anvil, that single blow being a secret signal bringing the aid of his fellow artisans.[87]

It is also significant that Matthieu presents the obscure Flemish seers, the unseen sources of the rumor of the assassination, as some sort of occult practitioners, modern-day versions of Apollonius of Tyana and Caius Cornelius. Not coincidentally, Apollonius's telepathy had been explained in terms of meteorological catoptrics; according to Pietro Pomponazzi, it was through an analogy with a mirror that Bishop Pietro Barozzi had explained the magus's knowledge of distant events. Of the means used by Caius Cornelius to witness the remote battle of Pharsalia there is no discussion; the crucial point is that he saw whatever he saw without ever leaving Padua. So, too, Galileo had used an optical device—*a quodam Belgam elaboratum*, "developed by a certain Netherlander," as he put it in the *Starry Messenger*—to see the distant heavens and to give news of them without leaving Padua.[88]

It seems likely, then, that Matthieu intended to portray these shadowy seers of Flanders, the original sources of the unhappy rumors of the king's death, in terms compatible with the first users of the telescope. Like those

Netherlanders, they see remote events, have some connection to a profession then associated with the manufacture of such instruments, and make observations that will eventually become the most compelling of news reports. Just as through its reliance on the telescope the *Starry Messenger* offered, or purported to offer, a less ambiguous image of the heavens than did contemporaneous almanacs, so the simplicity of the seers' rumor about the assassination confirmed and clarified the astrological prognostications that had preceded them. And Matthieu's depiction of a singularly well-informed Flemish goldsmith in the queen's orbit—a sanitized version of the sinister visionaries in Flanders, sources of the advance rumors of the king's death—has its real-life analogue in the foreign goldsmiths, engravers, and painters who provided rulers not just with art objects, but also with diplomatic insight, political news, and espionage in the early modern period.[89]

All this is articulated about the central conceit of Henri IV's blindness or shortsightedness, as if the monarch's disregard for available astronomical information, his illusory vision of neutral and enemy territory, and his eventual appearance in foreign news reports were the correlates of the scant attention paid to Pierre Jeannin's initial disclosures concerning the Dutch telescope. That Matthieu believed that there was the slightest causal connection between these sorts of blindness appears extremely unlikely. But it is more than credible that he, like other royal historiographers of the period, sought in late 1610 or early 1611 to enhance his arguments through reference to technological instruments and processes dear to the increasingly influential Jeannin, and sufficiently remote from the obscure maps and grand designs of Minister Sully.[90] Such a rhetorical strategy would explain the tension between Matthieu's presentation of the Netherlandish visionaries as both dabblers in the occult and possessors of a valuable optical secret, one that had earlier been within the grasp of the French king and his fallen adviser.

Vox populi, vox dei

Nicolas Pasquier, son of the celebrated humanist, historian, and jurist Étienne Pasquier, took up the story of the assassination in a letter to one of the late king's councilors, and published it in 1623 in an epistolary collection.[91] It is improbable that this letter was actually written and sent to its addressee shortly after Ravaillac's trial; it is more likely a meditation on the murder composed a decade after the fact. As Charles Sorel observed in his

Bibliothèque françoise of 1664, "some of [Pasquier's] letters are genuine missives or familiar letters, such that one would send to friends; the others are
consolations, remonstrances, and warnings."[92] The news of Henri's death
seems to be of this latter sort. The first letter in the first of ten books of
missives in Pasquier's epistolary collection, it is by far the longest and the
most artful of those that appear in that volume of almost 1,000 pages.

Pasquier was to some extent indebted to Matthieu's account. Like the
royal historiographer, for example, he insisted on the king's fatal focus on
maps: the letter began by noting that the incident took place as Henri was
planning to go from the Louvre to the Arsenal and from there to the lodgings
of Monsieur d'Escures. Pasquier also emphasized the blindness of the victim
and his companions. The king "not being able to read without his glasses,"
had handed a letter to the duc d'Espernon, and was speaking optimistically
about his planned foray when his carriage was blocked in a busy street. As
the impetuous monarch stuck his head out of the carriage to observe the
reasons for the delay, only to return to the discussion of his eventual easy
passage through the Ardennes, "this cursed, execrable [Ravaillac] saw him,
and at that moment a diabolic vision came to him, saying, 'Go and strike
hard, you will find them all blind.'" Pasquier reported that the assassin had
acted so promptly that "no one of the [king's] company saw him give the
two blows."[93]

Where Matthieu had contrasted Henri's cartographic image of his campaign on the outskirts of the Spanish Netherlands to the telescopic vision
someone in Flanders had of him, however, Pasquier substituted the opposition of the military map and the astrological chart, both referred to as *cartes*.
The morning of the assassination, for example, the king was allegedly
reminded of the inauspicious prognostications made by La Brosse, whom he
called "an old fool," and he was presented with several letters to the same
effect, which he ignored. The middle-aged Escures, whom Pasquier describes
as "all gouty," arrived at the Louvre with his map, and enjoyed a better
reception than the astrologer and the letters. Rather improbably, the king
inquired the day's date, and being told it was the fifteenth of the month,
replied that he had escaped the danger of the fourteenth, and made plans to
leave the Louvre.

Pasquier suggested that the still uneasy king insisted on an outing
because he had already promised his retinue that "they would learn something from the map drawn up by Escures [*la carte qu'Escures en avait dressée*],
or rather, because of a necessity of celestial influences which were impossible,
or at the very least, quite difficult for anyone to resist." Put differently, the

astrological chart drawn up at the hour of the king's birth, and not the military map of the upcoming expedition, was the relevant document; the "old fool," not the middle-aged gout-ridden one, was the better adviser.[94]

In certain respects Pasquier's letter appears a *retardataire* version of Matthieu's chronicle. His insistence on the ultimate accuracy of the astrologers, for instance, seems old-fashioned when compared to the historiographer's gesture toward telescopic vision, as does his elaboration of the already copious catalogue of prognostications of the king's death. A doctor in Toulouse warned Henri that the stars were conspiring against him, Helvæus Rosselin, physician and astrologer in Alsace, had written prophetic verses on his account, and Catherine de Médicis and Nostradamus had foreseen his death during the twenty-first year of his reign. Pasquier himself had encountered an almanac six months before the assassination where the fatal day was marked, *Dies illa, dies iræ, calamitatis & miseriæ*. A series of poetic prophecies written a century earlier, for the period 1590 through 1613, was off by one year, Pasquier admitted, but substantially correct in its enigmatic depiction of international politics. Choice verses by Nostradamus and his imitators appeared in other almanacs for that year, and had circulated among the people. Two *mathématiciens*, one from Besançon, the other a Scotsman, having drawn up the horoscope of the king in 1608, predicted with apparent accuracy that the king would be killed in two years; as regards Sully, they foresaw that "someone powerful who holds one of the premier positions in France, stripped of his charges, would be as miserable as he once was happy." Pasquier completed his account of the overwhelming astrological consensus about the king's fate by explaining how the German Johannes Rudolphus Camerarius had misreckoned the date of the assassination, and by noting that one of Henri's own councilors had warned him of a dangerous illness that might kill him between the winter and summer solstices of 1609–1610.[95]

Pasquier's spectacular catalogue of prognostications is part of a larger and more pertinent argument about foreknowledge and communication, and it is this issue that is most crucial here. Near the beginning of his letter about Henri's death, he stated that "just as we never see fire without smoke, nor a body without shadow, so the death of a great Prince will always be predicted to us a long time before it happens, either by the voice of the people, or by divine inspiration, which God reveals to some of His servants, or by astrologers."[96]

All three sorts of reports were within divine control. Pasquier mentioned several different pious individuals who had had visions of the assassination and tried to warn the king, and he blamed Henri's powerful Jesuit confessor,

Father Pierre Coton, for keeping the most insistent and credible of these messengers at bay.[97] This was the smallest class of persons having advance notice of the assassination, the risk of being mistaken for a confederate of the killer, or a raving lunatic, or both, being significant, as contemporary documents make clear. A somewhat larger group of seers would be the many astrologers whose predictions Pasquier compiled and discussed. But it is the largest and most inclusive group, the people, that is of the most interest in Pasquier's argument.[98]

Adapting the proverb *vox populi, vox dei*[99] to his own purposes, Pasquier presented the rumors preceding the king's death as a popular exposé of what God knew to be true, and in this he differs significantly from Matthieu, who emphasized the diabolical nature of such reports. Often understood to refer to the political voice, or choice, of the populace, or to common and thus incontrovertible opinions, on this occasion *vox populi, vox dei* represented for Pasquier the oracular quality of hearsay tales: "A popular rumor is the voice of God." Pasquier referred to the accuracy and divine origin of such rumors elsewhere in his letters; in discussing the signs preceding the disgrace and assassination of Marie de Médicis' Florentine adviser Concino Concini in 1617, he remarked that the people's "voice is compared to God's, because we often see that popular opinion brings out amazing prognostications, and from this it seems that there is some secret and hidden virtue in [the masses] that makes them sense from afar the good and the evil that will befall the greatest men."[100]

Directly after the assassination of Henri IV, in his *Most Humble Counsel to the Queen and Regent of France for the Conservation of the State during the Minority of her Son*, Pasquier had offered a somewhat different explanation of the *vox populi*. In this work of 1610 he argued that "the best and most sound advice will come to you from the common voice of the people. It is, as they say, the voice of God, from which you should take great instruction in your affairs." In order to assure the queen that in the case of long-term political strategies and options, as opposed to short-term predictions of unavoidable events, the *vox populi* was not strictly a matter of popular opinion, he went on to define "the people" as the clergy, the nobility, and the third estate.[101] A related understanding of *vox populi, vox dei* occurs in Traiano Boccalini's caustic meditation on advance rumors; these were associated with both a compelling popular political will—as if one common term for journalistic reports, *avviso* or *avis* or the early modern English *advices*, were best translated here as "advice"—and with the typical venues for news and gossip. Thus Boccalini noted in a discussion that carefully suppresses any connection between *vox populi* and divine directives that "often what princes ought to do is treated as if it had already been done. The death of the duke of Guise [at the hands of King Henri III in 1588] was publicized in Rome long before it actually happened. And before he decided upon anything, Pope Paul III would take counsel from trusted men about what was being said in *i Banchi* [the epicenter for news in Rome], and he found that frequently it was correct."[102]

In 1632 Théophraste Renaudot—the founder of the *Gazette* and thus presumably somewhat skeptical of the *vox populi*—would offer rather bland support of "the common opinion that rumors, for all their uncertainty, are images of future events."[103] In the case of Henri's death, popular knowledge of the coming disaster seemed for the most part geographically determined, as the advance reports mentioned by Pasquier were largely confined to the Spanish Netherlands:

Fifteen days before this piteous spectacle, a merchant from Douai wrote to a friend of his in Rouen and asked him to send word if it were true that the king had been killed. The letter was shown at [Ravaillac's trial]. In this same period, a similar rumor circulated in Lille in Flanders, as I learned from a merchant from that place. Another merchant from Antwerp wrote to a certain Flemish merchant staying in Paris, a friend of mine who lent me an excerpt from the letter, which I quote to you here. These are his words: "It is much noted that they were talking about the death of the king here twelve days before it happened. No one thought much of it at the time, but at last it turned out to be true. We are all astonished that such news circulated in this city. It seems that some people knew that this thing was meant to be." One of the leading men of Cambrai said of the king, eight or ten days before this sorrowful event, "this old man has great plans, but he will not get far." [Jean de Thumery, Seigneur de] Boissise, Councilor of State and at that time Ambassador to the Marquis of Brandenburg and the Duke of Neuburg, told me that a rumor went around Antwerp, Bois-le-Duc, and Maastricht, that the king had been killed ten days before it happened, and that in Cologne, at the same time, letters from Antwerp were read in public in which they sent news of the King's death, even though it had not yet taken place. [Philippe] Galland, rector

of the College of Boncourt, upon returning from Arras, told me that
they had word of the death of the king in that city eight days before
it happened. It often happens that news precedes the event.[104]

Some of these stories about advance notice of the assassination are contradicted by the principals' correspondence. In Pasquier's account, for
instance, the ambassador Boissise appears to have signaled the presence of
such rumors in Antwerp, Bois-le-Duc, Maastricht, and Cologne, around May
4, 1610, making his tale a hyperbolic version of that offered by Jean Hotman
de Villiers, then in his company in Dusseldorf, about letters from Antwerp
and Brussels describing the king's death a mere two days before it actually
took place. And yet both Hotman and Boissise seemed unfamiliar with these
advance rumors seven days *after* the assassination.[105] Nor do Hotman and
Boissise appear any better informed later that day. "I have yours of the 10th
[of May]," Hotman wrote William Trumbull, the English *chargé d'affaires* at
Brussels,

> We have since had from one of your secretaries to [Archduke] Leo-
> pold, by name Fleckhammer, the news of the murder. Hearing
> nothing from you nor [Matthieu Brûlart, French ambassador to the
> Spanish Netherlands], we hope it is not true, and [that it is] put
> forward to help on the treaty which Leopold hopes to make with
> these princes. But Boissise and I and all who rely on the king's help
> are alarmed.[106]

Both Boissise and Hotman received some of their information from the
press. The former, according to Pasquier, alleged that "in Cologne, [ten days
before the assassination], letters from Antwerp were read in public," suggesting that the privately circulated rumor had entered the public domain via the
newsletter. Hotman, for his part, was one of several persons to allege that the
killer or his family were Flemish, rather than French.[107] Just as Henry Wotton, the English ambassador in Venice, reported that "the French king had
been killed in his coach by a Walon,"[108] Hotman noted that "the Gazettes
say the assassin had a wife at Brussels," and he then proceeded to air the story
of letters—possibly newsletters—from Brussels and Antwerp dated May 12
and describing the king's death.[109] And it is the relationship of the advance
reports—Pasquier's divinely inspired *vox populi*—to contemporary newsletters that merits further examination here.

From Douai, April 30

Nicolas Pasquier was clearly an avid newsreader; his stories about advance notice of the assassination are followed, for instance, by a description of a celestial event of 1608 in Ravaillac's native Angoulême drawn from the annual newsletter *Mercure François*,[110] and several of the "consolations, remonstrances, and warnings" that figure in his epistolary collection appear based on journalistic accounts of current events. He was not, however, openly interested in the processes characteristic of this emerging industry. In contrast to letter writers such as Jean-Louis Guez de Balzac, Nicolas Claude Fabri de Peiresc, François Malherbe, and Guy Patin, who alluded frequently, if scornfully, to corantos, gazettes, and *avvisi*, Pasquier had nothing to say about the commercial source of any item of news, and he referred instead to prominent individuals, some of whom he named, and upon whose word he relied. He noted neither how any given story differed from or conformed to popular opinion, nor how recent it was, nor what purpose it might serve, and he made no requests for such information from his putative addressees. Some of this studied indifference to journalism is due to his unwillingness to present himself as an individual who was forced to seek out the very kind of information that came to him unsolicited. It is also an effect of the moralizing epistolary genre to which he aspired: his letters are often meditations upon a single theme, differing very little from essays, and affecting none of the variegated tone and shifts in tenor characteristic of his newsmongering peers.

In a letter of his famous father one finds an interesting if somewhat dated observation about the production of news. Writing decades earlier to a friend, Étienne Pasquier had closed his letter with a meditation on the popular generation of rumors and reports, a subject low enough to require, by way of compensation, a lofty allusion to Virgil's alleged mode of poetic composition, where unpolished texts, like baby bears, were licked into shape:

> And don't think, however, that in receiving this that you will have much news. I realize today in fact what imagination alone had suggested to me in the past, that news stories are born within our Palace [of Justice] with custom, and that they derive their birth, their growth, their progress, and their end from the increase or decrease of this custom. You might think by chance that I am joking, but it is true, and it is perhaps not wrong to derive the cause of this, if you consider that the abundance of business causes the amassing of

crowds of people, and crowds are not only the mother of news, but
more than this, like the bear, the ones who lick them, or better,
cuddle them like babies, arranging them in all the different ways
that one might desire. From this it happens that at the Exchange in
Lyon, on the Rialto in Venice, at the *banchi* in Rome, there is never
a shortage of news. And from this it happens that within our Palace
one lacks no more for news than one does for lawsuits. Indeed I can
even say—because it is true—that these are related things. And
when the Palace is without lawsuits, it is also without news, and you
would find no better index of a lack of lawsuits, than if you were
running short of news.[111]

Father and son appear to differ somewhat in their impressions of news.
Étienne Pasquier's belief that "the people" generate news from the raw material of lawsuits or business does not have a divine correlate: there is no indication here that this *vox populi* has anything to do with the *vox dei*, and the
apparent malleability and multiplicity of popular rumors would militate
against any such connection.[112] His sense that the habitual confluence of
large groups of people interested in maintaining their fiscal wellbeing created
situations and events suitable for discussion—news, in a word—was certainly
correct for the latter part of the sixteenth century, and represents a departure
from an older view in which the newsworthy occurrence was relatively rare,
unrelated to social and civic customs, and divinely ordained. Étienne's
emphasis on the inevitability of news almost certainly arises from the appearance of serial publications, and his insistence on the extreme variability of
reports arranged in "all the different ways one might desire" may be related
to the presence of competing journalistic organizations. But while places such
as the Exchanges in Lyon, Venice, and Rome, or the Palace of Justice in Paris
remained important venues for seeking out written and oral reports of current
events, by the close of the second decade of the seventeenth century the places
where news was born, to use Étienne's metaphor, had less to do with the
presence of crowds searching for profit or justice in urban centers than with
military activity in the more remote reaches of the Holy Roman Empire, for
the Thirty Years' War obliged a sometimes unwilling public to focus on a
variety of places, peoples, and issues at a great remove from them.
    It is in light of certain formal traits of early modern newsletters that we
should reexamine Nicolas Pasquier's account of the advance reports of
Henri's death circulating in the Spanish Netherlands in 1610. The fact that

his letter collection was not published until 1623 meant that his readers would have already come into contact with journalistic materials such as the French-language weekly newsletters published in Amsterdam and distributed in Paris from 1620.[113] The format of the *Courant d'Italie & d'Almaigne, &c.* is illuminating in comparison to Pasquier's presentation of the advance reports. There were generally about three weeks between the date of the last recent event recorded in Van Hilten's *Courant* and its publication date, and the stories usually, but not always, ran from the oldest to the freshest news. Where multiple reports arrived from one location in a two-week period, they were arranged by dateline, rather than grouped under a single place of origin. Thus, for example, the datelines of the *Courant* of September 12, 1620 were

> *From Rome, August 23,*
> *From Venice, August 28,*
> *From Vienna, August 26,*
> *From Prague, August 31,*
> *From Cologne, September 5,*
> *From Frankfurt, September 3,*
> *From Cologne, September 8,*
> *From the Camp of the Prince of Orange, September 9.*[114]

Stories from a particular city often concern the presence of reports and rumors circulating within that area, rather than actual events and activities occurring there. Thus the report from Prague of August 31, 1620, records the recent arrival of a messenger from England, speculates upon the monies sent to the newly crowned king of Hungary by the king of Denmark, and describes Cossack attacks in Moravia and Hungary.[115] Prague itself appears in this period, just two months before the Battle of White Mountain, the calm eye of the storm; the great imperial city is portrayed as a sort of *entrepôt* for news and diplomatic activity, but swept clean by conscriptions, and defended only by its bourgeoisie.[116]

The *Courant* published by Van Hilten almost a year later, on September 4, 1621, reflects the increased turbulence and expanded horizon of the Thirty Years' War. Its datelines, slightly disordered and some almost four weeks old, run as follows:

> *From Rome, August 7,*
> *From Venice, August 13,*

From Vienna, August 18,
From Prague, August 19,
From Vienna, August 19,
From Linz, August 23,
From Hamburg, August 25,
From Frankfurt, August 29,
From Bunten in Switzerland, August 11,
From the city of Berne, August 12,
From Spier, August 20,
From Udenheym, August 20,
From Berchstraten, August 25,
From Cologne, August 31,
From Wesel, September 1.[117]

The same pattern of second- or third-hand reportage obtains, especially in items sent from Italy; the article from Rome, for instance, concerns news from Messina via Palermo, while that from Venice involves confirmation from Genoa of an execution in Spain, and a report from Milan about the allegiance of Catholic cantons in Switzerland. A stop-press article, finally, includes a brief synopsis of the latest activities of the troops of the prince of Orange in the Rhineland, several rumors having no provenance more specific than "from France" about the cessation of hostilities between Catholic and Huguenot forces in that country and the onset of the plague in the camp of Louis XIII. The *Courant* of September 4, 1621, finishes on a note of uncertainty—a device commonly used to sustain interest in subsequent issues of the serial—by noting that "There is a rumor [in Hamburg] as well, that the king of Sweden laid siege to the city of Riga in Livonia."[118]

It is not surprising that the *Courant*, published in the United Provinces, occasionally contained criticism of French policy, which was then cautiously pro-Spanish. Such views were expressed in the item of August 11 from Bunten in Switzerland, which maintained that the French ambassador was trying to lead the Swiss "back under the yoke of Spain," and called for a levy of troops.[119] Observations of this sort would not necessarily have displeased all French readers of the *Courant*; France was at this point far too divided by conflicts between Catholic and Huguenot nobles, and by the factions within those groups to have achieved a coherent policy toward European antagonists in the Thirty Years' War. But this very divisiveness, and the fact that most of the news about France in the 1620s concerned its alarming return to the civil wars of the latter decades of the previous century, would have signaled to all readers the precarious position of their country.[120]

What is especially important here is the way in which the *Courant* of the early 1620s compares to Nicolas Pasquier's presentation of the rumors preceding the death of Henri IV in May 1610. Were the format of Pasquier's reports adjusted slightly, his letter would conform closely to that of a newsletter of the 1620s, with its general tendency toward chronological ordering, its reportage of emerging rumors as well as of actual events, and its several meandering returns to news centers of importance. For a reader familiar with the *Courant* and other such serials, Pasquier's account of the advance reports of the royal assassination a decade earlier might be read as follows:

> *From Douai, April 30,*
> *From Lille, April 30,*
> *From Antwerp, May 2,*
> *From Cambrai, May 6,*
> *From Antwerp, May 4,*
> *From Bois-le-Duc, May 4,*
> *From Maastricht, May 4,*
> *From Cologne, May 4,*
> *From Arras, May 6.*

Much of what Pasquier records about the rumors was said to have originally appeared in epistolary form, a detail which would have increased the similarity between his account and that of publications like the *Courant*, where individual news items were often introduced as materials gathered from letters. The appearance of reports from Antwerp on both May 2 and 4, and the fact that the story from Cologne concerns the public reading of some sort of letters from Antwerp, are likewise typical of the repetitions in the format of the *Courant*.

From Pasquino to Coranto

Pasquier's substitution of journalistic features for the shadowy telescopic imagery Pierre Matthieu had used to portray advance reports is especially interesting, the news publication and the optical instrument often appearing as doubles of each other in this period. This focus on the particulars of the coranto also conforms to his quite pronounced desire to imitate his father, and in this connection I return to a rhetorical strategy of Étienne Pasquier, one echoed in updated form in the work of his son. In his *Jesuits' Catechism* of 1602, Étienne had used a dialogic form punctuated by commentary to bring the usual charges against the society, particularly those regarding its role in promoting France's civil wars, and in one of the more effective moments in the work, he had introduced the Roman statues Pasquino and Marforio as speakers.[121] While he appears to have been inspired by a Jesuit enemy's trivial association of *Pasquier* and *Pasquino*,[122] the device allowed him to present, in the artless tones he assigned to the talking statues, a wide variety of damning texts, some authored by members of the society, and others by their numerous enemies within and outside Roman Catholicism.

The following year, in *The Hunt for Pasquino the Fox, Discovered and Trapped in his Lair*, the Jesuit Louis Richeome made much of the conceit: a figure emerges "looking from afar like a lawyer with a great register, and a little closer like a porter at the Palace of Justice, having stuff like bunches of almanacs, songs, and other little books beneath his arm, but being now close at hand, he was instantly recognizable as the buffoon Pasquino."[123] Father Richeome's Pasquino cheerfully describes the Roman statue for which he is named as a pathetic and passive thing, plastered with satires, invectives, libels, and memoirs sent from North, South, East, and West, and crumbling beneath the burden.[124]

And it's just the same for me. For the same reasons I gather memoirs and investigations from all over the place, and I've got so much to say that I'm practically bursting. My work is just stuffed with it, since I filled it with everything I could find, or whatever people gave me from here and there, not just in France, but also in Scotland, England, and elsewhere. The Roman Pasquino puts his goods out in the Piazza Navona, but me, I sell my stories in the Palace of Justice, in the courts, in crowds, in marketplaces, and at the fairs of France. Because between me and him, there's this difference: he, acting like a weighty man, doesn't budge from his spot, and has people consult him at home, and being composed of more solid and pressing material, he doesn't write anything, but just bears up under the stuff of others. But me, being of lighter material, and fed and lifted by the wind, I go from place to place, from house to house, learning and carrying news, which I season with just a pinch of gravity, and bring to light at the expense of my enemies.[125]

This unflattering portrait of his father was crucial to Nicolas Pasquier, and his counter strategy in the account of the royal assassination was to offer information chiefly associated with a meddling and murderous Society of Jesus, but to present it in a relatively neutral format more reminiscent of the emergent coranto than of the colorful Roman Pasquino, and thus less available to attacks from the Jesuits. Far from the indiscriminate newsmonger portrayed by Richeome—part beast of burden, part packrat, absorbing and dispersing rumors and reports from every direction, and all with a clear bias against the Jesuits—the anonymous "people" of Flanders, underwritten by the faceless and comparatively featureless format of the coranto, offer information that appears specific in its focus, corroborated by accounts available in nearby locales, and studiedly indifferent to the notorious role in the royal assassination ascribed by many to the Order.

This treatment of the advance reports is also remarkable for the capabilities it does *not* assign to "the people," for while Matthieu's account implied that a chain of associates stretched from his diabolic seers through the Flemish populace to the correspondents of the queen's Flemish goldsmith, Nicolas Pasquier emphasized no particular professional or social attributes in those who knew of the king's death before May 14, but merely gestured toward a few merchants, a leading man of Cambrai, and the spokesmen they and others like them found in an ambassador and a prominent college rector from Arras. This failure to connect specific interests or vocations with the advance reports, or rather the willingness to disperse such particularizing details within the larger public implied by the coranto, distinguishes him from the last two instances I shall examine, the accounts offered by Michel de Marolles in 1656, and by Pierre Gassendi in 1641.

Secrets of State

Though James Howell attempted to make the mythical Florentine astrologer's meditation authoritative and urgent, such premonitions figure elsewhere as no more than a hazy recollection. The astrological forewarning of Henri's death, while the first noteworthy event recorded in Michel de Marolles' *Mémoires* of 1656, appears shrouded in superstition and mild confusion. Marolles, a scholar, genealogist, translator, connoisseur, and collector,[126] cast his child-hood in Touraine as an unlikely idyll spent with a gentle mother, a fable-telling aunt, picturesque peasants, and amiable local priests, but punctuated with the acquisition at nine years of age of a clerical benefice, with visits from an ambitious father on furlough from the court of Henri IV, and with brief encounters with some of the most powerful men in France.[127] No character better captures the young Marolles' suspension between these two worlds than that of Father Marc Durand, an aged, cheerful, and inconsequential priest, author of a poem about Mary Magdalene whose aesthetic mediocrity the memoirist would only later discover, and "a great fan of news, such that his friends, which [Father Durand] had in great number, acquainted him with items from everywhere, in order to please him."[128]

Marolles, his mother and aunt, the agreeable peasants, and Father Dur-and eventually get news of the event which signals the close of an era, and yet it is the acquisition of the information, rather than the incident itself, which is at stake in this account:

> Thus the end of the Reign of Henri IV, and the end of so much good, and the beginning of an infinite number of evils came when an enraged Fury took the life of this Great Prince. I think I saw a funereal prognostication of this, when, on the evening of the day on which he was killed, a great light shining in the middle of the night made the whole countryside appear as if it were on fire. I saw it when we were getting ready to go to bed, and those who saw it with me were seized with a certain terror, but it didn't last long. And though many people thought it was nothing but a flash of lightning, for all its size, when we knew the following day of the fatal event, my mother—a little credulous about stories people were telling about prodigious events—did not fail to explain this apparition as a sure sign of the misfortune that had happened.[129]

Marolles' peculiar diction about a prognostication that appears only after the assassination is worth noting; the luminous sky signals not the approach of the king's death, but rather the approach of the *news* of such. The distinc-tion may appear slight, for the news itself obviously depends upon and derives from the event, and follows it, in this case, by a mere day's delay, and yet it is clearly crucial that emphasis has been shifted from the murder of the king of France to the state of mind of his subjects, as if the incident had not happened until it was fully known. The vignette of their penumbral knowledge of the assassination and eventual acquaintance with the fact recalls, in more somber tones, Marolles' depiction of Father Durand, the provincial priest who was somehow animated and even pleased by news, much of it quite dated, and some of it surely unpleasant, from near and remote regions. It is the moment of reception of information, more than the moment and tenor of the event itself, that is of note.

Marolles' eagerness to ascribe faith in prodigious signs to his mother and to the storytellers she encountered, and to distance himself, at fifty-six, from the child who saw the flare and who may or may not have held such beliefs is part of that larger pattern that pits the sleepy and sentimentalized backwater he would soon leave against the inevitable rigors of foreign travel, education, and the establishment of a career. That the mature Marolles would explicitly reject astrology later in his *Mémoires* is not surprising: he did so in the course of a melancholy elegy of the philosopher Pierre Gassendi, a close friend whose death in October 1655 he recorded near the end of his work.[130] And it is Gassendi's account of Henri's assassination, presented in his celebrated biography of Nicolas-Claude Fabri de Peiresc, that best informs Marolles' treatment of the event in his autobiography.[131] In his *Life of Peiresc*, first published in 1641, reissued in 1651 and 1655, and in English translation in 1657, Gassendi took pains to dissociate astrological prognostications and astronomical observation, and yet he, like Howell, hinted at an obscure connection between the news of the king's death and Galileo's *Starry Messenger*.

Gassendi related that in early 1610, Peiresc had received a recent Spanish almanac predicting Henri's death, had alerted the president of the Parliament of Aix and a prominent astronomer, but that the king himself had scorned the warning. "Even the king's ambassadors, and namely his agent in Venice, that best of men, Jean Bochart de Champigny, had already warned the king of it. And this was also proved by the fact that every sailor from Marseilles who had been in Spain in the two months [prior to the assassination] reported that a rumor had been broadcast there that the French king had just been or was just about to be killed with a sword."[132]

Gassendi's narrative further states that the Spanish astrologer Hieronymus Ollerius, having hit more or less by chance in 1609 on the king's imminent death, sought to profit from his lucky guess, and that whatever hints he had of the event derived not from the stars, and still less from evil spirits, but from knowledge of a plot hatched in Spain. No man, it appears, is a prophet in his own country. By 1611 Ollerius was under investigation by the Inquisition in Spain for his practices as a prognosticator, but in Gassendi's account, the circulation of a similar rumor about the assassination in Italy, conveyed to an indifferent Henri through an ambassadorial conduit, serves to undermine the astrological basis of the Spaniard's prediction.[133] Where the tenor of Marolles' account is tentative and nostalgic, that of Gassendi is knowing and ironic. His insistence, for example, that the archbishop's courier announced to Peiresc the "unheard of and most sorrowful death" of the king is undercut by the subsequent story of advance notice of the assassination and, more succinctly, by the verb *renuntiare* or "to retell." While Gassendi's use of *infandum adeo parricidium*, the "truly unspeakable treason," to describe an act of which so many were allegedly speaking clearly participates in the same corrosive rhetoric, it also suggests the paradox that would, at least for Henri, be the case: despite the endless advance reports of his death, the event only existed as news when the king did not.

Thus while Marolles and his fellow villagers enjoyed on the evening of May 14 a funereal prognostication of their future sorrow—the incident coming into being, for them, when the news reached them—matters were clearly otherwise for Henri. No narrative forerunner of the assassination could substitute for, much less deter, the incident itself. At the very most, Henri's knowledge of the news of his death was in that brief interval between the stabbing and his death—a moment in which he allegedly uttered no more than the obvious and the obviously erroneous, "I am wounded; it's nothing!"—a pared-down version of what every French sailor returning from Spain had known for months, that the king of France had just died, or was just about to die, of a sword wound.

Gassendi's account is also valuable for its allusion to the role allegedly played, albeit to little effect, by the French ambassador in Venice, Jean Bochart de Champigny. There is no independent evidence either to confirm or to deny the suggestion that Bochart took such action in early 1610; a powerful political figure who would later become *premier président* of the Parliament of Paris, and a man of Gassendi's personal acquaintance, Bochart had died about a decade before the story appeared in the biography of Peiresc.[134] Unlike the Venetian theologian Paolo Sarpi, who had nothing but contempt for Bochart and referred to him as "that best of Jesuits,"[135] Gassendi appears to have been sincere, if hyperbolic, in his admiration for this

"best of men;" that the episode was intended at once to cast doubt on astrological predictions of the assassination and to portray a devoted and alert
servant of the crown is certain. It is noteworthy in this regard that Marolles'
version explicitly names a Carthusian colleague of the newsmongering Father
Durand, "Father Bochart, brother of the *premier président* [Jean Bochart] de
Champigny," as one of the prominent men encountered in the course of an
otherwise uneventful childhood, as if to suggest a muted and much attenu-
ated version of the explicit connections that Gassendi had drawn between
diplomatic activity and foreknowledge of the assassination.[136]

But it is above all significant that Gassendi located the site of such foreknowledge in Venice. He returned to the issue of what else could then be
learned in Venice, and how, in the very next paragraph of his biography.

> Peiresc then received a letter written [in Venice] by [Lorenzo] Pigno-
> ria on the third day of that month, alerting him to the fact that
> Galileo, using the recently invented telescope, had discovered certain
> great and marvelous sights in the heavens, and above all four new
> planets, which he had named the "Medicean stars," wandering
> about Jupiter. As is known, early in 1609, when he was handling
> different sorts of lenses to test out the effects, Jacobus Metius of
> Alkmaar fell by chance on the combination of a convex and a con-
> cave glass, by means of which—especially once a tube has been
> added—small things seem great, and distant things, near to the
> observer. Thus the first discovery of the telescope must be granted
> to [Metius], though Giambattista della Porta had already publicized
> a somewhat similar one. In truth, on the basis of just the rumor that
> had come to him of the invention, Galileo himself began to think
> first about the cause of the telescope's effects, and then about the
> means of making it, such that after various efforts and attempts, he
> came at last to the construction of one that was most exact.[137]

What Gassendi had to say about the invention of the telescope conforms,
more or less, to the version promulgated four years earlier by René Descartes
in his *Dioptrique*, the upshot of which was the not entirely accurate depiction
of Jacob Metius as unlearned, and of his discovery as a matter of chance, "to
the disgrace of our sciences."[138] It is possible that Gassendi was following the
views of Peiresc himself, who insisted in a metaphor about newsgathering in

1626, in a letter known to his biographer, that Metius was the *vray inventeur primitif* of the telescope.[139]

Gassendi's description of Metius's habit of testing out different combinations of lenses as something that resulted by mere chance in the instrument that was the Dutch telescope is puzzling; so, too, is his insistence on the date of 1609. All other advocates for Metius's claim as *primus inventor* had declared that his discovery dated to 1606, 1607, or 1608; Metius himself, in the deposition made in October 1608 to Cornelis d'Aerssen, secretary of the States-General, alleged that he had been experimenting with glasses for about two years. Descartes' *Dioptrique* of 1637 described the event as one that had occurred "around thirty years ago," and the brother of the protagonist, in a work published in 1614, stated that the discovery had been made "about six years" earlier.[140] In view of the fact that publications describing the device dated from late September 1608, Gassendi's chronology did little, or rather nothing at all, to support such a claim.

It is likely that Gassendi's source for this trivial error is Girolamo Sirtori, whose treatise on the telescope, published in 1618, includes this same mistake, surely typographical in origin. More crucially, Sirtori's description of his own quest to learn the secret of the device he examined at a silversmith's in Milan in May 1609 is the story of a frustrating trip from there to Venice, and then a more successful one to Spain, where he encountered an old man whose rusted lens-making tools appeared to predate by some decades the Dutch discovery. Sirtori identified his Spanish acquaintance as "a brother of Roget of Burgundy, who at one time lived near Barcelona, a man of great diligence, who first introduced and established the art [of lens-making] in Spain." A prominent family of lens makers by this name did, in fact, live in Barcelona at this time, and there are ambiguous references to *olleres de llarga vista*, "glasses for seeing far," in wills from that city dating to 1593, 1608, and 1613. While the Rogets were not actually from Burgundy—a capacious term that might have referred to an area either in the Spanish Netherlands or in France—the accidental resemblance of these protagonists to the queen's goldsmith, the Flemish Nicolas Roger, is worth note.[141]

Gassendi's story is thus most remarkable for the insignificant error it appropriates from Sirtori, and for the significant argument it suppresses. Consider its structural similarities to the one which immediately preceded it, the tale of the advance reports of Henri's assassination. As Gassendi would have it, Metius, like Ollerius of Barcelona, but unlike Roget of Barcelona, had arrived at a crucial bit of information in 1609 by chance, and had sought to profit from it; less interested and more accurate versions of both the plot and the invention would soon emerge from Venice through the efforts of Bochart and Galileo. That Bochart's knowledge of the plan to assassinate Henri did not derive from Ollerius's *Almanac* and presumably predated it—he "had already warned the king of it"—means, among other things, that it is *not* the exact counterpart to Galileo's acquaintance with "just the rumor" from the Netherlands, a detail which Gassendi may have attempted to minimize by his manipulation of the overall cast and chronology of the episode. The eventual appearance of the "real" news of Henri's death and the Medici Stars in May 1610 offered a kind of retroactive legitimacy to Bochart and Galileo, though it seems clear that in both cases such prescience appeared suspect. The end result of the two events, in any case, was a new and out of place Medici regime on earth, in the guise of the unpopular regency of Marie de Médicis, and a rather more manageable one in the heavens, conveniently restricted to the orbit about Jupiter.

A kind of background noise competes with Gassendi's story: the effort to involve Venice as the locus of foresight in the narrative of the political plot against Henri IV without somehow inculpating the diplomatic personnel was clearly a strenuous one. The king's ambassadors, singularly unpersuasive in this most crucial of instances, are nonetheless called *regii oratores*, and the "very best" of them, a man who might just as well have done nothing, is portrayed as *Venetiis agens*, variously rendered as "an agent in Venice," or someone "acting in Venice." That the loyal Bochart's admonition was ratified by sailors returning to Marseilles from Spain is troubling; contemporary documents routinely note that criminals were sent to row in his majesty's galleys at Marseilles.[142] This is not to construct a conspiracy theory about the royal assassination—though many flourished in this period—and still less to question Gassendi's faith in Bochart. It does suggest, however, that the king's agent in Venice figures above all as a means of suturing together two narratives that Gassendi and his peers saw as surely, if obscurely, associated. For Bochart actually did have some connection to the latter story, that of the telescope's emergence in the Veneto; he examined one such instrument, a rival to Galileo's version, just as the latter sought to gain a monopoly for its development.

Extant documents suggest that Galileo heard a rumor of the Dutch invention sometime before the end of 1608, but that the then common supposition that telescopic devices involved a lens and mirror combination delayed his manufacture and perfection of the refracting telescope.[143] By late spring 1609, however, he might have received either an accurate description of the instrument—an event which he would present, somewhat misleadingly, in the *Starry Messenger* as Jacques Badovere's confirmation of the rumor from The Hague—or an actual prototype from France, as is related in a letter written in late August 1609 by Giovanni Bartoli, secretary to the Tuscan resident in Venice.[144] As Peiresc's friend Lorenzo Pignoria and Bartoli both reported, a "foreigner" had appeared that July in Venice with a telescopic device;[145] Galileo was ready to meet the challenge, and he, too, soon demonstrated his instrument to the *Signoria*.[146] Though Bartoli stated that it was rumored that the importunate newcomer was told by Paolo Sarpi that he was wasting his time in Venice, another rival, a Frenchman with a spyglass, appears to have immediately emerged.[147]

Bartoli's letters of September 5, September 26, October 3, October 17, October 24, October 31, and November 7, 1609, describe his efforts to procure a telescope for the Tuscan secretary of state.[148] Because Galileo was twenty-five miles away in Padua, under the pressing obligation to manufacture twelve telescopes for the Venetian Senate, and unable to share the particulars of the design with others, Bartoli turned to the Frenchman. Though Bartoli gradually came to regard these instruments as a costly prop for some joke—*per riuscire una burla*[149]—he reported in mid-October that the telescope he had tried to buy was destined for the "Ambassador from France," Jean Bochart de Champigny.[150]

Given the great antipathy between Paolo Sarpi and Bochart—the French diplomat having publicly and dramatically exposed, with the aid of Pierre Coton, the Venetian theologian's philo-Protestantism in mid-September 1609[151]—and the fact that Sarpi had tried to protect Galileo's monopoly on the production of telescopes in Venice, it is reasonable to assume that the French instrument maker had benefited from this hostility when he sought Bochart as a customer. Bartoli, for his part, portrayed the vendor rather ironically as the blustering representative of some specialized French knowledge, and his arcane but eminently available information as a burlesque version of a state secret: "Galileo's *secreto* or telescope is now being publicly sold by a certain Frenchman, who turns them out here as a secret of France, and not of Galileo [*come secreto di Francia, non del Galilei*]; and perhaps it isn't the same thing, this one being worth so little money."[152]

Though Gassendi's aim as Peiresc's biographer is to demonstrate his subject's proximity to the protagonists in the Venetian drama—a claim borne out by Peiresc's rapid acquisition of the *Starry Messenger* and several tele-scopes[153]—what appears most critical to the grand narratives of the spring of 1610 is the constant association of the royal assassination in France with some aspect of the no less complicated story of the telescope's development in the Netherlands and in Italy. Though the French would have to claim François Ravaillac as one of their own, the rumors of advance notice of the murder in cities and countries far removed from Paris provided some way of implying that the disastrous but foreseeable plot originated elsewhere, and was exe-cuted when, by an unlucky chance, the services of a traitorous local individual were secured. Conversely, the frequent suggestion that the French had, if not the Dutch or Galilean telescope, crucial components or a reasonable facsimile of it, converts the actual inventions into a series of lucky guesses that just happened to have taken place outside of France. Thus it is not at all surpris-ing to find, in the small world of the European intelligentsia and the even smaller web in which writers as diverse as Howell, Matthieu, and Gassendi tried to fit them, that Bochart's secretary in Venice was Jean Baptiste Duval, the man who had a few years earlier provided Marie de Médicis with an Italian translation of the treatise on the telescopic properties of concave mir-rors, and who sought out such glasses for her in Italy in January 1609, pre-cisely when her short-sighted spouse brushed aside the marvelous Dutch device proffered by Pierre Jeannin.

None of this implies an actual connection between trade secrets in optics and *arcana imperii*, though Galileo's close friend Traiano Boccalini, among others, had gestured toward the sometime convergence of technical and polit-ical insights into state secrets in his satirical *News from Parnassus*. And Boc-calini's suggestion that understanding of optics—roughly, the proper sort of perspective—led to or coexisted with timely political knowledge was echoed, albeit in more hesitant tones, by his friend Angelo Grillo. In a letter written shortly after the invention of the telescope, Grillo had described those armed with Galileo's glass as civil servants entrusted with the secrets and correspon-dence of rulers, "secretaries of the moon and of the stars," and had stated that the instrument had opened a "new school of almost indecent curiosity," as if the device would inevitably fall into the wrong hands.[154] Narratives such as those offered by Matthieu and Gassendi, where France's political loss is presented as a missed opportunity in technology, appear to exploit this impression: the chance discovery that allowed "some Netherlander" to see remarkable distances was reconfigured as the sinister view of the approaching murder from Flanders or from Spain. The accounts of Pasquier and of Marolles, while differing greatly in content and in tone, share a movement away from the technological determinism implied by the conflation of the epics of the Dutch invention and the French assassination. Their emphasis on the political incident as a rapidly and widely disseminated story shifts the focus to the emergence of a vast and news-hungry public, Pasquier's coranto-readers and Marolles' rural rumormongers and stargazers. And yet the erasure of the optical motif is incomplete, for they are consumers of events well beyond their immediate horizon, people for whom the telescopic view is a natural perspective.

CHAPTER 3

Galileo *Gazzettante*

In late 1604, physician and future canon Leonardo Tedeschi sent Galileo Galilei a treatise on the New Star, a subject often connected with news because of the traditional association of a related phenomenon, the comet, with plague, famine, war, and regime change. Leonardo avoided all such speculations. His tract began with the pious Aristotelian dictum that it was better and more agreeable to have even superficial and imperfect knowledge of superior and noble matters than fuller and more solid acquaintance with things here below, and he maintained that the new star, far from generating "killings, plagues, famines, horrible winds and earthquakes," had been distinguished by stasis, "a very tranquil and constant serenity in the air," windless and most temperate weather.[1]

Leonardo's brother Giandomenico, an aspiring newswriter, evidently preferred more excitement, and the subject of this chapter is the depiction of the hazardous aspects of his sometime profession in terms of contemporaneous developments in astronomy and in optics. Those *novellisti* portrayed in such fashion generally had no real connection with either field, and this would appear to be true of Giandomenico, whose interests lay closer to statecraft and to letters, who was an active member of the Philharmonic Academy of Verona, and whose professional aim was to produce lively accounts of current political and cultural events.[2] In commenting in 1614 on a skirmish subsequent to the recent publication of the *Dictionary of the Italian Language*, a pamphlet war between supporters and detractors of the Tuscan dialect, for instance, Giandomenico cheerfully predicted, despite his own involvement in this conflict, that "what had begun with pens would finish with daggers."[3] It did not, but squabbles over the *Dictionary* or "that insolent and awful book" generated interesting rumors for the next few years.[4] In this same period the Venetian Republic awarded Giandomenico the prestigious title of "Knight of San Marco," and he began to use his connections to prominent individuals to compose a gazette of some sort in Verona.[5]

Two of the men he approached for this venture, Guido Bentivoglio, the papal nuncio in Brussels, and Angelo Grillo, a Benedictine priest and poet then in Mantua and Venice, had a certain involvement with Galileo and strong interests in telescopic exploration, and the association of the aims and fates of journalists and astronomers was a pronounced feature of their perspective. As I will suggest below, the wild success of Traiano Boccalini, and the implicit pairing of his *News from Parnassus* with Galileo's *Starry Messenger* and *Letters on Sunspots* provided both a template and, in time, a cautionary tale for those who, like Giandomenico, were determined to convert piecemeal revelations about current events into status and profit. Thus while we have almost no material from Tedeschi's rather brief career as a journalist, the kinds of concerns raised about both Galileo and Boccalini, and above all the persistent coupling of their activities, offer some context for the aspirations and risks of the most ambitious newswriters in those years.

"What will you not manage, finally, with your stubborn importunity?" wrote Bentivoglio to Giandomenico in 1614. "Here, in spite of my many tasks, and above all in spite of decorum, is a long letter from me, but I really shouldn't be responding to your trifling gazettes from Verona with my heroic news from Flanders." Bentivoglio did supply a great deal of information about current events, ending the same letter with a brusque gesture to the hierarchy he had momentarily abandoned: "I'll return to my role as *nuncius*, and I leave you to yours as *gazzettante*."[6] Angelo Grillo seems to have been more hesitant; while many of his own letters had already appeared in print, these slightly dated and very circumspect missives did not offer much in the way of news. His reply to Giandomenico suggests, nonetheless, the offer of covert collaboration: the newswriter might publish what Grillo disclosed. "In closing your letter, you call for me to stroll about the public piazza with you. I don't go out, Sir, because it is a hazardous season. But I consent."[7]

Letters of this period comment on Giandomenico's tendency to talk too freely, on his intractable desire to meddle in political affairs, and on his strong interest in Tacitus, then considered suspect, and in Niccolò Machiavelli, whose prohibited writings circulated in clandestine fashion.[8] Between 1616 and 1618, these habits, in connection with his gazettes, led to his imprisonment in Venice. The most dramatic change in Giandomenico's behavior during and subsequent to his incarceration—and one noted with some embarrassment by his patrons and friends, though perhaps encouraged by his respectable brother, Canon Leonardo Tedeschi—was a ceaseless flow of sonnets, composed without books, ink, or paper, evidently a sort of substitution for his "trifling gazettes."[9] "What do you have to say about Signor Tedeschi?" Bentivoglio asked another clergyman, the canon Paolo Gualdo, in 1618. "What do you think about that dark and gloomy Parnassus where this new swan of ours was suddenly born?"[10] As this chapter will suggest, another equally crucial influence, for Tedeschi and those monitoring the emerging news industry was that overly bright Parnassus where Traiano Boccalini had first focused on the telescope.

Spyglasses

Consider, by way of background, the early association of optical instruments with reportage in the correspondence of Tedeschi's confidant Grillo. In writing to several friends and a relative about the telescope very soon after its invention, Grillo referred to the device as "a secretary of the moon," and as "a most excellent spy."[11] He insisted around 1610 or 1611, not very accurately, on the ordinariness of the brand-new telescope, and drew on the familiar trope of friendship as an uncommon magnifier of virtues when he sent lenses for assembly into two of the Dutch instruments to Tomaso Arigucci, a scholarly acquaintance in Perugia.[12] But his contemporaneous allusions to the same device were more interesting and significant when they concerned another friend, Francesco Maria Vialardi, a translator, agent and adventurer whose newsletters for a very elite clientele were complemented by his extensive travels, his activities as an informer, his role as a "reader of mathematics" in a fledgling academy in Rome, and his bout in prison for possession of incriminating texts.[13] On receiving a letter after several years' silence from Vialardi, Grillo responded with celebrated lines from the *Aeneid*,

> *Verane te facies, verus mihi nuncius affers?* "Do you present yourself to me as a real form, a true messenger?" Is this or is this not Your Lordship? In truth I am somewhat doubtful. And why shouldn't I be, after I have been gone from Rome for such a long time, since you, who know how to find men who are off the map, and even to spy out what is being done on the moon, didn't bother to look for

me around Padua, or here in San Benedetto in Mantua, where I
have been for three years?[14]

Some of the differences in valence in Grillo's letters are a function of
social context. In writing to Arigucci in Perugia, Grillo implicitly taxed this
friend with a certain belatedness strongly associated with provinciality; it
began with the observation that the latter's missive had taken seventy days to
reach him, turned to assurances that the telescope was already old news to
those around Mantua, and concluded with the generous gift of four lenses
sent under separate cover for eventual assembly as a pair of instruments. The
reply to the inextant letter from the better-informed Vialardi is, on the other
hand, probably a reaction to the agent's claim to have used a high-quality
telescope in Rome; such an opportunity would have been provided by Cardinal Ferdinando Gonzaga, the recipient of Vialardi's newsletters. This patron
enjoyed close ties both to the astronomer Giovanni Antonio Magini, who
was observing the moon's craters in June 1610, and to Cardinal Scipione
Borghese, who had in that same month replaced an early and inferior telescope from the Netherlands with a superior one made by Galileo.[15] In early
1612, moreover, Vialardi would inform Gonzaga about charges that the
astronomer had not discovered Jupiter's satellites, but rather had appropriated them from a much earlier book by a Dutchman. While the tenor of
this news is difficult to judge, it was probably reported as an absurd rather
than likely allegation, for the assertion seems modeled on the more familiar
one of his theft of the telescope, and is undercut by the significant differences
between the instrument from the Netherlands and that provided by Galileo.[16]
     Thus while what Grillo had to say about the telescope and its revelations
in the letter to Arigucci appears to be drawn from the recently published
Starry Messenger, his reply to Vialardi suggests that here it was the agent who
had taken on the role of *nuncius* or messenger, and Grillo who was the tardy
recipient of news. Just as Arigucci was, at least in Grillo's letters, the very
image of belated knowledge, so Vialardi appears the icon of current information in his correspondence.

News from Parnassus

What made Vialardi so valuable to his patrons, and so intriguing to Grillo,
was the fact that he disclosed his findings to a very restricted public, patrons

who in their turn offered access to novel inventions such as the telescope.[17] The motif of the spyglass would be deployed in somewhat different terms as it and its revelations became more familiar; such changes involve less emphasis on the solitary tasks of spying out and ensuring a limited and strategic diffusion of timely information, and considerably more attention to the unsettling possibility of widespread access to such knowledge.[18]

In tandem with this development was a shift of focus from the currency of the news to the more enduring outlook that seemed to accompany each perishable item. Put differently, it was much less the impact of particular telescopic observations—especially since these were notoriously open to appropriation and reinterpretation by proponents of other world systems— than the suggestion of a long-term agenda directed to a broad public.[19] Though that agenda fell squarely within the ambit of natural philosophy, the steady transition from news of particular findings to the recognizable orientation of a Copernican was soon conceived as a political and cultural matter.[20] As if to suggest that the newsworthiness of Galileo's first discoveries was their least important feature, and their still obscure connection to political culture the most important one, the initial pairing of the author of the *Starry Messenger* with the satirical newswriter Traiano Boccalini increasingly included a secondary couple, Machiavelli and Tacitus. In this sense, the very real forward momentum of a rapidly developing field was accompanied by a retrograde motion that did nothing to compromise or to question those observations, but served instead to signal the larger backdrop against which such developments occurred.

Boccalini's extraordinarily popular *News from Parnassus*, a series of almost three hundred satirical bulletins, was published in Venice in installments in 1612, 1613, and 1614, and quickly translated to most European vernaculars. Though begun around 1605 and originally offered in the summer of 1609 in a restricted format and in manuscript to Cardinal Scipione Borghese, who was both the recipient of Boccalini's genuine newsletters and, thanks to the efforts of Guido Bentivoglio, one of the first men in Italy to obtain a Dutch telescope, the earliest version of the *News from Parnassus* failed to draw on any optical motifs.[21]

This emphasis appeared in fall 1612, for the first printed *ragguaglio* or "bulletin" opened with an elaborate discussion of the sorts of glasses courtiers use. Early modern rulers, particularly in Italy, did supply their courtiers with glasses, but Boccalini's descriptions would have been recognizable adaptations of the kinds of claims routinely made for pretelescopic devices and eventually for the Dutch instrument itself.²² Thus while several mid- to late sixteenth-century authors had described various optical arrangements or "perspectives" that allowed the viewer to light up dark and relatively distant places, or to see "whatever is being done in yonder bedroom," or to recognize friends from afar, Boccalini's fictive newswriter portrayed glasses that "illuminated those salacious men, who in the furor of their libidinous ways find themselves short sighted, such that they cannot distinguish honor from vituperation, nor a friend from an enemy, nor a relative from a stranger."²³ He further praised glasses whose value lay in their ability to obscure, rather than to reveal, as for example those that allowed courtiers to engage in escapism by focusing only on the remote, or those that exaggerated the size and availability of the rewards promised them. This latter sort of eyewear was conflated with "those glasses lately invented in the Netherlands, purchased at great expense by grand personages, and distributed to courtiers."²⁴

The dismissive treatment of optical aids was to some extent prompted by a literary antecedent. In 1582 the comic Perugian poet Cesare Caporali had published a *Voyage to Parnassus* and *Reports from Parnassus*; the former text would inspire Miguel de Cervantes's *Viaje del Parnaso* in 1614, while a detail in the latter suggests that spectacles induced an ethical blindness among courtiers.²⁵ A hideous statue described by Caporali's newsmonger wears "a pair of glasses / Of extravagant and unheard of effect," the frames made of "bone of envy," the lenses fogging up around virtue, such that its wearer saw no farther than a hand span.²⁶ Boccalini's skeptical reaction to the telescope appears to have at least one historical analogue: in Tuscany, the grand duchess Cristina gave two pairs of spectacles with lenses of English glass to a visitor to the Medici court, Monsignor Bonifazio Vannozzi, in the spring of 1610, and in the summer of that year the grand duke presented the same man with a telescope. Vannozzi was delighted with the eyeglasses, or at least with their exotic origin, but he promptly gave the spyglass away: he compared the *moderno occhiale* unfavorably to some legendary telescopic mirrors, claiming that the former had some value because of the Medici provenance, but adding that Galileo's achievement had already been realized by Archimedes, and described by Giambattista della Porta.²⁷ And in 1627, even after nearly two decades of crucial astronomical discoveries, the unflattering association of the glasses that misled courtiers with "those glasses lately invented in the Netherlands" would be reiterated and regionalized in the Italian context: the moralist and cleric Secondo Lancellotti observed that "Galileo's *occhiale lungo*, which so many people use, and which makes objects appear larger than they are, might well be an emblem of Envy."[28]

Boccalini's close friendship with Galileo, Grillo, and the poet and glass-maker Girolamo Magagnati perhaps accounts for his familiarity with conventions surrounding pretelescopic instruments and for his relatively cool reception of the Dutch device. It is worth noting, in fact, that in 1601 Grillo had written a letter commenting on courtiers' reliance on eyeglasses that rendered remote objects misleadingly close, and that Magagnati, in celebrating the satellites of Jupiter or Medici stars in May 1610, referred to observation as "spying," but made no mention of the device on which this discovery depended, though he twice compared the grand duke of Tuscany to a component in those earlier pre-telescopic instruments, the concave mirror.[29]

It is also possible that in 1609 or 1610, years in which Boccalini undertook several missions for Cardinal Borghese and then settled in Rome, he used the rare but relatively inferior telescope Bentivoglio had sent from Brussels to this patron just six months after the invention emerged in The Hague. Given that a letter which Boccalini had written sometime between 1596 and 1604 had alluded favorably to *l'occhialon politico della longa vista*—roughly, a strong lens used to correct political myopia—it would appear that his initial encounter with the Dutch instrument was somehow problematic or anticlimactic.[30] Only near the end of the first installment of *News from Parnassus* did the telescope become explicitly associated with the air of political perspicacity Boccalini sought to convey. To exploit a metaphor that would have been less striking in the earlier versions of Boccalini's work, the *News from Parnassus* offered a particular political and cultural orientation, or a "perspective."[31]

The nascent optical and astronomical metaphors are clearly an elaboration of Boccalini's passing depiction of history, in his commentaries on Tacitus, as a primitive visual aid, "a bright flare for illuminating the road ahead," and (in keeping with Tacitus's own resistance to astrology) as the only legitimate sort of prognostication.[32] Thus Boccalini's reporter revealed that several powerful monarchs had lodged complaints with Apollo, the ruler of Parnassus, against Tacitus, charging that "with the seditious material of his *Annals* and *Histories* he had manufactured certain glasses" that allowed simple people to see the machinations of their sovereigns. Out of affection for Tacitus, his "first minister and greatest chronicler," Apollo reached a compromise: the fewer such spectacles allotted to pernicious rulers, the better, and their use was to be limited to political elites such as secretaries and counselors.[33] A subsequent news bulletin both reinforced the connection between the ancient texts and the telescope, and suggested that distribution of the latter was not, in fact, so restricted, for Apollo gave a benighted literary figure "a pair of those excellent glasses lately made in the furnace of Tacitus the political [thinker]" in order that he might describe the world with greater accuracy.[34]

A crucial reference to optical devices appeared toward the end of the first installment of the *News from Parnassus*, in the course of the eloquent defense offered by the most celebrated of secretaries, Niccolò Machiavelli, in support of his own works, long banned in Parnassus, as they were, though only in theory, in the Venetian Republic. Boccalini's Machiavelli claimed that the troubling political precepts in his writings, far from being his own inventions, were simply "taken from the actions of certain Princes" whose names he would have been happy to supply to Apollo, and whose unhappy legacy he had merely put into print. "And I don't know why," he continued, "it is good to adore the original as if it were holy, and to burn the copy as if execrable. And why should I be so persecuted, while the study of History, which is not just permitted, but even commended, is notorious for having the power to convert all those who attend to it with the *occhiale politico* into so many Machiavellis?"[35]

The artful Florentine went on to associate political discernment of the sort he and Tacitus seemed to supply with the study of the natural world. Rather than an activity so absorbing that it screened its practitioners from an unpleasant political reality, or from the detrimental activities of a particular patron, in Boccalini's account sustained natural observation seemed a precondition for proper political judgment.[36] "It is a blessing that people are not as simple as many imagine, as if those same men who with their supreme intelligence were able to search out the most recondite secrets of Nature somehow lacked the ability to perceive the true goals behind the doings of their Princes, even when great artifice was used to disguise them."[37]

Unlike Tacitus, Machiavelli is treated without clemency in Parnassus, though the implied equivalence between the two authors, and the facile substitution of the *occhiale politico* for the sort of discernment that was both the approach to and the product of such readings suggest a ready redundancy for the Florentine secretary. But even as he disappeared from the text, the uncanny emphasis on optical devices remained. The bulletin ended with the judges' conclusion that "the attempt to turn simple people into wary ones, and to try to make those moles, whom Nature had so prudently created blind, see something of the light, was a wish to set the whole world on fire."[38] In phrasing Machiavelli's project in terms of visual and thermal effects accomplished with either a mirror or a lens, both of which were components of two early and relatively unsatisfactory pretelescopic instruments, Boccalini did not necessarily intend a reference to Galileo, despite the latter's familiarity with such devices. A more likely candidate, if such specificity is warranted, would be Galileo's close associate, Fra Paolo Sarpi, who was known throughout Venice and beyond both for his political astuteness and for his investigations of the natural world, including that of optics.[39]

Little of this mattered, perhaps, in the aftermath of the *Starry Messenger*: to refer to an optical instrument, to gesture to "the most recondite secrets of Nature," and to allude to the doings of Princes, particularly in connection with another famous observer of the Medici regime, was to all appearances to speak of Galileo. The spectral presence of the astronomer in the *News from Parnassus* would become increasingly pronounced in the coming decades.

Something New Under the Sun

The first suggestion that Galileo's *Starry Messenger* and Boccalini's *News from Parnassus* shared any common denominator appeared in 1611 in a letter appended to a vernacular translation of Tacitus and concerned, in large part, with the ongoing quarrel over language. In defense of Sienese terms judged low and obscure by Tuscan purists, Adriano Politi, already a discreet promoter of the *Starry Messenger*, implied that the very currency of these words made them as elevated and bright as stars.[40] Such familiar lexical items, he added, were clearer than the faint stars "of the sixth magnitude, or those that astronomers call 'nebulous,' and less obscure than those of the sort that [Cesare] Caporali goes looking for with an astrolabe in his courtiers' soup."[41] The allusions are to two impressions of stars rendered obsolete by the recent emergence of the *Starry Messenger* and by the *News from Parnassus*, then about to make the transition from manuscript circulation to print: what those without telescopes had had to say of the so-called nebulous stars, and what Caporali had recounted in his *Reports from Parnassus* about courtiers forced to use astrolabes to seek out star-shaped pasta in greasy and unappetizing broth looked remote and farfetched when the works of Galileo and Boccalini appeared.[42]

For his part, Galileo was initially treated by skeptics as a reporter of sorts, and his *Starry Messenger* as an overblown bit of news. Thus Giambattista della

Porta, who had maintained privately in 1609 both that Galileo's device had been poached from one of his publications, and that it was a *coglionaria* or "load of balls," wrote in triumphant tones to the astronomer himself in 1614 that he and an ingenious collaborator were developing a new and much more powerful instrument, and that "God willing, with it we will spy out the doings [in the Empyrean], and will compose an *Empyrean Messenger*."[43] And even as friendly an onlooker as Lorenzo Pignoria would suggest to Paolo Gualdo in January 1611 that in his *Starry Messenger* the astronomer had merely proffered an invention, rather than a credible account of celestial novelties, and that the only certain news was that of his recent illness: "Galileo has fallen sick again, and this comes from a reliable source. In sum, going around carefully searching out the secrets of the heavens was always a risky business, especially if he has told some tall tales."[44]

What is less predictable than Galileo's popular connection with journalism is Boccalini's association with developments in astronomy. One of his bids for patronage, a short letter written in October 1612 to Francesco Maria della Rovere, duke of Urbino, alludes to the latest in astronomical news, Galileo's quarrel with the Jesuit Christoph Scheiner over the nature of sunspots. In sending his recently published *News from Parnassus* to the duke, Boccalini wrote, "If I were content for [my bulletins] to be seen only by my peers, I know perfectly well that I would be making the mistake of showing paintings to cobblers in order to get their views on color."[45] The remark evokes Pliny's well-known story of the classical painter Apelles, who hid behind his canvases in order to have the candid views of passersby, one of whom was a cobbler. This man correctly criticized Apelles' depiction of sandals, but soon overstepped the mark in commenting on the way in which legs had been rendered, giving rise to the proverbial recommendation that the cobbler stick to footwear.

Boccalini's allusion is not quite accurate—the point of Pliny's story is that Apelles is anything but the peer of the cobbler, and the issue of color does not arise in their exchange—but the deformity may be due to his interest in the tale's deployment in the context of the solar debate. The Jesuit Scheiner had published his tracts in early 1612 as "Apelles hiding behind the canvas," a pseudonym much derided by Galileo's supporters, and perhaps by the Italian astronomer himself, whose references to *il finto Apelle* could be read as either "the masked Apelles" or "the fake Apelles."[46] Galileo was inevitably portrayed as "the real Apelles," and his manuscript responses to his rival were particularly well known in the Veneto, where Angelo Grillo managed to discuss them, and even have them read by the conservative scholar Andrea Chiocco to the Philharmonic Academy in Verona in the spring or summer of 1612 without making clear whether he himself subscribed to such views, where another solar observer took on the artsy pseudonym of "Protogenes" and entered the discussion in this same season, and where the time-consuming task of copying Galileo's text and images of the sunspots in September 1612 meant that the letter intended for *il finto Apelle* reached its addressee nearly two months after its composition.[47] Boccalini's ambiguous image of the ancient artist in October 1612 suggests at the very least that he followed something of this quarrel, and that his sympathies lay with "the real Apelles."

Boccalini's gratuitous depiction of an eclipsed and cloud-covered Apollo, trailing a "stinking fog," in the third volume of the *News from Parnassus*, might likewise be a passing allusion to the quarrel between Galileo and Scheiner.[48] In this same volume he also elaborated the association of Apelles with Tacitus, whose candid chronicle of political misdeeds was portrayed as an optical instrument.[49] The Roman historian, described by Apollo as "my political Apelles," was characterized by his realistic representations of the protagonists of his works: "In the canvases of his *Annals*, with the marvelous paintbrush of his pen, Tacitus depicted from life."[50]

No News

Boccalini's own fate, and especially its narrative afterlife, seemed to reinforce cautionary tales concerning news that was overly candid and indiscriminately distributed. His sudden death at the height of his success in late 1613 was popularly assumed to have been the result of poisoning by Spanish agents. By the mid-seventeenth century, he was said to have succumbed to a fatal nocturnal beating with a sandbag, a picturesque form of the harassment to which Euclid had been subjected in the *News from Parnassus* for having exposed, through the publication of a geometrical theorem, the furtive pursuit of others' wealth as the focal point of all princes and all private citizens.[51]

Galileo, too, was occasionally described as a victim of Spanish repression; thus Rudolph II's librarian, Martin Hasdale, writing to the astronomer from the imperial court in Prague soon after the publication of the *Starry Messenger*, related that "the Spanish believe that the book of Your Lordship must be suppressed, as they allege it is harmful to religion, which is the cloak under which they legitimize every sort of thuggery." The claim was made in a private letter, and perhaps undercut by the fact that Hasdale then described the Augsburg humanist Marcus Welser as "wholly Spanish, and not very friendly to Venice." While the perception of Welser as an enemy of that republic, and eventually as the author of a polemical anti-Venetian tract of 1612 was (and still is) common, the correlative suggestion that he favored the suppression of Galileo's works would have lost force after his quite open support of the *Letters on Sunspots*.[52]

It is their odd relationship to that crucial feature of news, timeliness, that best served to decrease the substantive differences between the publications of Galileo and Boccalini. As startling as the disclosures of the *Starry Messenger* were, the long-term thrust of the pamphlet was to deprive astrologers, in particular, of their traditional role as forecasters of coming events, an eventuality encoded in one of a series of provisional titles, *Astronomical Announcement to Students of the Heavens*, and readily recognized by several onlookers just before the work emerged from the press. This shift from the traditional preeminence of that sort of forecaster to the newfound importance of the astronomical reporter, however, was only temporary. Galileo's early emphasis on priority and on exclusivity, so pronounced in the *Starry Messenger* and in the two anagrammatic disclosures concerning the apparent shape of Saturn and the moon-like phases of Venus, was already slightly diminished during the dispute over the sunspots in 1612.[53] Here, for instance, he proved uncharacteristically forthcoming about the observational arrangement that would in theory allow any reader equipped with a telescope and a dark room to project and trace the solar markings, almost certainly because he believed, to his subsequent regret, that after his *Letters on Sunspots* these particular phenomena could offer little more in the way of novelties.[54] His gesture to the misinterpreted sunspots of the distant past and his insistence on their ceaseless movement about the solar body for the foreseeable future seemed likewise to evacuate the issue of its newsworthiness, as did the pace with which he took up this study in late 1611 and terminated it in early 1613.[55]

It is true that the relative complexity and length of the *Letters on Sunspots*, and its extensive preprint censorship, precluded a publication as extraordinarily rapid as that of the *Starry Messenger*, and that Galileo's burdens at the Medici court, advancing age, declining eyesight, and ill health would have made him increasingly less able to undertake lengthy astronomical observations, and finally that further telescopic discoveries were unlikely, given the optical power of even the best of his instruments.[56] But the most significant factor was Galileo's wish, as ill advised as it turned out to be, to present not a breathless series of unconnected telescopic novelties, but rather a whole scientific program, a full-fledged orientation to the Copernican world system. Thus while much of the authority of that presentation rested on the newsworthy quality of Galileo's earlier findings, the emphasis had inevitably shifted from the timeliness and exclusivity of particular discoveries and disclosures, all of which were becoming both increasingly dated and available for replication and refinement by other viewers, to the long-term and influential outlook that put his observations, and by implication those of others, into perspective.[57] The fact that his difficulties with the Roman Inquisition began with the *Letters on Sunspots*, and involved both that work's treatment of Scripture and its Copernican agenda, likewise indicates that institution's concern with something other than solar ephemera and priority claims.[58]

Boccalini's bulletins were, by comparison, *never* newsworthy. The *News from Parnassus* did refer, here and there, to current events, but the various topics taken up by that strange assembly of mythological characters, of ancient, medieval, and early modern writers, and of still living personages would have served as a poor substitute for the newssheets of the period. The fact that Boccalini did not have a strong political agenda, moreover, further deprived his work of the topical fervor that animated the private correspondence of many of his peers.[59] What the *News from Parnassus* did offer, and with some reliability, was a kind of orientation, a sort of posture for the audience to adopt in the face of political and cultural developments yet to occur. As would be the case in Galileo's later writings, especially the *Assayer* and the *Dialogue Concerning the Two Chief World Systems*, this perspective was that of the well-educated, skeptical, and debonair reader, one well disposed to both the Veneto and Tuscany, somewhat anticlerical but not openly irreligious, relatively intolerant of intellectual or political oppression, and when there was no other remedy for such evils, appreciative of that last resort, the caustic style. This orientation, more than any other attribute, made Boccalini's news bulletins appeal to Italian, English, French, Spanish, and German readers for many decades after the original publication.

The notion that Galileo and Boccalini offered readers something closer to a particular perspective than to a series of novel announcements has its analogue, moreover, in the way many early modern thinkers assessed Tacitus, often understood as the presentable double of Machiavelli.[60] Such writers favored and sometimes imitated a curt, slightly corrosive style, emphasized Tacitus's ability to identify the private motives behind public events, and in some cases reduced his texts to a set of precepts, variously portable to different historical contexts.[61] His work offered a *ciencia politica*, as the Spanish scholar Baltasar Álamos de Barrientos put it in 1614 in his influential translation and commentary on Tacitus, a system explicitly modeled on the practices of Hippocrates and Ptolemy, whose close observation of particulars, he argued, had produced aphorisms of predictive value for the well-being of the body politic.[62]

This piecemeal approach acquired its critics: John Milton observed that such scholars "cut Tacitus into slivers and steaks."[63] It also generated a significant textual apparatus, as the emphasis on the timeless currency of Tacitus meant that translators needed at once to deploy an anachronistic vocabulary, and to justify those tendencies. Discussions of these maneuvers were especially protracted in Italy, where they merged with the contemporaneous debate over Tuscan cultural hegemony. The Sienese scholar Politi, who in his adaptation of the Florentine *Dictionary of the Italian Language* had defined the new word *gazzetta* as "a brief commentary on public events," or "ephemerides," in 1613, insisted on the prominent place that reportage had enjoyed even in imperial Rome some five years later in a Venetian edition of Tacitus's works. [64] No longer portrayed exclusively in terms of the recent past of civic affairs or the near future of astrological prognostication, the *gazzetta* was to be surveyed critically, and with that detached and skeptical perspective elsewhere associated with Boccalini and Galileo:

> The practice of writing news for pay, by persons who make a particular profession of it, and who maintain themselves in this business as if by a trade, is something that arose during my days at the court of Rome, and in other celebrated cities of Europe. Thirty years ago it was practiced only by a few secretaries of princes, and by statesmen for use by their own patrons, and by their important friends. And the word *gazzetta* is well suited to [Tacitus's] material because the *menanti*, as the practitioners of this craft are called in Rome, tend like *gazze* [magpies], in order to fill up a page, to make a great clamor by writing up a lot of chattering, and sometimes absurdities, in order to appear that they deserve the fee that they request.[65]

For Politi, the empty chatter of the early modern newswriter had its analogue in contemporary historiography, likewise a venal pursuit. Apart from the Florentine statesman and historian Francesco Guicciardini, who would have been "a true friend of the true, had the printing press not refashioned his pen," Politi argued, "few authors would be found, who at least in their descriptions of princes' deeds, did not have their eyes more on the rewards for their adulation than on writing, or not suppressing, the truth."[66] This assertion, first published in Politi's translation of 1604, appears to anticipate the perspective of Boccalini's Machiavelli, who had stated in defense of his own censored works that princes interested in manipulating their subjects needed to render them oafish and uncultured by "banning *buone lettere* [fine literature], which is what turns blind minds into so many Arguses."[67] Only the threat of lasting revelations of the sort "our writer" had made, Politi continued, would check the all powerful; "the fear of seeing the still living stain, and the infamy of his own misdeeds" was the sole deterrent.

That newsless present inhabited by Boccalini and by the later Galileo thus emerges in Politi's work as well: the epoch appears animated chiefly by the skeptical gaze of Tacitus and his followers. Gazettes are filled with trivialities, and most contemporaries' histories are mendacious hagiographies of still-living rulers; more crucially, the vivid disclosures of "our writer" prevent the most powerful, "entirely without superiors, and with no restraints but reason and the desire Nature has given to all for a good reputation after death," from acting on the worst of their impulses.[68] What is significant in this argument, otherwise a transparent fantasy about the continued relevance and coercive power of Tacitus, is the covert allusion to the unmentionable Machiavelli. "Macchiavello," as it was often written, is the name princes fear in the "still-living stain" or *macchia* of their misdeeds, and the compromised Guicciardini, described as *vero amico del vero*, would better have been portrayed as *vero amico del Macchiavello*.

Burlesque Mysteries

One of the most interesting meditations on the fate of Boccalini came some thirty years after his death when Cardinal Bentivoglio offered a conspicuous pairing of the writer with Galileo in his *Memoirs*. In describing his own life as a young man in Rome around 1600, Bentivoglio noted that his study of history was accompanied by some prior knowledge of geography, and he added that he had already had private lessons on a traditional cosmological text, the *Sphere* of Sacrobosco, with Galileo in Padua several years earlier.[69] Gesturing to Galileo's condemnation in 1633, and to his own role as an

unwilling participant in that process, Bentivoglio observed that the astronomer's difficulties "had been brought on himself for having wanted to publish
in print his new opinions about the movement of the earth."[70]

A peculiar feature of his comments—the sudden retrograde movement
from Rome in 1600 to Padua around 1595—is explained by the cardinal's
immediate turn to Boccalini, whose teaching is presented as an extension of
that of Galileo. The satirist was introduced as his geography teacher in Rome,
as "the great anatomist and dissector of Tacitus," and as one who had "transferred his soul, so to speak, into that imagined King Apollo of his." Bentivoglio further stated that this Tacitean doctrine permeated the "imaginary
gazzettante" of the *News from Parnassus*, "as costly as these burlesque mysteries were to him."[71]

Bentivoglio's discussion ended with bland remarks about the dangers run
by great minds unescorted by judgment and integrity, but what is crucial
here, first of all, is his determination to see Galileo and Boccalini in terms of
each other. In this reading, Galileo, like the satirist, was characterized by an
improvident willingness to publicize matters best left unprinted; Boccalini,
like the astronomer, distinguished himself with a perspective more properly
solar than terrestrial, and exposed such "burlesque mysteries" at his own
peril. There is no suggestion that Bentivoglio necessarily disagreed with the
views of either man: indeed, his positive emphasis on the private character of
his lessons from them militates against such an interpretation. In his view,
however, both had adopted the more public role of *gazzettante* with disastrous consequences.

More intriguingly, this depiction of Boccalini bears a family likeness to
the self-portrait of another more famous sitter, Machiavelli. In his now-
famous letter to Francesco Vettori, which circulated in the first half of the
seventeenth century, if at all, only as a manuscript copy of a lost original, the
exiled Machiavelli described his activities shortly after the fall of the Florentine Republic, stating that a day of rustic pastimes was inevitably followed by
an evening where "clothed in royal and curial robes," he entered "the ancient
courts of ancient men" to converse with them. [72] The letter is remarkable for
its insistence on ritual; apart from the fastidious garments and the predictable
commerce with classical authors, Machiavelli presented himself as the ultimate consumer of texts—"I feed on the food that is mine alone"—and as
one whose rapport with such writers was so strong that he temporarily
assumed their identities and abandoned his own: "I transfer myself entirely
into them."[73]

Bentivoglio's portrait of Boccalini deploys these features, albeit in slightly more disturbing fashion. The startling graphic detail of *il grande anatomista e minuzzatore di Tacito*—the great anatomist and dissector, or rather the butcher boy of Tacitus—registers only after the subsequent disclosure that Boccalini, thus fortified with the ancient author, "transferred his soul" into the fiction that was "King Apollo."[74] Whereas Machiavelli consumed Livy in order to "transfer [himself] entirely" into him and to prepare his *Discourses* on that author, the suggestion here is that Boccalini digested Tacitus so that he might practice a sort of ventriloquism—literally, speaking from the stomach—in his guise as the fake Apollo. Cardinal Bentivoglio's unease with these details, entirely understandable, is signaled by his addition of "so to speak," as if to sanitize the whole portrait, or to convert it into a palatable vignette of the humanist reader and writer, a move somewhat undone by Milton's allusion to commentators' tendency to "cut Tacitus into slivers and steaks." Bentivoglio's repeated references to the *News from Parnassus* as "burlesque *mysteries*" effectively serve to narrow the gap between Boccalini's practices and those of a cult.

There is nothing quite so strange in Angelo Grillo's assessment of Boccalini and Galileo, though this author also associated the two figures, and made a slight gesture to the likes of Tacitus and Machiavelli. Of the latter pair, Grillo offered little explicit commentary, apart from warning the newswriter Giandomenico Tedeschi that Machiavelli was a "monster" and "truly a stain" or *macchia veramente*, and that it was Tacitus's style, and the great variety of things he treated, rather than the substance of his precepts, that made him appealing.[75] He appears to have taxed both Tacitus and Machiavelli with a kind of physical cruelty, as if the harsh practices they described were necessarily converted into corporal punishment. Writing to the brother of a recidivist monk, Grillo defended the disciplinary measures of his own order by contrasting them with these two writers:

> Your Lordship is not well informed. In religion, we advance with pastoral piety, not with despotic power. We govern ourselves with the precepts of the Gospel, and with the regulations of the Rule of Saint Benedict, not with the maxims of Cornelius Tacitus, much less of Machiavelli. And the rod of the abbey is more apt to guide than to punish. Your Lordship's brother saw this several times, but he failed to benefit from it. And therefore one must try a more robust medicine, in order to avoid the iron and the fire.[76]

Grillo's conflation of the astronomer and the *gazzettante* was managed in less troubling terms. The association itself should come as no surprise, for Grillo was closely involved in two simultaneous exercises, the debate over the sunspots and the publication and manuscript circulation of *News from Parnassus*. Ever cautious, he appeared unwilling to condemn Christoph Scheiner, the "masked Apelles," for his interpretation of the sunspots as regularly shaped and imperishable "solar stars" rather than as the nebulous masses on the sun's surface minutely measured, described, projected, and drawn by Galileo. Writing to the intermediary between Scheiner and Galileo, German humanist Marcus Welser, in the fall of 1612, Grillo sought to minimize the crucial differences in these hypotheses by deftly transferring the clouds discussed by Galileo to his adversary's obscured identity. "I thank you," he wrote,

> for the second discourses of our unknown Apelles, who is now so
> well known for his famous disguise, that he looks to me more like
> an Apollo, than an Apelles. That is, an Apollo perhaps more visible,
> and more agreeable, under that veil or thin cloud of modesty, than
> the other Apollo, whenever he is entirely uncovered, and denuded,
> since his overwhelming visibility renders him almost invisible, and
> unmanageable, and unwelcome to our sight.[77]

Such circumspection is typical of Grillo's published correspondence; several other letters with speculation concerning either natural philosophy or political events break off with hasty references to the dangers of such pursuits.[78] The need to obscure one's arguments, or to veil the object of one's criticism—to correct, in other words, for the excessive detail provided by Tacitus's *occhiale* and by Apelles' portraits—seemed to Grillo most crucial in the case of Boccalini. After the latter's unexpected death in November 1613, Grillo saw to his burial, and sent a poignant description of his late friend to the aspiring newswriter Tedeschi, who likewise knew Boccalini, or his works, well enough to mourn him.

Grillo's assessment of the risks Boccalini took in the *News from Parnassus* was perhaps influenced by the rumors surrounding the writer's fate, and doubtless part of a general effort to rein in Tedeschi's journalistic efforts. Given that the Veronese nobleman was jailed just two years later, such advice was evidently insufficient. More significantly, Grillo's depiction of Boccalini was clearly inflected by the popular image of the telescopic observer, who in seeing all, reports all to a curious readership. "He showed me many of his writings," Grillo confided,

> where, as was his habit, he criticized this age through censorship of the past, and scourged the vices of the living by castigating those of the dead. I told him my opinion, as a friend, and in many instances, I urged him to adopt a less pungent style, and to offer less detail, so that readers might not go from the general to the specific and from the specific to the individual. Because in certain passages in his *News*, you see someone who instead of being masked has over his face a veil that is so fine and so transparent that not just the courtly world and speculative types who have the eyes of lynxes but even the shortsighted can recognize him and call him by his own name.[79]

This assessment implies that Boccalini provided a type of understanding that was not limited to insiders, the members of the court and sharp-eyed natural philosophers, but also to those who could normally discern nothing but the closest objects. The pairing of a political and scientific elite, the fact that these men are associated at once with mirrors and with telescopic vision—*'l mondo cortigiano, & specolativo, c'ha gl'occhi lincei*—and the suggestion that such perception now was within the grasp of the shortsighted appear no more than the logical consequence of Boccalini's own depiction of Machiavelli in the *News from Parnassus*, which likewise made the attentive observation of nature a precondition of political perception, and which further alleged that all those who tended to the reading of history, thanks to the Tacitean *occhiale politico*, would in time become "so many Machiavellis."

That Grillo's portrait of Boccalini inevitably involves not just Galileo, but also Machiavelli and Tacitus, is further implied by the hint of violence he associated with these latter authors: the *gazzettante* is equipped with a scourge to correct the vices of his contemporaries. The move from the punitive to the pungent is, moreover, predictable, given the insistence of Machiavelli himself and subsequently of Bentivoglio on the figure of digestion; several months earlier, as if to shift from the standard rhetoric of incorporation to the more disconcerting gesture to ritual meals, Grillo himself had called the second volume of the *News from Parnassus*, "a magic aliment," and had claimed, "I don't know if I devoured it, or if it consumed me."[80] That Grillo had in mind the Florentine secretary, and especially his forbidden works, while describing Boccalini to Tedeschi is indubitable, for in noting that the study of politics was the particular passion of his late friend, he added, "it's delicate food for great men, and the true and living book for Princes."[81]

The foregoing suggests that the double depiction of Boccalini and Galileo signaled, at least in the case of well-informed onlookers such as Bentivoglio, Grillo, and perhaps Politi, some fundamental similarity with Tacitus and Machiavelli. While Tacitus is very occasionally invoked in connection with Galileo,[82] the axiomatic link with Machiavelli is more surprising, and seems an artifact of Boccalini's *News from Parnassus*. For while both the Florentine secretary and the astronomer shared a covert interest in Lucretius, and both were evidently known to their closest friends as talented impromptu performers of verse and translation, the connection of Machiavelli with Galileo is not routine.[83] And though the eventual condemnation of Galileo's *Dialogue* in 1633 meant that his name would be durably associated with that of Machiavelli, whose works were likewise banned, there is only the slightest, and most casual, indication that he knew or cared at all about obtaining such texts.[84] Just as the *Letters on Sunspots* were being put through the press in early 1613, Giambattista Amadori, a Florentine physician, wrote to Galileo's close friend Lodovico Cigoli in Rome, asking for the medical works of four early modern authors, at least one of which had been prohibited for confessional reasons. Seemingly aware of the burden of his request, Amadori confided that Galileo had informed him that Cigoli had frequent contact with powerful men in Rome "in the circle of [Cardinal Scipione] Borghese," and it was clearly these acquaintances who were meant to facilitate the matter. In a postscript, Amadori boldly sought two more banned authors, neither connected with medical matters, but both Florentines, Giovanni Boccaccio and Niccolò Machiavelli.[85]

The Harried Messenger

The part played by Galileo was further scrutinized in the work of Alessandro Tassoni, who followed Boccalini's *News from Parnassus* closely in 1614–1615, who shared his taste for satirical disclosures, and who was sometimes thought to have authored a continuation of that author's work.[86] Like his friend Politi, Tassoni appears to have been generally aware of recent astronomical discoveries and conjectures, and his closest associates, the Paduan canons Albertino Barisoni and Paolo Gualdo, were both warmly acquainted with Galileo as well.[87] His interest in natural philosophy was sufficiently pronounced that later scholars have occasionally maintained that he, like Galileo, had been a member of the first scientific society, the Academy of the Lynxes.[88] The tenor of Tassoni's rejection of the authority accorded Aristotle by his early modern followers, moreover, strongly resembles the sort of doubt raised time and again by Galileo. Nowhere was this likeness so pronounced as in 1613, when both writers addressed harsh criticism of their recent works. In his *Letters on Sunspots* Galileo mocked the "strict defenders of every Peripatetic trifle" and their habit of collocating disparate Aristotelian texts rather than observing the heavens, suggested that the motivations of his Italian detractors had nothing to do with a desire for truth, and ridiculed their understanding of the terms "generation" and "corruption," while Tassoni accused a Bolognese Peripatetic, along with his colleagues, of willfully misreading the ancient philosopher as a Christian, of treating him not just like a prophet but like a candidate for sainthood, of offering absurd interpretations of the word "generation," and above all of being "salaried by Aristotle."[89] Tassoni's closing remarks in this private letter of 1613 could have been written by Galileo:

> Your Lordship should neither be scandalized nor moved to anger because I am not always in agreement with Aristotle; I find his doctrine beautiful and intelligent, but I want to say new things, and since this is my goal, I ask my friends their opinion not so they can tell me what I have said against Aristotle, but so that they can correct me in the event that I have made errors. You and your lot, because you are Aristotle's hired hands, are obliged to defend his doctrine, right or wrong. But I am not his dependent.[90]

For his part, Galileo owned the 1608 edition of Tassoni's encyclopedic *Pensieri*, and his annotations on the verse of Petrarch draw repeatedly on Tassoni's commentary on that poet, published in 1609.[91] Given that Tassoni had been, like Galileo, a student in Pisa around 1582, and that his recollection of that period involved less what he learned than the names of wines he had especially enjoyed, his participation in a carnivalesque mock battle, and his particular friendships with Florentines, their relationship might even date to that early epoch.[92]

While Tassoni's letters of 1617–1618 record his numerous attempts to obtain a Galilaean telescope from Barisoni, the instrument disappointed him, and was in any case destined for terrestrial, rather than celestial, observations.[93] He was much more conscious of the impact that various contributions to astronomy would have *outside* the discipline, when the news originally offered to specialists had been converted to cultural prestige or to a more generalized intellectual authority. Thus in seeking a prominent patron in late 1612, he gestured to the sudden rise of Galileo, noting in a letter to Cosimo II, grand duke of Tuscany, that even "the smallest stars were treated with great pomp" at his court.[94] It is possible that one of the matters addressed in the work Tassoni sent along with this letter—"Whether the Earth Moves"—prompted a celebrated Scripture-based objection to heliocentrism at the Medici court in late 1613, and was the impetus for Galileo's *Letter to the Grand Duchess Cristina*: that ruler, like Tassoni, argued that the miracle at the battle with the Gibeonites, where Joshua bade the sun to stand still, could have no meaning in a Copernican context.[95]

The astute Tassoni also recognized the import of the condemnation of Copernicanism in 1616 within a day of that decision, and quickly appended the same discussion "against the movement of the earth" to the next edition of his *Pensieri*.[96] He followed the chatter about Pope Paul V's interest in one of the comets of 1618 more closely than the phenomenon itself, claiming that he lacked the means to measure with any precision its place in the firmament, and reporting after several weeks that no one discussed it any more, once they had seen that the pontiff laughed off their astrological predictions.[97]

In these tendencies, to be sure, Tassoni appears to differ little from his contemporaries, but what is most interesting about him is a series of covert gestures to Galileo's discoveries. While in his *Pensieri* he rejected Copernicanism with familiar physical arguments about the supposed effect of a mobile earth on birds in flight, projectiles, and objects dropped from towers, he also assented to telescopic observations of an opaque and mountainous moon, of a spotted sun, of the innumerable stars in the Milky Way, and of superlunary comets.[98] Galileo is not named in the course of any of these references, but merely figures in a list of exemplary sixteenth- and seventeenth-century natural philosophers and astronomers, and as part of the ballast that allows the Moderns to outweigh, and to outdo, the Ancients.[99] That said, a possible source of Tassoni's knowledge, if not his friends Barisoni and Gualdo, was the "Preface to the Reader" in Galileo's *Letters on Sunspots*, written in 1613 by Angelo de' Filiis, secretary of the Academy of the Lynxes. This text offered a breathless list of the astronomer's recent discoveries, with special insistence on their newsworthy qualities:

And thus Signor Galilei shows us innumerable squadrons of fixed stars, scattered throughout the entire firmament, many, once dim and indistinct, in the Galaxy and in the nebular [stars]; he has found Jupiter's royal retinue, and he has discovered that the Moon has a mountainous and varied surface. All this he has made known and transmitted to everyone in his *Astronomical Announcement [Starry Messenger]*. Astonishment arose quickly, novelty in the heavens being the last of our expectations. Pursuing his enterprise still further, Galilei reveals the new triform Venus to be an imitator of the Moon.[100]

Despite Tassoni's apparent erasure of Galileo's name from his discussion of telescopic discoveries, the astronomer does appear, after a fashion, in a context reminiscent of Boccalini's satirical council of the gods at Parnassus. This was in Tassoni's *Secchia Rapita*, a mock epic begun around 1611, published in abridged fashion in Paris in 1621 to avoid censorship, and in an integral edition in Venice in 1630. The poem depicts a centuries-old conflict between Modena and Bologna, a quarrel whose origin was said (wrongly) to have been the theft of a wooden bucket. The single most important classical precedent for this sort of burlesque was the pseudo-Homeric *Battle of the Frogs and Mice*, a work whose translation to Italian, incidentally, Galileo himself had undertaken in summer 1604, at the point where Barisoni was still a student of his.[101]

In canto II of the *Secchia Rapita*, various deities gather to confer about the emergent squabble. Tassoni presents them in a register that consistently begins with soaring promise, only to founder, stanza by stanza, on the most picayune of details. An instance typical in its lyrical first line—*Non comparve la vergine Diana*—shifts quickly to studied banality,

> The maiden Diana did not appear,
> For on rising she'd gone to the woods,
> To do her washing in a fountain
> In the marshes of Tuscany,
> Not to return until the North Star
> Was making its way across the dark sky;
> Her mother hastened with her excuses,
> Knitting a stocking all the while.[102]

The stanza would have had a familiar ring to Tassoni's audience, the absent goddess of the moon, lingering over her laundry, being a burlesque version of the poolside maiden portrayed in one of the most celebrated of Petrarch's poems, the madrigal *Non al suo amante più Dïana piacque*. But its prosaic particulars, such as the suggestion that Diana was around during the day, but would only be visible once evening had advanced, that her garb might have been spotted and speckled with dirt, and that she had some special connection with Tuscany, would have had much greater currency in the aftermath of the publication of the *Starry Messenger*. What was typical to the point of banality in lyric—that proper names such as "Diana" could refer at once to a deity, to a celestial body, to an idealized woman, and perhaps to a flesh-and-blood individual of the poet's acquaintance—emerges as both comic and potentially unsettling in Tassoni's verse, and provides the poet with the occasional opportunity to convert the epic of the stolen bucket into a commentary on more recent news and newsmakers.

It is thus notable that Tassoni's representation of Mercury has the same minimal level of astronomical information, allied with a deft reference to that deity's function as a messenger.

> With Jupiter's cap and with his *occhiali*
> Mercury followed next, and in his hand he held
> A satchel, where he gathered the prayers
> And queries of mortal men,
> And arranged them in two vials
> That his father had in his cabinet,
> And which with great focus and care
> He signed twice a day.
> Finally Jupiter arrived, in royal guise,
> With those stars that they found on his head. . . .[103]

Tassoni's verses convey nothing about the planet Mercury, emphasizing instead a messenger with "glasses" or rather telescopes, a sort of harried and filial factotum, created, in effect, by Jupiter. This figure is transparently Galileian. It would have in all likelihood been recognizable to many in Tassoni's original audience, for at least two other authors had identified the astronomer with Mercury directly after the publication of the *Starry Messenger*.[104] And Tassoni accomplishes with great finesse what his rival Giambattista Marino would manage only with the labored ventriloquism of canto X of the *Adone*, where

Mercury, "messenger of Jupiter," paraphrases the findings of the *Starry Messenger*, predicts the invention of the telescope, and at last mentions Galileo by name.[105]

Tassoni's representation also differs from the latent genealogies evoked by those envious of Galileo's success, who tended to suggest either that his observations of celestial phenomena were his creations, or that what he claimed to have invented actually came from elsewhere, and was nothing but his illegitimate offspring. Tassoni implied instead, in an astute if very reductive reading of the genesis and impact of Galileo's first telescopic news, that the messenger owed his very existence to Jupiter. Though the detail about Jupiter's hat is oddly reminiscent of Galileo's description, in the *Starry Messenger* and in subsequent works, of the way in which the telescope stripped bright celestial objects of irradiation or their "coiffure," it appears more likely that Tassoni merely intended to evoke the cap or crown with which the deity was distinguished when he was represented as a planet.[106] Such emphasis would heighten, at least momentarily, the relevance of the subsequent verses, where the Medici stars at last make their brief appearance.

This sort of opportunism—Tassoni's readiness to introduce, where it suited him, passing allusions to Galileo's latest discoveries—suggests that he assumed that his unmarked references would be legible to his initial audience, precisely because of the enormous impact of those celestial disclosures. (It must be noted that this initial comprehension, if it took place, was not to last: around 1760 Voltaire allegedly described Tassoni to Giacomo Casanova as the author of the "only tragicomic poem which Italy possesses," "a wit and a learned genius as poet," but "not learned for ridiculing the Copernican system."[107]) Though these brief vignettes have no connection to the business about the bucket, Tassoni seems also to have concluded that at least one longer episode might be treated as a burlesque version of Galileo's recent pronouncements. Canto VIII begins with a review of grotesquely inadequate troops—and here Galileo's close friend Paolo Gualdo makes an anachronistic appearance—before returning to negotiations over the wretched bucket. To entertain the visiting ambassadors, a blind harpist commences a tale about Endymion and that "beautiful goddess of the first heaven," variously identified as the huntress Artemis, Diana, or Cynthia. The suave story soon turns lascivious, and is interrupted by the lady who requested it, for the upshot is that the pure goddess of the moon regrets her chastity, and roundly curses the fruitless years spent in the solitary pursuit of wild beasts.[108]

This tale can naturally have no ending, for its real importance is its brazen manipulation of the short formulation of one of Galileo's telescopic discoveries, the realization that Venus had phases like those of the moon. Galileo's phrase *Cynthiae figuras aemulatur mater amorum*, conventionally translated "the Mother of Love imitates the shapes of Cynthia," but more suggestively rendered "the Mother of Loves envies Cynthia's manner," has here been reduced to the basic premise of the sometime similarity between the two celestial bodies. Until the blind seer is censored, the chaste moon seems to have been intent on emulating Venus's amorous style.

The reversal of the Galileian formula—where the moon might be said to have phases like those of Venus—is in its way more startling than the translation of the phrase to the licentious mode of romping goddesses, for it implies, quite absurdly, that the planet's varying appearances were somehow more familiar to terrestrial observers than those of their own satellite. Such a suggestion erases the telescopic discovery of Venus's phases. And yet it paradoxically depends for its ultimate legibility upon knowledge of both this revelation, and of Galileo's one-line description of the finding in late 1610, the news of which he remarked to his rival Scheiner had "been communicated to so many that [by mid-1612] it had become famous," even though he had not bothered to put it in print.[109]

While one might easily regard Tassoni's tendency to minimize or to burlesque the astronomical content of Galileo's latest discoveries as nothing more than the sorry index of their rivalry, the backdrop of Boccalini's *News from Parnassus* provides another and more illuminating reading. The insistent presentation of the *nuncius* as one who, to his own peril, disclosed too much to all onlookers is tacitly acknowledged in the *Secchia Rapita*. Rather than doubling such figures through the association of the astronomer and the newswriter, and magnifying the challenge to authority that they posed, Tassoni simply reduced Galileo to a harried messenger, or as Grillo would have it, covered him with "a veil that is so fine, and so transparent, that not just the courtly world, and speculative types, who have the eyes of lynxes, but even the shortsighted can recognize him." And as if to hide them in plain sight, he converted the extraordinary discoveries of the very recent past— those *occhiali*, the moon's spots, Jupiter's satellites, and Venus's phases—to things stranded between the eternal present of the mythological and the absolute banality of the quotidian. These piecemeal and picayune items, shorn of their original context, would have appeared remote indeed from the coherent cosmological framework where Galileo was seeking to place them.

Celestial Assemblies

Given the original association of political perspicacity with optical instrumentation in Boccalini's *News from Parnassus*, it is to be expected that some
writers would present the telescope as a device for revelations and prophecies.
A typical instance is the anonymous *Report of the General Council of Parnassus
in which the Origin of the War in Italy is Debated, as Well as the Remedies to
Bring about Peace*, composed in the fall of 1614, and thus a scant year after
Boccalini's death. In the midst of a raucous argument, Apollo strikes the
discussants with lightning, and when the smoke clears only the Palace of
Spain remains standing, though a "stinking smoke" pours from it. As Bocca-
lini's imitator tells it,

> The *virtuosi* were not fooled, however, because they finally found
> certain tiny holes, and taking up the "Flemish spectacles," which
> make distant things seem nearby, they recognized that Apollo's
> anger was working its way underground through these holes, which
> had seemed almost invisible, and that it was going to strike at the
> very foundations [of the palace], which had originally appeared
> more solid than marble.[110]

The *Report* goes on to predict on the basis of this telescopic observation
the imminent fall of Spain, and the correlative rebirth of Italy. The overall
strategies—the hyperbolic premise, the emphasis on the utility and availability of the Dutch telescope, and the seamless conversion of the bogus and
unreal news of Apollo's tantrum into the long-term promise of something
potentially true and relevant, the evacuation of Spanish influence from the
Italian peninsula—are roughly modeled on Boccalini's bulletins.

What is crucial here is the *Report*'s appropriation of current discussions
of the moon's telescopic appearance immediately after Galileo published the
Starry Messenger, a strategy that would suggest popular as well as elite association of the astronomer with the *gazzettante*. The notion that solar rays
worked their way through a labyrinth of underground channels in a perfectly
spherical moon, emerging occasionally as bright spots beyond the boundary
between dark and light on the lunar globe, was proposed in 1612, for instance,
by the well-established scholar Giulio Cesare Lagalla, and was intended as an
alternative to Galileo's insistence on the sun's play over a rocky surface.[111]
While we are unable to judge whether the anonymous newswriter regarded

this sort of theory as a viable physical model or as just one more comic element in his satire, at the very least we can conclude that his reassignment of the sun's action to Apollo and the porous lunar substance to the Palace of Spain is evidence of some knowledge of the latest discussions in astronomy as well as in politics, and of an expectation of a similar degree of awareness in his audience.

Conversely, natural philosophers sometimes seized on the genre of the Parnassian bulletin. The most striking instance appeared in 1619: the anonymous *Celestial Assembly Concerning the New Comet* exploited Boccalini's model to render not a Copernican but rather a Tychonic explanation of the recent phenomena. This work privileged the same rhetorical strategies as the *News from Parnassus*—direct observation by qualified members of the intelligentsia, mockery of traditional authorities, and a fast-paced procession of speakers—and ended with a combative gesture, and an insistence on extraordinary visual realism.

> The Tychonians were thus holding forth very heatedly about how the Comet had predicted the ruin of the Peripatetic and Ptolemaic view, like a massive tree, as the Scythian once said to the Great Macedonian, which had grown for years and years, but was in a single hour struck down and uprooted.[112] And so they made an emblem—having gotten Caravaggio to paint it—of a falling tree with the motto "One Hour." And under the cover of night, they stuck it on Aristotle's portal like a coat of arms, and it seemed to everyone a foul pasquinade at his expense.[113]

Despite their shared antipathy for the Aristotelian position, and their professed faith in observational data, the Tychonians and Copernicans had crucial differences, as they coveted the same readership. It appears that the canon Leonardo Tedeschi, the conservative scholar Andrea Chiocco, and perhaps even Alessandro Tassoni were at least as interested in the Tychonic position as in the Copernican one.[114] Thus while the loquacious "Tycho" of the *Celestial Assembly* gestured favorably to Galileo, pretending even to have recent news about the telescopic appearance of the comet from him, and making a passing swipe at his rivals' interpretation of the sunspots, the general tendencies of the pamphlet would have seemed a brazen bid to recruit those who had originally been the Pisan's followers, particularly in the ambit of educated nonspecialists, whose expectations of polished, caustic, and lively

Italian prose would have been high.[115] For his part, Galileo obtained the *Celestial Assembly* almost immediately after its publication, and he and his associate Mario Guiducci set about planning a riposte to that work—as usual, a parade of speakers around Parnassus—before abandoning the constraints of that genre for the *Assayer*, which was both bolder and more detailed, but less timely in its emergence.[116] The emphasis, after all, had shifted from the newsworthiness of a particular issue to the accomplishment of longer-term goals.

Human Seraphim

Where does the obscure Giandomenico Tedeschi figure in this context? The one-time newswriter's close acquaintance with Galileo's associates Angelo Grillo, Paolo Gualdo, and Cardinal Bentivoglio is undoubtedly significant, as is the fact that he was also a friend and a protégé of Andrea Chiocco, the conservative Veronese academician to whom Grillo had sent several summaries of the quarrel over the sunspots and of the astronomer's other discoveries in 1611–1612.[117] And though the debate over Tuscan in 1613 did not end, as Tedeschi had predicted, "with daggers," it is noteworthy that several of the most prominent advocates of the dialect—Chiocco, Welser, Filippo Salviati, and Galileo himself—were also involved in investigations of the sunspots in that same year, and that in early 1610 their greatest antagonist, Paolo Beni, had been familiar enough with the claims of the *Starry Messenger* to have sent to a correspondent in Naples an illustrated overview of that pamphlet before it emerged from the press.[118] Tassoni, typically elusive, seems to have objected more to the canonization of several minor writers of the twelfth and thirteenth centuries than to the implicit claim of Tuscan hegemony.[119]

The position of Adriano Politi is yet more difficult to define, but most striking in its frequent alignment of linguistic issues with astronomical discoveries. His peculiar suggestion of 1611 that terms whose currency rendered them lofty and clear were more stellar than outmoded astronomical and poetical references to stars anticipates Galileo's own caustic discussion of the range of figurative meanings assigned by some astronomers and poets to the same word in the *Letters on Sunspots*.[120] His observation that the brilliant prose of Boccaccio and poetry of Petrarch was a chance matter of individual genius, rather than the "particular prerogative of Florence" is echoed, somewhat altered in tone, in Galileo's assertion that the appearance of the dark marks on the solar globe was not "a particular privilege of Florence," but a configuration visible to all.[121] Like Galileo's *Letters on Sunspots*, Politi's abridged compendium of the Tuscan faction's *Dictionary of the Italian Language* emerged from the Roman press of Giacomo Mascardi in 1613; its prefatory claim that because of recent erudite works in the Tuscan vernacular, "even without Latin or Greek, any unimpaired mind might become a worthy speaker, philosopher, and mathematician" seems an allusion to the cultural impact of the grand duke of Tuscany's celebrated philosopher and mathematician. In the summer of 1614, Politi's compendium struck the Paduan Lorenzo Pignoria as a well-priced imitation of the "insolent and awful" tome issued by the Tuscan faction, but throughout the fall the scholar was rumored to have been jailed because of false material—Sienese proverbs—boldly inserted in the work. Only late in the year did Pignoria report that the book was being sold without resistance in Venice, and that little by little the Tuscan academicians would discard their objections.[122] In short, Politi appears distinguished by his close and sympathetic observation of Galileo's astronomical activities, but also by a slight resistance to the correlative claims others in Galileo's orbit made for linguistic hegemony.

While the conflict over language does not map perfectly onto the quarrel over the solar phenomena, Tedeschi and other onlookers, as for example the Tuscan poet Alessandro Allegri, would have noticed that the general agenda of the pro-Tuscan faction appeared commensurate with Galileo's ever more influential scientific program.[123] Just as the supporters of Tuscan presented their dialect as an appropriate lingua franca for all of Italy, insisting at once on several centuries of a cultural pedigree provided by Dante, Petrarch, and Boccaccio, on the lively currency of a vocabulary drawn from contemporary merchants and artisans, *and* on the portability of that hybrid idiom to qualified speakers well outside of the region, so Galileo appeared to present his telescopic discoveries as the product of a particular sort of education, experience and perspective, but also as a series of insights increasingly available to credentialed observers entirely remote from him.[124]

Grillo's detailed description of Boccalini in his Galileian guise was certainly legible to Tedeschi, as was the veiled, but almost axiomatic, reference to Machiavelli in the allusion to the "delicate food for great men, and the true and living book for Princes" in that same portrait. To judge from the sole fact of Tedeschi's imprisonment, the cautionary tale appears not to have had sufficient impact: within several years he was detained in what Bentivoglio would term the "dark and gloomy Parnassus" where his poetic career began, and where his early tendencies as a newswriter and imitator of Bocca-lini presumably underwent some modification.

It is therefore especially interesting that among the scant traces of Tede-schi after his liberation from jail is at least one effort to associate himself in print with a work of natural philosophy characterized by its apparent resis-tance to Galileo's telescopic discoveries. This involved a prefatory sonnet for the *Astronomical and Geometrical Harmony*, an important treatise on sundials begun around 1617 by a fellow citizen of Verona, the Capuchin friar Teofilo Bruni, and published posthumously in 1621 and 1622.

While the issue of the earth's movement was irrelevant to Bruni's work, the Church's condemnation of the Copernican hypothesis in March 1616 seems to have prompted that author to include, in the final pages of his book, a discussion of the lunar globe as viewed through a telescope, and an implicit rejection of Galileo's explanation of its telescopic appearance. Bruni's deci-sion to insist that the rough lunar surface was filled with a crystalline matter that preserved that globe's spherical perfection, and to avoid mention of Gali-leo's name was surely a way of suppressing the unsettling novelties which that "imaginary *gazzettante*" had divulged, and above all the world system in which he subsequently sought to inscribe such phenomena.

Finally we will reveal the true explanation of the moonspots, which has been hidden from philosophers for so many centuries, and has lately been obtained by us, and I believe by others as well, through use of the telescope. With [this instrument] we have clearly under-stood that the body of the moon is not smooth or polished, as phi-losophers imagined it had to be in order to receive light and shine, but rather it is even more mountainous, and dotted with caverns, than is the Earth, and something like a sponge. Within these cav-erns, the light is somewhat dimmed, just like the secondary light that we perceive in the corners of rooms, or in wells, and this is what makes us see those moonspots. We are not arguing, however, that these caverns are empty, but rather that they are filled with that subtle matter of the heavens, which is unable to receive light. . . . Anyone can perceive this imperfection that we find in the five- or six-day-old moon by looking at it through the telescope, because one can see in the region bordered by light that it resembles a saw, or a piece of unevenly melted ice with many thick and dark holes, like a sponge. And depending on one's position sometimes one will see a few isolated areas of the moon, and [rings] that can be nothing other than mountains and caverns where the light cannot enter.[125]

What is interesting about the verses Tedeschi wrote to accompany this work is the ambiguity of his portrait of the late Bruni, and by implication of Galileo himself. The poem opens rather conventionally with an octave in praise of the Creator's handiwork, and as is also conventional, reserves its more complicated arguments for the concluding sestet. While the context of the sonnet, its pious tone, and its early reference to the Earth's weight and apparent immobility would all establish Tedeschi's credentials as Bruni's confederate, the sestet itself falls somewhat short of the glorious paean one would expect.

> But measuring the vastness, and knowing
> The place, the size, and the movement of each sphere,
> Dividing the hours of the Sun's path,
> This is the work of a man, who in a lonely and remote cloister
> Follows the ranks of the Human Seraphim,
> Fixing his eyes on this immensity.[126]

The sheer weirdness of the image—the slightly comic contrast between the bright and vast heavens and the lowly Capuchin cloister where Bruni labored—is perhaps no more than an unintentional effect of Tedeschi's lyrical effort. The "Human Seraphim" in whose wake Bruni walked was, of course, the Seraphic Father and founder of his order, Saint Francis of Assisi. But the verse also presents the scholar as the distinctly inferior follower of a more contemporary visitor to the celestial realms, and the remoteness and isolation of the cell where he worked as the very emblem, if not the cause, of his benighted state. In this reading, "the sun's path," and its minute division into days and hours would likewise be an index of that futile Ptolemaic labor which the lyrics were meant to celebrate. The man who so outdistanced Bruni, and in effect his antagonist, would naturally be the author of the *Starry Messenger*, whose telescopic discoveries seemed to his enemies a brazen attempt to appropriate a role reserved for the highest ranks of the angelic hierarchy.

The possibility that the "Human Seraphim" serves as a covert reference to the astronomer, particularly in light of the most celebrated of subsequent literary elaborations, is worth some consideration. The shift in Galileo's image from that *nuncius* with the spyglass to the "imaginary *gazzettante*" who disclosed too much too fast, to the harried messenger of Tassoni's mock epic, and finally to an angel poised for a disastrous fall mirrors what seemed to many an inevitable course. Given that his manuscript notes show that in composing the *Starry Messenger*, Galileo understood *nuncius* as the Latin translation of the Greek *angelos*, and that he had considered alternatives closer in meaning to "informer," "auctioneer," and "crier," the potential for such a trajectory had been there from the start.[127] It anticipates John Milton's association of the astronomer with Lucifer, once the brightest of the Seraphim, in *Paradise Lost.* In the great English epic, for instance, the shield carried by the "Superior Fiend" is compared to "the Moon, whose Orb / Through Optic Glass the Tuscan Artist views / At evening," while his presence on the Sun recalls the solar projections of the real and false Apelles: "There lands the Fiend, a spot like which perhaps / Astronomer in the Sun's lucent Orbe / Through his glaz'd Optic Tube yet never saw."[128] Significantly, however, it is less Lucifer's fall than the angelic rank he once occupied that underpins his connection with Galileo, for when Raphael makes his way as a messenger from the Creator to the short-term residents of Eden, his perspective is portrayed "As when by night the Glass / Of Galileo, less assur'd observes / Imagin'd Lands and Regions in the Moon." Like Galileo, Raphael is thus both the bearer of a celestial message and angelic: "At once on th'Eastern cliff of Paradise / He lights, and to his proper shape returns / A Seraph wingd."[129]

The note of dubiety associated with astronomical observation in Milton's account, a function of the temporary attribution of seraphic powers to mortal men, is likewise crucial.[130] It is clear that in his *Areopagitica: A speech of Mr. John Milton for the liberty of unlicensed printing to the Parliament of England*, the English scholar believed that the astronomer's fate was due generally to the Roman Catholic Church's unwillingness to allow open investigation of such matters, and to the specific collaboration of individuals like Friar Teofilo Bruni: "[In Italy] I found and visited the famous Galileo, grown old, a prisoner to the Inquisition, for thinking in astronomy otherwise than the Franciscan and Dominican licensers thought."[131] While Galileo counted at least one Franciscan, the Capuchin friar Valeriano Magni, among the most fervent of his supporters, it is worth noting that an anti-Copernican manuscript circulating in the Veneto shortly after Galileo's condemnation and written by a Capuchin of Verona was attributed, rightly or wrongly, to Teofilo Bruni, who had died over a decade earlier.[132]

What Milton makes explicit—that some saw in the novel disclosures of those starry messengers and human seraphim matter for censorship, rather than issues for debate and discussion—is anticipated in the views of Bentivoglio, Grillo, and Tassoni, and signaled by their various treatments of the role of the *gazzettante*. That such gestures became so coded, that they involved a reflexive nod to a vaguely defined but sinister political landscape, and that they tended to insist less and less on the newsworthy quality of particular discoveries, and more on the long-term impact of Galileo's program, all imply an early recognition of the way in which this sort of science had "rearrange[d] the rest of intellectual space."[133] Such matters are, at best, inchoate and latent in Tedeschi's verse. What the latter really believed about Galileo, and whether he saw the astronomer as an "imaginary *gazzettante*" poised for inevitable punishment, is and will in all likelihood remain unknown to us. But there is the chance that in that "dark and gloomy Parnassus" from which he had recently emerged, or in the afterimage left by Boccalini's bulletins, the author of the trivial news from Verona had learned to present his subject with a veil so fine that it both did, and did not, prevent recognition.

CHAPTER 4

Cameras That Don't Lie

The early impression of the telescope and its precursors as analogues of the newsletter would undergo various revisions as both the optical instrument and the journalistic medium developed. This chapter examines the fiction of a modified version of the telescope—one adapted for use in the *camera obscura*—as the ideal and objective transmitter of news, a notion of particular appeal to ambassadors abroad. The text that best exemplifies this fantasy, Henry Wotton's letter to Francis Bacon describing the use Johannes Kepler made of such a device in 1620, while well known to historians of science and of art, has been analyzed neither in terms of its historical context, the grim aftermath of the defeat of Protestant forces at the Battle of White Mountain, nor with particular regard to its author's own varied career as an ambassador.[1]

In this reading, Wotton's discussion of the *camera obscura* centers on the fundamental relationship of optical issues to the problematic reportage of foreign affairs, and is informed by some knowledge of meditations on both the dark room and pinhole apertures made in this period by Protestant and Catholic astronomers. Their observations were in turn often characterized by references to the great confessional conflict that would be the Thirty Years' War, such that the ideal of an objective machine for recording neutral data about either the natural or the political world seemed in 1619–1620 at once the most useful and the most suspect of notions. In this and the following chapter, therefore, I will address that implicit transfer of faith from the *tubus emmissitius*, or "prying glass," to the *tubus immissitius*,[2] where visual data were projected through the telescope into a dark room, as through the eye onto the retina, in terms of the problem of accurate reportage and subsequent interpretation of current events.

Broadly speaking, these two chapters will examine a number of early modern references to the *camera obscura* in terms of recent adaptations of

Jürgen Habermas's notion of the public sphere. While Habermas's public sphere was originally associated with the rise of the bourgeoisie, with late eighteenth-century culture, and with an increasingly rational, purposive, and oppositional discussion of political issues previously under the exclusive control of the state, scholars of early modernity have refined, disputed, and backdated this model, particularly in connection with emergent news forms. Thus they have insisted more on the regularity and reliability of the networks that made the diffusion of news possible from the late sixteenth century onwards; on the political importance of news from the early seventeenth century; on the fact that news was routinely exchanged, discussed, and assessed within absolutist regimes and by individuals nominally extraneous to the political process; on the overwhelming importance of war in the creation of the discursive arena; and on a recognizable rhetorical continuum, animated above all by a pellucid "plain style," between the private letter of Renaissance humanism and the widely distributed newsletter of the 1600s.[3]

Among historians of the Enlightenment and beyond, the Habermasian public sphere, by contrast, has been subjected less to this temporal adjustment than to a marked emphasis on spatialization: the original settings of coffee houses and salons in which rational political discourse took place have multiplied to include other sites, other modes of communication, and above all, other "publics," typically those not included under the rubric of bourgeois citizenry. A critique of and corrective to those tendencies insists, however, on their paradoxical nature: to accentuate the social particularity of a given group is to relegate it to the margins of that public, and to present its viewpoint not as a rational perspective available to all, but rather as the corollary of narrow desires. This critical account emphasizes, moreover, the connection between the presumed rationality of discourse within the public sphere and the seeming abstraction from specific social interests and biases: within the powerful fiction of the public sphere, the claim to a universal and disinterested view is predicated on the successful masking of particularity. Thus the crucial move is the illusory presentation of a given sector of the public, its diverse composition and shifting interests, as a unified collective subject embodying a stable, legible, and rational perspective, in sum, a natural perspective. Finally, those historians not reliant on the Habermasian model to explain the emergence of the French Revolution and its eventual transition to the Terror nonetheless have recourse to a similar lexicon to describe the spontaneous establishment of a powerful, ultimately ungovernable and irrational collective, an oversized projection of a massed body, the phantasmic tribunal of public opinion, a figment called into being and dispersed from one day to the next.[4]

While the backdating of the Habermasian public sphere is of obvious interest to my discussion of early modern news, the critique of its recent appropriation by historians of the Enlightenment and beyond is also striking for the resonance of the terms it deploys. These are almost invariably optical in tenor, for in that reading the unified public is a temporary "phantasm," an "illusion," a "projection" conjured up from the equally fictive space of the sphere. The attempted naturalization of public opinion, in other words, extends the lexicon of projection, once limited to whatever might be fleetingly seen in the dark room, to include the viewer as well; the emphasis shifts from the unstable tenor of "opinion" to the legitimizing force conveyed by "public." The availability of the optical idiom is no accident, for these terms evoke the issues of central concern to those who gestured to the *camera obscura* as an idealized transmitter of news. Thus while early users of the telescope acknowledged significant discrepancies in the instruments, eyesight, abilities, and intellectual agendas of those who took up that apparatus, the fact that the *camera obscura* often accommodated more than one individual, and that it allowed direct transcription of terrestrial and celestial objects led to the fictions of an idealized, abstracted, and competent viewer who happened to inhabit the dark room, and of a permanent and unequivocal record of the shifting phenomena under scrutiny. Because the *camera obscura* was sometimes taken as a model for vision, and because the eyes of dead animals and humans were occasionally examined in this period to see what was on their retinas, the unreal scenario of the disembodied eye was perhaps all the more persuasive.

Though the public sphere and the *camera obscura* both have an indubitable connection to specific physical sites—coffee houses, salons, reading rooms, closed chambers with minute apertures—they do not function as such unless particular behavior takes place there, that is, unless the designated protagonists, in examining whatever lies beyond their immediate setting, articulate views that appear free of the bias and error normally associated with the term "opinion." While both the discursive and the optical undertakings are characterized by a tension between the dynamic horizon of events and the alleged stability of a perspective, or between an affirmation of objectivity and an acknowledgement of the theatrical or even uncanny nature of the exercise, early descriptions of the *camera obscura* often emphasized these tendencies more forcefully. They insisted at once on the geometrical accuracy of transcriptions performed in the dark room and on the roles of knowing charlatans and of their ignorant dupes, and in a slightly later period, on the liveliness of the image and on the inanimate eye.[5] Like the eventual fiction of the disembodied rational public, these earlier illusions, more dramatic and openly coercive, typically involved a hint of violence: the observed object was a reanimated death mask, or the scene of carnage, or the instrument itself was lifeless.

The Curious Composure of the Eye

Henry Wotton's interest in optical instruments and his professional concern with the problem of obtaining and relaying accurate foreign news are characteristic of many who moved in diplomatic circles in the early modern period. His acquaintance with the *Starry Messenger* in March 1610, just as it emerged from the press in Venice, where he was serving as ambassador, inspired him to present it as the most exotic piece of foreign reportage.[6] Wotton initially appeared more convinced by the optical instrument, one of which he promptly sent to his ruler, than by the astronomical information it offered, noting cautiously that Galileo "runneth a fortune to be either exceeding famous or exceeding ridiculous," but by mid-October 1610 he was confident enough about those claims to figure as the dedicatee of a Scotsman's attack on Martin Horky's *Brief Excursion Against the Starry Messenger*.[7]

Wotton's pronouncements on more general sorts of rumors and news are frequent, and suggest a wary but resigned reliance on such reports. In August 1608, a Venetian writer related that the English diplomat had "confessed that not all that Ambassadors might say was gospel truth, their food is news, and among the things reported to them some are true but sometimes they have to swallow flies."[8] In 1614, after being blamed for a diplomatic debacle in the Netherlands, Wotton wrote to a colleague with what seems studied insouciance that he was unfazed by the popular outcry against him among the Dutch, for "in truthe the condicion of publike servants weare most miserable if they were bound to be troubled with vulgar voices and with the gatheringe upp of all the breath and rumour that is soe easilye spent, especially in such sensitive times and in States of this composition, whoe, having bravely purchased their freedome, are content (it seemes) that some of it shall appear even in their discourses."[9] That same bruising freedom of discourse, however, served him better some months later, when he relayed various reports picked up during a visit to prosperous Amsterdam, "a magazen of rumours as well as of other commodities."[10]

In his later life as provost of Eton, he had considerable enthusiasm for the science of optics, which he pursued with his protégé John Beale. Their projects included the construction of a telescope around 1628, and Beale and his brother Richard went on to fashion a polemoscope for the nearsighted Gustavus Adolphus of Sweden, a device for reading at night without a candle, spyglasses for use on a tempestuous sea, and some version of the *camera obscura*.[11] But it also appears that the study of optics had interested Wotton in his own youth; he commented on the similarity of the recreations of his charges at Eton and those that were his at the same age.[12] In what follows, therefore, I will examine Wotton's early interest in optics, and an unfortunate diplomatic incident that figures as an unhappy precedent to his encounter with Kepler's *camera obscura*. The dark room, in other words, serves as an idealized corrective to the problematic transfer of information between a semiprivate civil space and the public domain.

In his early biography of his close friend and disciple, Izaak Walton offered an account of the young man's simultaneous introduction to both the science of optics and the celebrated Italian legal scholar and Protestant refugee Alberico Gentili, recounting that the latter was particularly impressed with three lectures that Wotton had produced involving the eye's structure, the traditional debate over intromission and extramission, and finally the "blessing and benefit of Seeing."[13] These discourses on the eye were likely to have been made around 1587, and on the face of it they represent three very different aspects of optics.[14] Gentili's alleged desire to impart mathematical knowledge to Wotton after the three lectures on the eye implies that the young Englishman's understanding of the mechanism of sight derived from the burgeoning anatomical tradition and was in some way opposed to, or negligent of, the geometrical treatment of light and the production of images.[15] This is consistent with the reports that Wotton enumerated the various parts of the eye and described their particular offices—for so the anatomists did with increasing precision throughout the latter half of the sixteenth century—as well as with the inconclusive speculation about ancient theories of extramission and intromission, and the emphasis on the utility of vision, which also figure frequently in the early modern discussions and depictions of the parts of the body.

The crystalline humor was a subject of particular interest to anatomists. It was described by Andreas Vesalius in 1543 as lenticular in shape and endowed with the ability to magnify.[16] Because it was also regarded by many as the principal instrument of sight, succeeding writers within the anatomical tradition often insisted upon this particular: in 1554, for example, Francesco Maurolico portrayed the crystalline humor as a biconvex lens, and the influential Realdo Colombo reported in 1559 that "if you were to remove the crystalline humor from its natural place, and bring it near alphabetic characters, they would seem larger and easier to read, and I suspect that the invention of eyeglasses arose from this."[17] Though he insisted that the retina alone was sensitive to visual stimuli and was thus the principal instrument of vision, in a 1583 work Swiss professor of medicine Felix Platter explained the crystalline as a focusing lens or "internal eyeglass" for that receptor.[18] In 1585 Jacques Guillemeau asserted that the crystalline magnified "anie thing written or imprinted" twofold, and that this observation led to the invention of spectacles.[19] André du Laurens repeated these remarks in a popular work of the 1590s, and in 1615 Helkiah Crooke rehearsed the same story and added that vision occurred when the lens-like crystalline enlarged images and presented them to the optic nerve for interpretation.[20]

This tradition likely figured in Wotton's supposed discussion of "the curious composure of the Eye, and . . . how of those very many, every humour and nerve performs its distinct office."[21] The impression of the relationship between a mechanism for interpretation—in this case, a single lens designed solely for the magnification of small letters—and the eye itself would have complemented the contemporaneous early modern effort to develop a device for reading "the least letters" from a great distance, which long predated the actual advent of the telescope.[22] Inevitably, as the mere feat of magnification appeared to contribute to rather than to resolve visual ambiguity, the quest for more accurate interpretive devices became the search for more faithful models of the eye, and the emphasis shifted from eyeglasses to telescopes to those passive receptors of outside images, the pinhole aperture and the *camera obscura*.

Wotton's second disputation on this subject, which concerned the ancient debate over the theories of extramission, where a visual ray was thought to emerge from the eye and travel to the object, and intromission, where an image of the perceived object entered the eye of the beholder, is most pertinent to the *camera obscura*.[23] Vesalius's successor Colombo had promptly declared in the mid-sixteenth century that because "the way in which vision took place is not easy to explain, and the case is still under judgment," "this is not the place for discussing the matter," and in this abstention he was followed by John Banister; others such as French doctor André du Laurens and English royal physician Helkiah Crooke apologized for pursuing questions that were "rather Philosophicall then belonging to the art of Anatomy," and reviewed the traditional arguments at length before deciding in favor of intromission.[24] Two years after Wotton's disputations Giambattista della Porta would claim in his *Natural Magick* that his experimentation with the *camera obscura* had resolved the question "that was antiently so discussed," but it is unlikely that Kepler was entirely sincere in 1604 when he congratulated the Neapolitan writer for his alleged closure to an issue that for many had long since been put to rest.[25]

While the progression from an interest in the eye's powers, especially as these related to magnification and the reading of "minute letters," to more portable abstractions about how vision took place anticipates Wotton's acquaintance with the telescope and the *camera obscura*, it is also representative of the strong interest in both early modern England and Italy in pretelescopic devices in this period, particularly as proponents of these instruments routinely claimed their utility in the contexts of defense, combat, and espionage.[26] Wotton's third lecture, blandly presented as the "blessing and benefit of Seeing," signals an eventual focus upon the usefulness of and the limits to visual data in the reportage and interpretation of foreign affairs. This hypothetical trajectory would coincide roughly with Wotton's career as an ambassador, a profession which, like his mastery of Italian, suggests the influence of Gentili.

Even here, however, the jurist's legacy is problematic. Gentili's *On Embassies*, published just as the student and mentor became acquainted, traced the history of ambassadors and meticulously outlined their legal rights and responsibilities in foreign territory. In regard to the specifics of the ambassador's duties, Gentili was at once high-minded and vague, particularly in areas dealing with the inevitable conflict between honesty and political efficacy. Thus he suggested that, contrary to the impressions of others, he chose not to believe that permanent embassies were actually involved in spying on their hosts, and he urged that legates under such suspicion be treated with a wary leniency.[27] Yet he returned to the issue throughout his work, noting that intelligence gathering was a probable, if not primary, goal of many embassies, adding that in his *Laws* Plato saw fit to include spies in his republic, and asking at last why Venetian ambassadors "fail to admit this [practice] openly, since it is their custom on returning from an embassy to omit mention of nothing which pertains to the state in which they have been serving as ambassadors?"[28]

In his various discussions of lying—a topic that would be of great perti-
nence to Wotton—Gentili was even more ambiguous. His *On Embassies*
stated categorically that under no circumstances should the ambassador
deceive his ruler, even when acting in that sovereign's best interests: "the
ambassador is an interpreter, but is the man who reports something at vari-
ance with the substance of the negotiations, an interpreter?"[29] But Gentili's
straightforward assessment of the ambassador as a diligent and linguistically
adept transcriber of foreign affairs was undercut by his 1599 *Disputation on
the Use of Lies*, for here the legal scholar recognized the validity of what he
called "official fraud" and "official lie," untruths that neither harmed nor
benefited any private individual, and whose ultimate goal was the public
good.[30] And to the obvious objection that the "official fraud" designed for the
public good of one state—and such the ambassador's lie would presumably
be—would often constitute the damage of another, Gentili would answer, as
in his influential *On the Just War* of 1598, that human notions of justice were
relative, and that what rendered the deception practiced by a particular state
legitimate was its leaders' desire to maintain both internal order and, more
crucially, an international balance of power.[31] What the ruler knew to be
useful in the long run to his subjects, in other words, took precedence over
the rigorous honesty practiced on a daily basis by the "interpreters" he had
placed in distant foreign embassies.

From an Album of 1604

Into such territory Wotton ventured in his mid-thirties, having begun his
study of Italian with such fervor a decade earlier that he played a minor part
in the peninsula's ongoing debates over language as the addressee of a Sienese
scholar's treatise on the sources of Tuscan.[32] Two critical moments of his
diplomatic career, involving his problematic professional roles as interpreter
or liar, converge with his interest in optics. I turn in this connection to
his biographer's account of the incident for which Wotton is perhaps best
remembered today, his definition of an ambassador, injudiciously recorded
in an *album amicorum* in 1604.

The *album* was originally a white board or wall where public notices
such as the *acta diurna* of ancient Rome were displayed, together with longer-
term announcements such as the annals of the *pontifex maximus*, edicts, and
in time, mere lists of current senators. In the early modern period, the *album*

referred to a book in which Germans, Dutchmen, Flemings, Scandinavians, and occasional representatives of other European nations collected the signatures of their friends and professors. These entries were often accompanied by original poetic compositions, or by biblical or classical verses celebrating piety, fellowship, learning, or other aspects of student life, and by drawings such as a miniature portrait of the signer, a lavish version of his coat of arms, a landscape of the place in which the friendship was established, or some individual cultural features of that locale.[33]

The ritual seemed particularly to outsiders to embody both boorishness and a palpable snobbery; the erudite Peiresc, for instance, referred casually to "one of those uncouth Germans who pester you to write in their albums," even as he commented minutely and admiringly on the way in which their arms were painted there.[34] The arrangement of the album reflected its remote past as either a kind of notice board chronicling a variety of public events, or as a more restricted record of the names of those who dominated civic life. The structure of some such books was merely chronological, recording, like a personal diary bizarrely authored not by its owner but by his acquaintances, the high points in one's contacts and travels. Other albums, by contrast, made explicit a social and civil hierarchy, for their first pages would be reserved for princely and noble signatories, the middle ones for prominently placed non-nobles and savants, and the final section for friends without such distinctions and for family members.[35]

Both aspects were mocked; some detractors commented on the youthful bouts of debauchery so openly commemorated there, while others complained of the costly pretentiousness of those pages displaying the arms of the owner's most powerful acquaintances.[36] The variation, unpredictability, and sharply restricted circulation of the contents of these booklets contributed to their status: in 1820 Goethe, having been sent a fine English sonnet long sequestered in this way and attributed to "W.S.," published a German translation whose romantic title, "From an Album of 1604," was intended to enhance lyrics almost certainly not written by Shakespeare.[37]

Translation and deferred publication also figured in Wotton's experience of the album, but it ended less well. In August 1604, he encountered the custom as he journeyed from his homeland to Venice for his first position as English ambassador to that republic, a post he had chosen, in Walton's words, so as to combine diplomacy with "study and a trial of natural experiments."[38] He stopped in Augsburg and enjoyed the company of friends and local notables, and as he, Walton, and the enemy who eventually exposed him all recounted, Wotton was asked to sign, "after the German custom," the *album amicorum* of a certain acquaintance, Christopher Fleckhammer, a Catholic patrician of that city. There he offered a candid assessment of his profession, writing "An Ambassador is an honest man sent to lye abroad for the Commonwealth," and adding his name and official title. Though the phrase was innocent enough in English—"lying abroad" meaning "residing overseas" as well as "prevaricating"—Wotton's decision to write the defini-tion in Latin admitted of no ambiguity: *Legatus est vir bonus, peregrē missus ad metiendum Reipublicae causâ*. As Walton observed in his biography, the witless remark "slept quietly among other Sentences in this Albo, almost eight years, till by accident it fell into the hands of *Jasper* [Caspar] *Scioppius*, a Romanist," who published it in a work of 1611 as a "Principle of that Religion professed by the [English] King, and his Ambassador Sir *Henry Wotton*, then at *Venice:* and in *Venice* it was presently after written in several Glass-windows, and spitefully declared to be Sir *Henry Wotton's*."[39]

"This coming to the knowledge of King *James*," Walton continued, "he apprehended it to be such an oversight," that the disgraced man was eventu-ally forced to write two apologies in December 1612.[40] The first was a private one in English to his ruler; the second and more famous one was in Latin and addressed to Marcus Welser, the most prominent citizen of Augsburg, and published in early 1613.[41] In the latter Wotton sarcastically thanked Sci-oppius for his "most indulgent interpretation" of the pun, opposed the habit of private friends who "usually write or draw idle as well as true things with equal security in an *album amicorum*" to the Jesuitical frauds then emanating from the pulpit, and described his enemy, a Catholic convert from Lutheran-ism and an émigré to Italy, as "a starving turncoat, a mud-caked quack of the Roman curia, who scribbles only so that he might sup," "a half-baked grammarian, untouched by acquaintance with more solid learning," and "the scummy by-product of a corpse bearer and a camp follower."[42] Though Wal-ton finished the story with a reassuring if unlikely suggestion of James's for-giveness and Wotton's increased standing at court, what is more crucial here is not the happy resolution of the episode, but rather its relationship to Wot-ton's status as a student of optics, and especially to the dark room incident of 1620.[43]

The story nuances Wotton's early connection with Gentili in curious fashion. Walton had noted that Gentili, though pleased by Wotton's three lectures on the eye, had been unable to teach the young man all that he knew of either mathematics or law, and that he concentrated instead on his considerable linguistic abilities. In that it was in part Wotton's talent with languages—his ill-judged readiness to translate an equivocal English pun about his mission in Italy into an uninspired and unambiguous Latin phrase for his German friends—that caused his difficulties, it would seem that Gentili had in some way communicated to the young polyglot both the pious and useless role of the ambassador as a faithful "interpreter" of foreign business *and* the ultimate utility and overriding importance of the "official lie" in the arena of international affairs. Wotton's disgrace, of course, lay in the fact that he had conflated the two views, naively treating the task of the "interpreter" as the unproblematic transmission of information about the commonwealth's aims abroad.

Consider in this connection the way in which Walton began his sad tale: "For eight years after Sir *Henry Wotton's* going into *Italy*, he stood fair and highly valued in the King's opinion, but at last became much clouded by an accident, which I shall proceed to relate."[44] The implied comparison is between Wotton and a pane of glass, and in this it is not unlike the more successful figure used by Thomas Fuller in his biography of the royal physician and great student of magnetism, William Gilbert; "He had (saith my informer) the clearness of Venice glass, without the brittleness thereof."[45] But apart from a possible sojourn in the Veneto in 1573, and a reference in Gilbert's posthumously published *De Mundo* to a perspective glass for observing the Milky Way, there is little in the Elizabethan scientist's life, and nothing whatsoever in this earliest of biographical accounts, to suggest a genuine association with either Venice or its celebrated glass industry, while there are all too many such connections in Walton's narrative.[46] Wotton's near-native Italian or, as his biographer put it, his "connaturalness" to that language,[47] his long residency in Venice, and his early acquaintance with Galileo's telescope rendered him not unlike that idealized image of fair and valuable Venetian glass, a neutral medium through which outside information might be plainly transmitted. The figure thus has a peculiar application to the ambassador; his appalling clarity about his lack of honesty found a fitting material equivalent in the crude message that clouded those clear Venetian windows.

The association of what Gentili called an "interpreter" with clouded glass, moreover, had a resonance that extended far beyond Wotton's linguistic exploit, for translation then, as now, was routinely evaluated in terms of its clarity, or seeming invisibility, as a medium. Not surprisingly, this clarity—

the interpreter's ability to diminish cultural differences between the original and translation by presenting a text notable not for its difficulty and exoticism but for the apparent naturalness and familiarity conveyed by fluency—involved a kind of surreptitious domestication, an effacement of the translator so total that the reader believed the work at hand to be both the original *and* the recognizable product of his own age and culture.[48] A telling extension of the emphasis on apparent lucidity and the covert importation of meaning was the allusion to the telescope, in two English poems celebrating translations in the 1630s, as the ideal emblem of the process: it made what had been remote and indistinct suddenly accessible and deceptively clear, so much so, in fact, that the reader spied in the new text truths latent but not visible in the Italian original.[49] Wotton's effort, made a few years before the invention of the Dutch telescope and exposed shortly after it, quite evidently failed to conform to these canons of taste, for it did not efface but rather emphasized his disastrous role as "interpreter," rendered the original English word play on "lying abroad" alien and opaque, instead of familiar and witty, and magnified the ideological and confessional differences between him and his reader, the German Catholic Scioppius.

If the episode is construed as a kind of botched optical experiment, it is worth noting that Scioppius, the apparent victor, while not a student of optics himself, counted among his friends a number of devotees of that discipline, almost all of whom were associated with the recent invention of the letter-reading device that was the telescope.[50] The truculent Giambattista della Porta, who would claim the instrument as his own in 1609, for instance, had initially chosen Scioppius to bring the table of contents of his latest version of *Natural Magick*, the first book of which would concern optics, to Holy Roman Emperor Rudolph II in 1607.[51] In that same year Scioppius also made the acquaintance of Giovanni Antonio Magini, then one of Europe's leading astronomers and a student of catoptrics since at least 1602, and of Paolo Sarpi, whose investigations of both mirrors and glass lenses date to the late 1570s, and whose skill in optics was such that the telescope was sometimes said to be his invention.[52] Scioppius's close friends Antonio Persio, the Greek mathematician Johannes Demisianus, and the Germans Johannes Faber and Johannes Shreck S.J. were all early members of the Academy of the Lynxes—the first scientific academy, named for that sharp-eyed animal—and were all present at a banquet in Rome in April 1611 when Galileo offered a much publicized demonstration of the telescope, one which involved reading distant letters inscribed on a public building.[53]

Finally, there is the matter of Marcus Welser, the addressee of Wotton's public apology for the incident. He was another early member of the Academy of the Lynxes, and someone who relayed rumors about novel optical instruments, such as the one allegedly allowing an observer in the fall of 1610 "to read an open letter from a distance as great as a man might walk in an hour or more."[54] Wotton's decision to turn to this most prominent citizen of Augsburg was a wise one; Welser had been a close friend of Scioppius since 1598, and it might even have been through him that the latter had obtained the incriminating passage from Christopher Fleckhammer's *album amicorum*.[55] Though a fervent Catholic, Welser had a number of Protestant contacts, and it is possible that the English ambassador imagined that such relationships would guarantee him the widest and most sympathetic audience.

It is, moreover, likely that Wotton's choice of addressee in this publication was informed by knowledge of other and more celebrated missives then sent to Welser, the letters on the sunspots authored by Christoph Scheiner S.J., who wrote under the pseudonym "Apelles," and by his great rival Galileo. These date from late 1611 to late 1612; Galileo's three letters to Welser emerged from the press in collected form, in fact, at the same moment as Wotton's apology. Because Wotton was in Turin and then in Milan from mid-May to late June 1612—arriving there just as Galileo was sending manuscript versions of his first reply to Welser in Augsburg and to friends in Venice—because his contacts with the Venetian intelligentsia remained strong, and because Milan was the site of early sunspot study, he almost certainly had some notion that the addressee and apparent arbiter of this great debate was the Augsburg humanist.[56] And even if Wotton based his decision to direct his apology to Welser on no such information, it seems likely that some of its earliest readers would have associated it with those contemporaneous letters authored by Galileo and Scheiner.

Broken Church Windows

To this impression of the episode of the album as a failed optical experiment I would oppose Wotton's description of Kepler's *camera obscura*, arguing that the latter event constitutes a phantasm of the ambassador's accurate reportage of incidents observed while "lying abroad." That the *camera obscura* was understood as a kind of neutral transmitter of news was anticipated, in fact, by the treatment of pinhole images in both the second of Galileo's *Letters on*

Sunspots and Kepler's *An Unusual Phenomenon, or Mercury's Solar Transit* of 1609, where the technical issues of making and interpreting such observations took on a confessional cast. In writing to Welser in midsummer 1612, Galileo described a certain method of observing the solar phenomena without a telescope and the attendant risk to the eyes, and he insisted that this technique would reveal various novelties about a region generally considered inalterable and thus never newsworthy:

> I have then recognized the kindness of Nature, which thousands and thousands of years ago put in place the means of having some knowledge of these spots and through them, certain great consequences. For without other instruments, the image of the Sun and the spots is carried over great distances through each small aperture traversed by the solar rays, and imprinted on any surface held up to it. It is true that [such images] are by no means as well defined as those of the telescope; nevertheless, the larger ones are perceived distinctly enough. And when in a church Your Lordship sees the light of the Sun fall on the floor through some faraway broken pane of glass, hasten there with a large unfolded sheet of white paper, because you will discern the spots on it.[57] But I say further that Nature was likewise so benevolent that for our instruction she has from time to time speckled the Sun with a spot so large and dark that it was seen with the naked eye alone by a great number of people. But a false and inveterate idea, that the celestial bodies were exempt from all alterations and mutations, made them believe that such a spot was Mercury interposed between the Sun and us, and [this happened] not without shame of the astronomers of that age.[58]

The effect Galileo had in mind, that of the pinhole image, was produced by small fissures in stained glass windows, which had been in use in the Latin West for over a millennium: five windows in the Augsburg Cathedral, created around 1100, are now among the oldest known to us.[59] The revelation of solar phenomena depended, however, on the very recent knowledge of their existence. Galileo's meditation on the availability and frequent misinterpretation of pinhole observations such as this one, moreover, was followed by his explicit criticism of Kepler's recent misprision of an unusually large sunspot in 1607 for the passage of Mercury across the face of the sun.[60] The Italian astronomer's insistence on the broken church window as an acceptable tool for such observations—appropriate, perhaps, for one who worshiped in Florence, where stained glass, a comparative rarity in Italy, was relatively common—was also informed by Kepler's studied avoidance of religious locales in his *Unusual Phenomenon, or Mercury's Solar Transit* of 1609.

In that work, Kepler had described a series of pinhole observations of what Galileo subsequently showed to be sunspots rather than a transit of Mercury, but he had portrayed these experiments as if an ecclesiastical setting would have somehow thwarted them. This is not to say that Kepler never recognized the value of observations made through fissures in stained glass—he mentioned them in passing in his optical treatise of 1604, and would later discuss more fully such experimentation made in the famous Cathedral of Regensburg in 1613—but rather to suggest that the account most familiar to Galileo, and the subject of his meditation in the second of the *Letters on Sunspots*, manifests a real resistance to using this method. Thus Kepler had recounted that on the overcast day of his first view of the large spot on the solar face, he had been talking with a Jesuit priest about the amount of light required for these experiments, when

> around the fourth hour after noon I saw the clouds driven away and the sky tolerably clear, and I broke off the conversation with the Jesuit who had been discussing the matter with me, and retreated to my house for the business of this observation, under a high and wide roof equipped with a crack in a shingle. All of the solar outlines were imprinted [by a projection] about this size on a piece of paper. [Here Kepler included a drawing of the spotted solar body.] Wherever these solar outlines were in bright light, wherever the pinhole admitting the sun was so large that a full intersection of light rays could not take place, nothing could be made out, and I believe this argument goes against the Jesuit's.[61]

Kepler went on to verify this first observation immediately afterward "under someone else's roof," and then proceeded to the Old City of Prague to submit his findings to Holy Roman Emperor Rudolph II. On the way, inevitably, he encountered the Jesuit, and related his experience to him. "But this man, because he was devoting himself at that moment to prayer, bade the Sun to wait for him, and thereafter, by use of an aperture that was too large, he was able to make out nothing of those overly bright outlines, and, in a word, the ornamented Sun escaped his long attentiveness."[62] Kepler, for his part, went on to make a third set of observations that same day by devising an impromptu dark room outside of a place of worship: "on the steps that on the wide walkway of the Canons' chapel lead into the [Imperial] Armory, [two workmen and] I made a pinhole opening, and darkened the place as best we could, blocking the light with a door and curtains."[63]

Kepler's dark room techniques are clearly offset by those of his hapless Jesuit acquaintance, whose dawdling pace, inopportune moments of worship, and luminous observatory allowed him to see nothing of the sun. Whether Kepler intended to portray the Jesuit as no more than the inconsequential embodiment of poor scientific habits, or actually saw some connection between the priest's faith and his failure to conduct proper observations is impossible to judge. It is, however, worth noting that he had received a letter in January 1606 from Nicholas Serrarius, a prominent Jesuit in Munich, that bears a passing resemblance to his description of events in 1607, for this priest had written, "I could not, however, entirely neglect the great work of God, and so I interrupted my task from time to time, and going repeatedly to the little window in my study, I looked up at the eclipsed sun."[64]

It seems likely that in his *Letters on Sunspots* Galileo recast the incident in confessional terms. The allusion to the usefulness of broken church windows to amateur and professional astronomers directly precedes his criticism of Kepler's work, and it is clear that this setting—already suitable for such observations, designed for worship, and tolerably well lighted—would somehow justify Kepler's slow-moving and pious Jesuit target. This is not to suggest that Galileo had in mind a particular defense of any member of the Society of Jesus: he would learn within two months of finishing this letter that his antagonist "Apelles" was a Jesuit, and given that the mediator, Welser, was closely associated with the order, perhaps suspected as much already.[65] It is rather that his account might be construed as a response to the anti-Catholic strains in Kepler's story. More broadly, the motif of the broken church window is easily read as an emblem of the tensions that had animated European civilization since the Reformation.

The *Specularia* of Modern Statecraft

Windows figure elsewhere. Chief among the events leading to the pan-European conflict of the Thirty Years' War was the Defenestration of Prague, when Protestant leaders, provoked by a series of Counter-Reformation measures, entered the Chancellery of the Imperial Castle on May 23, 1618, and hurled two leaders of the Catholic nobility, Vilém Slavata and Jaroslav Martinic, from the windows of that building. The defenestrated men survived—an outcome attributed by their supporters to angelic intervention and by their enemies to the presence of a fresh dung heap beneath the castle windows—and the Protestant rebels soon formed an insurrectional directory, the first act of which was to demand the expulsion of the Jesuit order and its presumed instruments, Martinic and Slavata, from Bohemia. The directory also sought to establish support for its anti-Habsburg initiatives among both neighboring states and, with less success, more distant and powerful Protestant countries, most notably England and the United Provinces.[66]

The death of Holy Roman Emperor Matthias in March 1619, and the prospect of the election of the conservative Catholic Ferdinand, archduke of Austria and king of Bohemia, to the imperial office only served further to alarm the rebels. In August 1619 they deposed Ferdinand and shortly thereafter offered his crown to King James I's son-in-law, Calvinist elector Frederick V of the Palatinate, who unwisely accepted it. Despite his sympathy for the Protestant cause on the continent and his alarm for the fortunes of his daughter Elizabeth and his son-in-law, James recognized the invalidity of their enterprise, failed to see why dissatisfied nobles should be allowed to depose and elect monarchs at their pleasure, and understood that Frederick, soon called the "Winter King," was poorly equipped to sustain the inevitable Catholic assault. James's concern with the legalities of the issue was echoed in many of the publications about Bohemia available to English readers from 1619 through 1622, though the writers of these pamphlets were sometimes more optimistic in tenor about either Frederick's claims or his military potential.[67] James was also very unwilling to be drawn into the pan-European religious war many of his own subjects desired, and in this same period accelerated his unpopular search for a Catholic daughter-in-law as the best means of maintaining a delicate balance of power.[68]

James's apparent indifference to Frederick's plight gave rise to numerous pamphlets on the continent such as the *Letter of Wenceslas Meroschwa of Bohemia to Johannes Traut of Nuremberg Concerning the State of the Present War*, a document of May 1620 printed clandestinely first in Latin and subsequently in German, French, and Dutch translations.[69] Its original publisher was Sara Mang of Augsburg, whose late husband had printed Nicholas Trigault S.J.'s celebrated newsletters from the Far East in 1615, and one of Father Scheiner's treatises on the sun in that same year, and from whose press would soon emerge a sermon of the Jesuit confessor of Louis XIII celebrating the armed intervention leading to the re-Catholicization of Béarn in October 1620.[70]

Purportedly a private letter exchanged by two Protestant spectators of the growing unrest in the Holy Roman Empire and intercepted and published by Catholic authorities, this unlikely epistle examines the political options available to the imperial free cities.[71] Meroschwa recounted at the outset that his addressee had asked in a previous letter what course of action was to be adopted by such cities, particularly by Nuremberg, and had suggested in that missive that the choices were limited to allegiance to Frederick, or to Ferdinand, or neutrality.[72] For his part, Meroschwa responded that he saw no less than six options: Nuremberg might openly support Frederick or Ferdinand or remain neutral, but it also could simulate friendship with one or the other belligerents, while secretly aiding the enemy, or simulate neutrality and secretly aid one side or the other for financial gain.[73]

This wider range of choices was eventually reduced to the somber conclusion, surely designed to demoralize rebellious Protestant readers, that the sole plan worth pursuit was a brief period of simulated friendship with the Winter King—"in order that he not appear abandoned right away"—followed by genuine alliance with the inevitable victor, the Catholic emperor Ferdinand.[74] But what is most remarkable about the letter is the simile with which the six options are prefaced. "That tripartite division of yours," Meroschwa observed, "is derived from the academy and from those masters of ancient candor, but now just as modern astronomers discover with their telescopes new stars in the firmament, and new spots on the sun, so too modern statecraft has its *specularia* and its rules of optics, in which the other [hidden] members of this original division shine forth."[75]

Normally understood in classical Latin to mean window panes, typically composed of mica or selenite, the word *specularia* refers here to glass lenses: so the terms in the eventual German version, *Augenglaser* or "eye glasses," and especially in the Dutch and French translations, *buysen*, "tubes," and *tuiaux prospectifs*, "prospective tubes," and the context of telescopic discovery would suggest.[76] But the residual sense of the Latin word has a certain logic in terms of both the political moment of the letter itself, in the aftermath of the dramatic defenestration of Slavata and Martinic, and the recent exploration of sunspots through broken church windows.[77]

The most crucial element is the suggestion that with the telescopic era, a new set of political truths had been revealed to practitioners of statecraft.

The emphasis on the usefulness of the telescope has shifted from its actual utility as a military instrument—which must have been rather slight—to its value as a metaphorical analogue to political foresight and acuity. As the fake *Letter* works to make clear, however, increased knowledge about one's options did not, in fact, augment one's choices, but rather reduced them sharply. The sort of coercion in which translators specialized—their reliance on a pleasing clarity and fluency to convey something natural, familiar, and even inevitable in the texts they produced, a vision that exposed the utterly hidden truths of the original work as clearly as the telescope did new stars—is at work here.

This view of the telescope as the purveyor of excessive information finds an odd and belated analogue in the former soldier Johann Grimmelshausen's best novel, *Simplicissimus*, first published in 1668 and almost wholly concerned with the Thirty Years' War. *Simplicissimus* is relentless in its depiction of the visual aspects of warfare, but the telescope, the polemoscope, and the spy's traditional devices for "reading distant letters" have all been cast aside for some dubious acoustic machinery that functions unworkably well by day, and can thus be used only at night. As if to insist that what one saw with the telescope was already far too much, Simplicissimus told the reader,

> I devised an instrument wherewith if 'twas calm weather I could by night hear a trumpet blow three hours' march away, could hear a horse neigh or a dog bark at two hours' distance, and hear men's talk at three miles; which art I kept secret, and gained thereby great respect, for all it seemed incredible. Yet by day was this instrument, which I commonly kept with a perspective-glass in my breeches pouch, not so useful, even though 'twas in a quiet and lonely place: for with it one could not choose but hear every sound made by horses and cattle, yea, the smallest bird in the air and the frog in the water in all the country round, and all this could be as plainly heard as if one were in the midst of a market among men and beasts where all do make such noise that for the crying of one a man cannot understand another. 'Tis true I know there are folk who to this day will not believe this: but believe it or not, 'tis but the truth. With this instrument I can by night know any man that talks but so loud as his custom is, by his voice, though he be as far from me as where with a good perspective-glass one could by day know him by clothes. Yet I can blame no one if he believe not what I here write, for none

of those would believe me which saw with their own eyes how I
used the said instrument, and would say to them, "I hear cavalry,
for the horses are shod," or "I hear peasants coming, for the horses
are unshod," or "I hear waggoners, but 'tis only peasants; for I know
them by their talk." "Here come musqueteers, and so many, for I
hear the rattling of their bandoliers." "There is a village nearby, for
I hear the cocks crow and the dogs bark." "There goes a herd of
cattle; for I hear sheep bleat and cows low and pigs grunt," and so
forth.[78]

On the eve of the war that was for the first twenty-seven years of his
life all Grimmelshausen knew, Henry Wotton also manifested a pronounced
reluctance to mention optical instrumentation, as if it made available too
much information, most of it ambiguous and little of it pleasant, to ambassadors charged with the interpretation of foreign affairs. Faced with increasing
pressure to resolve the untenable situation in Bohemia, James I had chosen
Wotton, then returning from England to his old embassy post in Venice, as
his ambassador extraordinary to the emperor Ferdinand, and had instructed
him to send home accurate reports of what he found, to press for the rather
unlikely possibility of a truce, and above all to make clear that he had had no
foreknowledge of Frederick's and Elizabeth's unwarranted accession to the
Bohemian crown.[79]

It is difficult to gauge Wotton's sense of the merits of this unwelcome
mission. His detractors portray him as one ever ready to lie abroad for the
commonwealth; the waspish letter writer John Chamberlain commented in
July 1620 on his overconfidence, while the Venetian ambassador to England
reported "as far as I can make out he will be more of a spy than an ambassador," and noted that many at the English court felt that the wrong man had
been chosen for the undertaking. The fact that Wotton was simply passing
through Germany on his way to Venice, that he had enjoyed no great success
in previous negotiations, and above all that he was not "new, tender, and
easy to bend" seemed to work in his disfavor.[80]

Draft of a Landscape

When Wotton began his celebrated letter to Francis Bacon in December
1620, he had just received from the lord chancellor three copies of the *New*

Organon. As he told his addressee, he had had time only to read the first book of that work, but recognized "a certain congeniality . . . with your Lordship's studies," and regarded the "commerce of philosophical experiments" as "of all other, the most ingenuous traffic."[81] Toward the end of the first book of the *New Organon*, Wotton would have come across a passage of particular resonance. In his discussion devoted to the exchange of scientific information, Bacon had deplored the chaotic and haphazard fashion in which such data had always been collected, even by "Aristotle, so great a man himself, and supported by the wealth of so great a king [as Alexander]:"

> Men of learning, supine and easy-going as they are, have taken certain rumors of experience—as it were tales and airy fancies of it—on which to base or confirm their philosophy; yet, nonetheless, they have accorded them the weight of legitimate evidence. And just as if some kingdom or state were to govern its debates and affairs, not on the strength of letters and reports sent by ambassadors or trustworthy messengers, but on the gossip of the townsfolk and the streets— that is exactly the system by which experience is brought into philosophy.[82]

Bacon's criticism of the masses' meddling in questions of foreign policy is typical of an era in which the crown sought to combat the effects of printed newsbooks; a royal proclamation of December 1620, drafted by the lord chancellor, flatly declared that the "mysteries of state" were "no Theames or subjects fit for vulgar persons."[83] But the supposition that ambassadors would better be able to offer detailed and accurate reports of foreign affairs, the diplomatic equivalent of the meticulously gathered data on which scientific knowledge should rest, would have been far more problematic, especially in the eyes of Henry Wotton.

In addition to the sometime need to lie to a monarch unwilling to face the harsh realities of the Thirty Years' War, there was also the sheer difficulty of obtaining reliable information about crucial events, as Wotton had just found in his investigation of the disastrous defeat suffered by Protestant forces on November 8, 1620, at the Battle of White Mountain near Prague. While he was in Vienna, Wotton heard reports of the event around November 18, but hesitated to believe them in part because of their Catholic source, the two men defenestrated at Prague eighteen months earlier and then banished to Passau: "they lie like lieger-intelligencers for the Emperor, holding practice by letter with some of their inward party in Bohemia."[84] Though the "generalness of the rumour," a kind of *vox populi*, suggested to Wotton that while lying at Passau Martinic and Slavata had perhaps offered an accurate account of the battle, he concluded a letter to the secretary of state, Sir Robert Naunton, with an optical metaphor worthy of his close friend Paolo Sarpi, "I am willing to persuade myself that it will prove a thing multiplied and magnified beyond the truth, *alla solita* [as usual]," and with a request that neither His Majesty nor his addressee be further disquieted until he had more certain information.[85] Such information was evidently hard to come by; four days later, he referred again to the untrustworthiness of the report of those "who lie at Passauwe as lieger-intelligencers."[86] He ended his fruitless speculations with the assertion that "all things considered, we shall have the greater glory that do not believe it." Confirmation of Frederick's overwhelming losses, which some would attribute to sorcery, came to him still several days later.[87]

There was thus little in Wotton's immediate experience that conformed to "the letters and reports sent by ambassadors or trustworthy messengers" presented by Bacon as the ideal equivalent to the accurate and thorough collection of scientific facts. But while his view of most foreign reports as "multiplied and magnified beyond the truth" recalls the prevalent association of the telescope with the sorts of rumors published as news, he also imagined a modified version of that instrument as a peerless and objective receptor of data gathered from distant points. This was Kepler's *camera obscura*, described in the same letter to Bacon, at once the embodiment of a successful optical experiment, and a corrective to the disaster of the *album amicorum*.

> In [Kepler's] study I was much taken with the draft of a landscape on a piece of paper, methought masterly done: whereof inquiring the author, he bewrayed with a smile it was himself; adding, he had done it *non tanquam pictor, sed tanquam mathematicus* [less as a painter than as a mathematician]. This set me on fire. At last he told me how. He hath a little black tent (of what stuff is not much importing) which he can suddenly set up where he will in a field, and it is convertible (like a windmill) to all quarters at pleasure, capable of not much more than one man, as I conceive, and perhaps at no great ease; exactly close and dark, save at one hole, about an inch and a half in the diameter, to which he applies a long perspective trunk, with a convex glass fitted to the said hole, and the concave taken out at the other end, which extendeth to about the

middle of this erected tent, through which the visible radiations of all the objects without are intromitted, falling upon a piece of paper, which is accommodated to receive them; and so he traceth them with his pen in their natural appearance, turning his little tent round by degrees, till he hath designed the whole aspect of the field. This I have described to your Lordship, because I think there might be good use made of it for chorography: for otherwise, to make land-scapes by it were illiberal, though surely no painter can do them so precisely. Now from these artificial and natural curiosities, let me a little direct your Lordship to the contemplation of fortune.

Here, by a slight battle full of miserable errors (if I had leisure to set them down) all is reduced, or near the point. In the provinces there is nothing but of fluctuation and submission, the ordinary consequences of victory; wherein the triumphs of the field do not so much vex my soul, as the triumphs of the pulpit. For what noise will now the Jesuits disseminate in every corner.[88]

Kepler's representation of the field around him had all the features of an objective news report, and in this it would have differed markedly both from the "noise" or rumors of that other "field" then disseminated by the trium-phant Jesuits after the Battle of White Mountain, and from Wotton's own tardy, conjectural, and piecemeal accounts of that crucial episode. This is not to contrast Wotton's and Kepler's reactions to the struggle against Catholic oppression; it was in view of that very situation, in fact, that the ambassador had attempted to persuade the astronomer to emigrate to England, an invita-tion that, however flattering, seemed inconceivable to a man who defined himself as *Ego, Germanus.*[89] Like Wotton, Kepler is presented as an "author," and his impression of the bleak countryside around him was conceived as a "draft." Pinhole and dark room projections, moreover, were routinely said to be "inscribed" or "engraved" or "printed" in the early seventeenth cen-tury, as if to emphasize the smoothness of the transition to mechanically produced and legible images, and in this particular the markings in the *cam-era obscura* would have further resembled the diplomat's verbal impressions of the Austrian countryside.[90]

But what was inaccurate, late, and labored in Wotton's letters to Secre-tary Naunton appears the very emblem of precision, celerity, and seeming effortlessness in the record lately emerged from Kepler's *camera obscura.* Wot-ton's casual comparison of the optical instrument with the windmill involves a device familiar throughout Europe, but most closely associated with sixteenth-century Dutch innovations, and an arrangement where the heavy labor of grinding or papermaking appeared to be managed by unseen currents, the top of the mill pivoting "to all quarters" to benefit from the strongest breezes. Such a figure reinforces the impression of an easy and efficient mastery of an entire landscape.[91]

It also seems likely that Wotton's addressee, Francis Bacon, would have understood the optical device in just these terms. In Bacon's *Advancement of Learning* of 1623, he turned again to a meditation on the way in which Alexander the Great had supported the scientific research of Aristotle, and noted that if the wealthy Macedonian "placed so large a treasure at Aristotle's command, for the support of hunters, fowlers, fishers, and the like, in much more need do they stand of this beneficence who unfold the labyrinths of nature." The project would be costly, he warned, but "there will be no inroad made into the secrets of nature unless experiments, be they of Vulcan or Dædalus, furnace, engine, or any other kind, are allowed for; and therefore as the secretaries and spies of princes bring in bills for intelligence, so you must allow the spies and intelligence[r]s of nature to bring in their bills, or else you will be ignorant of many things worthy to be known."[92]

This comparison recasts that of the first book of the *New Organon* in slightly different fashion: there, the proper collection of scientific data was analogous to the trustworthy reports and letters issued by diplomats and messengers; here, the same task is depicted as an enterprise that was at once as costly and as valuable as the intelligence-gathering missions of the king's secretaries and spies. Whether Bacon imagined the specific case of one of those devices, the *camera obscura*, as the idealized analogue of Wotton's reportage of foreign affairs is difficult to know, but it is perhaps significant that the very first reference to the dark room published in England occurs in a subsequent book of *The Advancement of Learning*. It takes up an argument discussed at some length by Kepler, moreover, that of the optimal color of the screen onto which images of outside objects might be projected.[93]

Houses of Pleasure

The association between the *camera obscura* and the accurate transcription of news finds explicit development in a somewhat later document, but one which suggests that Wotton developed the conceit in letters dating from early

1621, probably during his negotiations with the Union of Protestant Princes of Germany. In 1633, during the second wave of press censorship in England, William Watts, the editor of a semi-annual English-language newsbook largely dependent upon previously published accounts of the Thirty Years' War, the *Swedish Intelligencer*, described his journalistic methods in terms of the same optical device:

> And thus, in this Island, have I done with this *forraine Story*; as in some *Houses of pleasure*, I have seene done with the *Landtskip* of a Countrey: where the *Hills* and *Woodes*, and *Houses*, have by *Perspective* and *Art Optick*, beene so brought thorow a small hole; that they have in little been reflected upon a *Paper*, or *polish't Stone*, in a *Study* or a *Dyning-roome*. And I have used the same *Art Perspective*: the Landtskip of these *Swedish Warres*, have been out of *Germany* brought home into my *Study*, which my *Paper* here reflects off againe. *In little*, I meane, and as I could; and though not in their iust *magnitude*; yet something towards their proportions.[94]

The dark room imagery is used to somewhat different effect here, but Wotton and Watts associate the idealized transmission of news with the same instrument. Watts's metaphor is also noteworthy in its implicit claim to celerity—as if no time at all had elapsed between the scenes played out in the German landscape and their depiction on his paper—for it masks what was a crucial factor in the English authorities' decision to tolerate the *Swedish Intelligencer*, the relatively long interval between event and publication. His emphasis on the miniaturized nature of his account and the faintly apologetic "as I could" are both gestures to the real constraints imposed by Charles I and his licenser, who demanded that this phase of the Thirty Years' War be recounted with some sympathy for the Protestant cause, but without fanfare or mention of England.[95]

His startling comparison between the "houses of pleasure" and his own workplace is more interesting still. In that the term then usually, but not always, indicated a place of debauchery, the simile and the substitution of studies and dining rooms for darkened bedchambers both emphasize and subsequently erase a conventional connection between the *camera obscura* and erotic activity, an association which animates, for instance, the last two stanzas of John Donne's contemporaneous "Canonization." More generally, the term may point to a pervasive unease about the transformation of individual endeavors such as letters or scholarship into profitable reading matter for an unregulated public. In this regard, Watts would have among his antecedents the young Jesuit Georg Stengel, who was happy to have helped Scheiner make observations of the sunspots in a *camera obscura* in late 1611 and early 1612, but chagrined to see them published by Welser, and who referred darkly in a private letter to other and yet unseen solar images, "that have not yet been exposed to public view," *quae nondum sunt prostitutae*.[96] The most plausible reading, however, where Watts's "houses of pleasure" are simply out-of-the-way summer palaces suitable for otiose optical experimentation, correctly identifies the other relevant feature in the circulation of the *Swedish Intelligencer:* it was more expensive to produce than a weekly newsbook, and it had a much lower and presumably wealthier readership.[97]

A sermon preached by Watts in 1637 is dedicated to Wotton, then provost of Eton College, and its insistence on preferment and benefits may signal a relationship of some standing.[98] Indeed, Wotton was perhaps in contact with Watts just a few weeks after he wrote the letter concerning the *camera obscura* to Francis Bacon, in January 1621. The chaplain, traveling in the company of Wotton's nephew and fellow ambassador Albertus Morton, had left England for Heilbronn on December 10/20, 1620, to offer James I's rather belated and limited support to the Union of Protestant Princes; Wotton, charged by the English monarch "to comfort and strengthen the United Princes and assure them of [the king's] good intentions" before going to Venice, proceeded from Vienna to Munich in search of documents allegedly damaging to the defeated Winter King.[99] Though it is clear that the Wotton and Watts could not have met in person at that point, it is possible that, always in very close contact, they exchanged letters in which the *camera obscura* was portrayed as the ideal transmitter of news.

While the image of the outside world produced by the scientist in his dark room represented to Wotton, Bacon, and Watts a fantastical corrective to the diplomat's own efforts to interpret "a thing multiplied and magnified beyond the truth," as if by a telescope, it also stands as a successful revision of the earlier episode of the *album amicorum*. There, an unnewsworthy remark recorded in "a Book of white Paper," "slept quietly"—as if sequestered in a dark chamber—for eight years before Scioppius "threw open the bookcase of friendship," allowing the diplomat's private observation to be inscribed in Venetian glass and exposed to the outside world; here, the timely data of the outside world was projected or "thrown," via the lens of the *camera obscura*, onto the piece of white paper designed to receive such impressions.[100] And while those who signed the *album amicorum* "in the German fashion" were accustomed, as Wotton noted in his apology to Welser, to "write or draw idle as well as true things," Kepler made it clear that he was proceeding "less as a painter than as a mathematician," and that this professional pursuit was one of accuracy and objectivity rather than leisurely pleasure.

Like Some Magistrate

The English ambassador had prefaced his discussion of the dark room with an observation at once troubling and revelatory: it was "more remarkable for the application than for the theory." Kepler's theoretical explanation of the *camera obscura* would have certainly involved a comparison of the optical instrument with the eye, for he had offered such analogies in his *Paralipomena in Vitellionem* of 1604 and repeated them more succinctly in his *Dioptrice* of 1611.[101] In the first instance, he was correcting della Porta's comparison of the sensory organ and the dark room: in 1589, in his alleged resolution of the dispute over extramission and intromission, the Italian natural philosopher had likened the pupil to the "opening of a window," which was conventional, but had also suggested that the anterior face of the crystalline humor, rather than the retina, served as the screen onto which intromitted images were projected. This arrangement effectively deprived the lens-like crystalline of its refractive power, and, as Kepler pointed out, was inconsistent with the optical properties of such lenses discussed elsewhere in the *Natural Magick*.[102]

It is probable, then, that something of what Kepler explained to Wotton about the actual function of the pupil—which was for him merely an empty and adjustable aperture, rather than a rigid glassy surface—added to Izaak Walton's impression of the optical experiment as the antitype of the Scioppius incident, where the window and what was etched on it figured far too prominently. More crucial for this context, however, is the issue that Kepler *failed* to resolve. As he stated in his optical treatise of 1604, the projected image, whether it fell onto the retina or onto the white paper in the dark room, would be both inverted and reversed.[103] In terms of his "draft" of the countryside around Linz, the problem of legibility was solved without difficulty: to correct the inversion, one simply rotated the paper 180 degrees, such that the details of the skyline ran along the upper border, and those of the foreground along the lower one. For the left-right reversal, one had only to hold the original drawing up to the light and to trace the outlines of the landscape onto the back of the paper.

But Kepler admitted he did not know how such fundamental adjustments were made within the eye of the observer; neither optical laws nor anatomical features explained them to him. As he stated in a memorable passage of his treatise of 1604, these issues, and the more general problem of interpreting visual data after it reached the retina, fell outside of the mathematical discipline of optics:

> How it is that the image or picture is united with the visual spirits that dwell in the retina and in the [optic] nerve, and whether it is escorted by these spirits within the caverns of the brain to that tribunal of the soul, its visual faculty, or whether the visual faculty, like some magistrate granted by the soul, descending from this superior court to this [optic] nerve and retina as if to an inferior one, meets the image on the way, all of this, I say, I leave to the discussion of natural philosophers.[104]

This division of labor must have held some interest for Wotton; though the puzzle in Kepler's experiment troubled him enough to prompt his reference to its modest theoretical basis, these delimitations would have also confirmed his impression of the optical instrument as an idealized device designed solely for the neutral transmission of information about the outside world, and not for the interpretation of such data. That task, as Kepler suggested in his allusions to legal summons to superior courts and to magistrates scurrying to inferior tribunals, was the business of high officials and low functionaries entirely unknown to him. This absolute disjunction between the passive intake and dispatch of graphic information that was seemingly as available as the landscape, and its subsequent oral evaluation by discussants remote from that region represents, of course, a radical difference from the endless cycle of lies, conjectures, and mistakes that were the lot of the ambassadorial "interpreter," and it is easy to see its appeal to Wotton in this troubled period of his career.

There are similar anxieties in the dispatches sent from Germany by James's favorite, Viscount Doncaster, to Secretary of State Robert Naunton in the same period, though these are expressed in aural rather than visual terms. Thus while Wotton wrote that he assumed reports of the Catholic victory at White Mountain would prove "a thing multiplied and magnified beyond the truth," Doncaster recounted that "this recent experience, I say, of the uncertainty of Fame's trompet upon the first sound, in matter of warre especially, made me then so wise as to attend a second before I gave his Ma^ty the alarme; and now repent me of it, hearing the niewes of Moravia confirmed from all parts." While Wotton seems to have envisioned the *camera obscura* as a recording device that would presumably pass the burden of interpreting accurate images of distant phenomena to other officials, Doncaster insisted, however disingenuously, on his own task as a simple recorder of sound, rather than as an active interpreter of speech: "Sir," he wrote to Secretary Naunton in July 1619, "you see how large and particular an accompt I have taken paynes to give you, not to the matter only, but of the very wordes, and almost syllables of all that is passed in my negotiation here." It is worth noting that the *result* of this proposed recording or transcript is nonetheless visual, and it takes the phantasmatic form of the optical image whose judgment is left to others: Doncaster hoped that "his Ma^ty may by the same light I have upon the place judge of a phancy wherewith both sleeping and waking I am troubled."[105]

That Wotton imagined an alternative to the diplomatic task before him—some means, at once natural and mechanical, of conveying relevant information about the political horizon to the scrutiny of other elite insiders—is not surprising. What is interesting about the proposed effacement of individual bias, or rather its subsumption into a geometrically defined perspective, is the way it anticipates similar maneuvers in the imagined public sphere, likewise animated by fractious individuals masquerading as a stable group whose particular viewpoint, in turn, is presented as a rational and sufficiently detailed description of the current landscape. Wotton's effort, of course, is to keep the monopoly and analysis of recent news entirely in-house, and in this he differs not at all from Bacon. Watts, in complicity with King Charles I, who resented "vulgar" access to continental news and sought to retain control of the discussion of English foreign policy, seems to have appropriated Wotton's metaphor, suggesting to a small sector of the public that this objectively transmitted news was for their consumption, even as censorship insisted upon its delayed broadcast and reduced format.

The most striking element of the dark room, however, and the subject of the following chapter, is its various emergence as an instrument more problematic than Wotton's idealized device. Thus the *camera obscura* does not merely record the progress of the Thirty Years' War, but also foresees the event, and describes its approach; elsewhere it excludes all information pertaining to current events, but exposes the enabling fictions of that spectral entity constituted by the rational analysis of just such information, the public sphere; still elsewhere, it figures not as a machine best deployed to gather and to scrutinize foreign reportage, but rather as a device for observing the burgeoning practice of news consumption itself.

CHAPTER 5

Cameras That Do

In the fall of 1611, a ponderous volume emerged from a press in the Lutheran city of Wittenberg, the second printed work in the telescopic era to mention sunspots. Unlike its predecessor, Frisian Johann Fabricius's *On the Spots Observed in the Sun*, published in the spring by the same press, this treatise alluded to the spots only in passing, and never mentioned the telescope at all.[1] For Ambrosius Rhodius, a professor of astronomy, the *camera obscura* was the instrument most worth attention, and much of the *Optica* is devoted to its revelations. His treatment of the question of the retinal image follows Kepler's discussion in close, even slavish fashion. Thus Rhodius, too, puzzled over the judicial faculty of the visual spirits, the "tribunal of the soul," the "laws of optics," the "ministry" and "office" assigned by some to other parts of the eye, the "low bench" of the optic nerve, and so forth, before concluding that the issue of interpretation was not in his bailiwick: "Here the Armory of Optics falls silent."[2]

The Armory of Optics was for the most part a noisy place. This chapter examines several depictions of the *camera obscura* that emphasize its connection to confessional conflict even as it is portrayed as a scientific instrument. The agreeable fiction of the camera that doesn't lie, articulated in December 1620 by Henry Wotton in his letter to Francis Bacon, is undercut by several cameras that do. News reporting, in these accounts, has nothing neutral about it.

Wars and Rumors of Wars

Though Rhodius suggested in the dedication of his treatise that the work was addressed neither to Jesuits nor to Calvinists—for it included the imprecation

"Begone, evil of the ill-omened Jesuit; begone, too, aimless crowd of the bald sect [*calvae sectae*]"—at least one member of the Society of Jesus, Christoph Scheiner, paid close attention to developments in astronomy and optics, as well as to current events.[3] The Swabian Scheiner was involved in the intelligence gathering and political meddling that Wotton portrayed as the disastrous specialty of the Society of Jesus.[4] Like most of his peers, Wotton regarded the archduke and future emperor Ferdinand II, to whom Scheiner's *Oculus* or treatise on the eye was dedicated, as "a prince wholly in the power of Jesuits."[5] Ferdinand's younger brother Archduke Leopold, whose letters of 1620 Wotton had described as "reeking" of another Jesuit adviser, received frequent doses of political counsel from Scheiner in 1620–1621, and regular missives regarding their lucrative traffic in sacred relics from 1625 to 1632.[6] Scheiner provided Leopold with occasional intelligence reports as well, detailing, for instance, the activities and demands of the English diplomats who succeeded Wotton in this crisis in Austria in July 1621, dismissing their objectives with the knowing remark, "But they're not Swabians, so you can get a lot more out of them," and portraying their efforts as "a thing well worth preventing, being most harmful to the Christian cause."[7]

Scheiner's long-term interest, the maintenance of older Jesuit colleges in Germany and the establishment of a lavish new one at Neisse, was moreover just the kind of undertaking deplored by Wotton in his reports to various English officials; the latter project was financed, in fact, by Ferdinand's and Leopold's brother, Archduke Carl of Austria, at Scheiner's suggestion.[8] In their correspondence of the early 1620s, Jesuit superior general Muzio Vitelleschi repeatedly admonished Scheiner not to overstep his role as confessor to Carl, not to dabble in the activities of the court at Vienna, not to recommend private individuals for civic offices, not to communicate except by secret code, and not to spend so extravagantly on clothing, food, travel, books, and optical instruments.[9]

But apart from the natural antipathy of Wotton and Scheiner, there is the strong possibility that Kepler actually discussed the latter's *Oculus* with his English visitor in the course of the conversation held in the astronomer's study. While Kepler or Scheiner might have been acquainted with a *camera obscura* that allowed the artist Hans Hauer to draw a landscape of Nuremberg in those years, this device appears to have received no theoretical treatment in print until 1636,[10] and it is more likely that Kepler sought to distinguish his dark room from one created by Scheiner, described to him by the mathematician of Archduke Maximilian around 1615, and treated at some length in the *Oculus*.[11] The phrase *non tanquam pictor, sed tanquam mathematicus*, "less as a painter than as a mathematician," in fact, might have been Kepler's way of differentiating his efforts from those of Scheiner, who had adopted the name of the ancient painter Apelles in the course of the sunspot controversy, who tended always to insist upon the applicability of such projections to all fields of the "science of drawing," and who would further develop such painterly metaphors in his *Rosa Ursina* of 1630.[12] Wotton's observation about the suitability of the dark room for chorography, in fact, has its echo in Scheiner's *Oculus*; the English ambassador's certitude that "to make landscapes by it were illiberal, though surely no painter can do them so precisely" is perhaps a grudging affirmation of the Jesuit author's unapologetic view of its use in painting. And Wotton's depiction of the entire project as "artificial and natural curiosities" may be reminiscent of Scheiner's tendency in the *Oculus* to describe dark room apertures without lenses as "Nature" and those with them as "Art."[13] In sum, the uneasiness that accompanies Wotton's presentation of the Keplerian device as a corrective to triumphal Jesuit sermons about the Catholic victory at White Mountain may be explained by his knowledge that Scheiner had apparently developed such an instrument independently of the imperial astronomer.

The dedication, addressed to the Catholic Ferdinand, and the frontispiece of Scheiner's *Oculus* are worth some scrutiny. Though the text itself was composed around 1617, the imprimatur of the work dates to February 5, 1619, and a poem honoring Scheiner to March 7, 1619, while the dedication is four months later, which suggests that the treatise was typeset in the interval, and that the front matter was added just as the *Oculus* was sent to booksellers.[14] The dedication bears the date June 12, 1619, just as the Bohemian rebels' siege of Vienna was being lifted, and shortly before their formal deposition of Ferdinand in favor of Frederick in August of that year.[15] It begins with a message of great political moment, but of little relevance to the technical particulars of the treatise: "Let he who reads Matthew 24 and Mark 13 understand."[16] These virtually identical chapters of the gospel concern Christ's warning to his disciples about the imminent destruction of the temple; as he reminded them, "there shall not be one stone upon another, that shall not be thrown down. . . . Take heed that no man deceive you. For many shall come in my name, saying, I am Christ; and shall deceive many. And ye shall hear of wars and rumors of wars: see that ye be not troubled: for all these things must come to pass, but the end is not yet. For nation shall rise against nation, and kingdom against kingdom: and there shall be famines, and pestilences, and earthquakes, in divers places. All these are the beginnings of sorrows."[17]

It is difficult to imagine who might have read these verses in 1619 and failed to understand them in terms of the inevitable conflagration that would be the Thirty Years' War. In view of Henry Wotton's doomed diplomatic efforts to avert this conflict and his difficulties in securing and relaying accurate reports of the decisive encounter at White Mountain, the peculiar promise "ye shall hear of wars and rumors of wars" would have seemed especially apposite. Scheiner's depiction of Ferdinand as "the eagle of the heavens," if banally martial and a familiar symbol of the Habsburg dynasty, becomes decidedly more sinister in the context of Matthew 24: 28, "For wheresoever the carcase is, there will the eagles be gathered together," which seemed to allude to the crucial aid the Catholic ruler would receive from the Spanish branch of his family.

Scheiner's task was first to suggest that the mandated allegiance to the apparent victor Christ also involved loyalty to Ferdinand, his representative in Bohemia, and then to relate that confessional issue to the more technical arguments of his *Oculus*. This was a difficult undertaking: his depictions of the way light behaved upon entering a minute opening in "a cubicle or heated room or other site so occluded" seem remote from the sound and fury of the battles so triumphantly promised in the dedication of the work, and his forthright confession that his anatomical model was the eye of a bull rather than that of a fresh human corpse a far cry from the mass carnage envisioned in Matthew 24 and Mark 13.[18] Yet the frontispiece, entitled the "Hieroglyphic Compendium of the Entire Work," is clearly devoted to both confessional and optical matters, as a cursory examination will show.[19]

The image concerns for the most part various types of dark rooms, with touches of anti-Galilean sentiment in its iconography. Thus the *camera obscura* on the lower right, as has been noted, features the inverted and reversed image projected by a Galilean or Dutch telescope, and the rubric *Non integer intrat*, "it does not enter unchanged," explicitly insists on the distortions of the Italian's observations of sunspots.[20] The peacock in the lower foreground, if an obvious reference to pride, also had a narrowly Galilean focus: in the *Sidereus Nuncius* the Italian astronomer had compared his visual impressions of the moonspots, subsequently much contested by observers within and outside the Jesuit order, to the eyes on that bird's tail.[21]

But what is more significant than the particular polemic between Galileo and Scheiner is the martial tenor that pervades the optical issues of this

FIGURE 2. Frontispiece, Christoph Scheiner, *Oculus*, 1619.

Reproduced with permission from Linda Hall Library of Science, Engineering & Technology.

frontispiece. Just above the dark room showing the defects of the Galilean telescope, for example, is one with two convex lenses, the configuration described by Kepler in his *Dioptrice* of 1611 and adopted by Scheiner for an upright projection of the sunspots in 1617.[22] The image of the man standing outside is neither inverted nor reversed inside the dark room, and the rubric *Post eversionem erigor*, which might be translated either "after an overthrow I am raised up again" or "after an inversion I am righted," presumably mirrors both Scheiner's sense of the eventual outcome of his conflict with Galileo and the optical characteristics of this sort of projection.

The triumphal tone of Scheiner's pronouncement has a particular applicability to the situation of the man to whom the work was dedicated, the Catholic Ferdinand, whose capital had just been assaulted by Protestant rebels, and whose title of "King of Bohemia" would be offered to James I's son-in-law, Frederick V of the Palatinate, within a matter of weeks. Ferdinand had indeed, to the immense surprise of most observers, been raised up again, and however insecure his estate in the summer of 1619 when the *Oculus* emerged from the press—when Moravia, Silesia, Upper and Lower Lusatia, and Upper Austria had joined the rebel forces—the aftermath of the Battle of White Mountain left little doubt that he was indeed as Scheiner had portrayed him, "The Sun and the All-seeing Eye," and, as the Eagle proclaims, a ruler with religious force, political legitimacy, and military backing.[23]

The other optical emblems of the frontispiece are characterized by the same air of prescience, one which makes the "good news" of the gospel the advance bulletin from which all other forms of wartime reportage devolved. On the lower left, for instance, the dull depiction of the conic pinhole projection is enlivened by a rubric of some political significance, *post angustias dilator*, whose translation "after restrictions I expand" applies equally well to Ferdinand's evolution from a captive of the besieged imperial city of Vienna and an impoverished ruler to one who, from the summer of 1619, could count on the military backing of Spain, the papacy, the Spanish Netherlands, Tuscany, and Lombardy, and who, to Superior General Vitelleschi's surprise, would offer a significant donation to Scheiner for the Jesuit church in Innsbruck in 1621.[24] Equally uninspired, from an optical standpoint, is the pronouncement beside the *camera obscura* just above, which states, "I am made dark so that I appear illuminated in this place"; the reading "I have been judged obscure in order that I might become illustrious here" is surely the one that most interested the dedicatee and his subjects. Both of these rubrics were also appropriate for the Jesuits of Bohemia, whose expulsion in June

1618 was followed two years later by their triumphal return and vigorous program to restore that kingdom to Catholicism.

The setting of the alleged victory is on the side of a light-colored elevation, a miniature version of the White Mountain where Ferdinand's forces would crush his rebellious Protestant subjects and their many allies after an hour's battle in November 1620. The aptness of this geographical detail is not very surprising, given that Prague had been invaded by Ferdinand's younger brother Leopold from White Mountain in 1611, but the message emanating from the temple-like structure at the summit to the distorting dark room with the Galilean telescope, "As soon as I have passed over the peaks I am borne to the lowest place," precisely embodied the fate of the Winter King's army.[25]

The foregoing associates Scheiner with the victorious Ferdinand, and, in a gesture that anticipates the Italian astronomer's status as a heretic, Galileo with the rebellious Protestant nobles who would suffer defeat at White Mountain. This overlay is emblematized in one of the more curious rubrics of the frontispiece, that just beyond the left hand of the man whose upright image is projected through the double convex lens. *Frustra ante oculos pennatorum*, "In vain before the eyes of birds," is a paraphrase of Proverbs 1: 17, the context of which is a father's advice to his son about the avoidance of evil companions. Predictably, the counsel offered in Proverbs is the sort that the young and fatherless Frederick V had failed to heed, despite such arguments from both his mother and his father-in-law, when approached by the insurgent Protestant noblemen who offered him the crown:

My son, if sinners entice thee, consent thou not.
If they say, Come with us, let us lay wait for blood, let us lurk privily
 for the innocent without cause:
Let us swallow them up alive as the grave; and whole, as those that
 go down into the pit,
We shall find all precious substance, we shall fill our houses with
 spoil:
Cast in thy lot among us; let us all have one purse:
My son, walk not thou in the way with them; refrain thy foot from
 their path:
For their feet run to evil, and make haste to shed blood.
Surely in vain the net is spread in the sight of any bird.

And they lay wait for their own blood; they lurk privily for their
own lives.
So are the ways of every one that is greedy of gain; which taketh
away the life of the owners thereof.[26]

In Scheiner's frontispiece the requisite nets, near the left foot of his
upright nobleman, are before the eyes (*ante oculos*) and thus apparent to the
triumphal Habsburg eagle, but well behind the eyes, sightless and otherwise,
of the vainglorious peacock of the foreground. By the fall of 1620, the dis-
gusted reaction of James's daughter, Elizabeth of Bohemia, to Ferdinand's
imperial election—"they have chosen here a blinde Emperour, for he hath
but one eye, and that not verie good"—appeared less accurate a description
than Scheiner's portrait of this all-seeing ruler.[27]

The conflation of the two rather different impulses of the frontispiece,
that uneasy combination of Scheiner's optical theories and his ruler's martial
undertakings, also explains the design's emphasis on the interdependence of the
eye and the hand. This sort of collaboration is boldly signaled by the ghastly
one-eyed fist in the center of the plate, by the frustrated individuals on the left
and right summits, and by the mottos that accompany them, "The hand sees
nothing without the eye," and "The eye is worth nothing without the hand."
While it is reasonable to see such observations as motivated by a general recognition of the nature of sensory illusion, and consistent with recent research on
this topic by Father Scheiner's fellow Jesuit François Aguilon, it is also true that
the insistence on the coordinated efforts of the eye and the hand can be read
as the Swabian scientist's attempt to suggest to a ruler and patron the relevance
of his field to the pressing military matters of the day. Given the regard many
early modern regents had for subjects possessed of optical knowledge, and the
supposed utility of the spyglass and related devices in warfare, Scheiner's suggestion would not have seemed farfetched.

The Story of a Dissection

A particular incident in the aftermath of the Battle of White Mountain also
seemed to affirm, in belated fashion, the heavy-handed alignment of Catholicism with perception, and heresy with blindness. In his *Oculus*, Scheiner had
freely acknowledged that he was following the work of Felix Platter and
Kepler in treating the retina as the sole instrument of sight, and the crystalline

as a mere lens for presenting images of perceived phenomena.[28] Such dependence on the work of the Protestant anatomist and astronomer meshes poorly with the confessional overtones of the frontispiece, but what may be more significant is Scheiner's silence about another anatomist, a man whose failure to recognize the true nature of the retina, and well-known opposition to the Society of Jesus would have made him the very emblem of unseeing heresy.

The two passages from Kepler's *Paralipomena* concerning the roles of the retina and crystalline cited by Scheiner contain not just applause for the work of Platter, but also discussion and eventual rejection of the theories of the Hungarian nobleman Johannes von Jessen.[29] A professor of medicine, the former physician of Rudolph II, and a close friend of Kepler, Jessen had treated the crystalline humor as percipient in his *Story of an Anatomy at Prague* of 1601, and he was obliged to find or to fabricate a means by which visual impressions were transferred through the retina to the optic nerve.[30] Scheiner attacked those kinds of arguments, maintaining that whichever part of the eye was granted visual power must have the ability "to offer incoming images a tranquil home, not the footpath of a journey."[31] But despite his respect for the anatomical tradition, his familiarity with other ocular anatomists of the period, and his insistence that the study of optics involved the contributions of physicians as well as those of theorists of vision, Scheiner avoided mentioning Jessen by name in the *Oculus*.[32]

It might be argued that the Jesuit's emphasis on his unfamiliarity with the human eye—"I have never had an abundance of excised human eyes available to me," he wrote, adding, "As I have said, an excised human eye has never passed through my hands"—takes issue with Jessen's fame as a dissector, for the latter's work is, as the title announces, the "story of a dissection solemnly administered by himself in Prague in the year 1600," the first public autopsy performed in that city.[33] But Jessen was quite well known to the Society of Jesus on other grounds, for he was rector of that Protestant stronghold, the Carolinum of Prague, which was described in a document written by Martinic and Slavata and forwarded by the Jesuits to Pope Gregory XV, as "the forge and the fountainhead of heresy and of all sedition," just as that order's college in the Old City of Prague, the Clementinum, was viewed as the focal point of Counter-Reformation efforts.[34] It was from the Carolinum that Protestants impatient with Catholic repression, regarded as the work of the Jesuits and of Martinic and Slavata, issued a list of their grievances two days prior to the celebrated defenestration of these two officials on May 23, 1618.[35]

In June of that year, Jessen had acted as a delegate to Hungary in search of anti-Habsburg support, and was subsequently imprisoned for six months. Ferdinand released this prominent figure as a show of good will, but in the spring of 1620, Jessen published his dissertation, originally written at the University of Padua, on the right of the populace to rebel against tyrants, and in the summer he approached the Hungarians again, offering them financial aid for their continued resistance to the emperor.[36] Jessen was also rumored to have been the author of an anti-Jesuit pamphlet published in Prague in this period, one which, under the fiction of a medical exam to cure Bohemia of its troubles, laid the blame for her maladies on the society, Martinic, and Slavata, and called for an improbable union of all German princes against the pope and his instruments.[37]

Those who followed "the wars and rumors of wars" circulating in weekly newsbooks of 1621 would not have been surprised to read early in that year that many of the restive noblemen of Bohemia had been imprisoned in the Castle of Prague for their involvement in the failed rebellion.[38] Inevitably, in an article of June 25, 1621, English newsreaders would learn that "Doctor Jessenius / lately Professor in the College of Caroline in the olde Cittie / first his tongue was cut out / then beheaded / quartered and hong uppon 4. stakes."[39] A rival coranto, making the most of advance publication of the types of torture and execution each of the rebels was to undergo, offered a slightly different version of his death: "*Iohn Iessenius* Doctor his tongue cut out, quartered alive, but grace given him, he is first to have his tongue cut out, then his head cut off, and his body quartered, and the quarters hangd before the gallowes gate, and his head set upon the Tower."[40] The emphasis of all reports on the various forms of torture duly inflicted on the leaders of the Bohemian Rebellion, while typical of the period and of such accounts, is rendered particularly gruesome in the case of the famous anatomist, onetime author of the "story of a dissection solemnly administered by himself in Prague in the year 1600."

Scheiner, of course, could not have known Jessen's fate as he was composing the *Oculus*, but it does seem reasonable to conclude in 1618–1619, when the physician and university rector distinguished himself in seditious activity, that he ran the risk of severe punishment. In the pamphlet attributed to him, Jessen himself predicted not the individual destiny of a martyr, but rather the widespread massacre of his coreligionists, a bloodbath like that of French Protestants in 1572.[41] It is probable that Scheiner's studied avoidance of Jessen's name and work in the *Oculus* was motivated by tact, or by genuine uncertainty about the rector's eventual fate, or by an unwillingness to mention a scholar whose political interests and religious beliefs were so clearly opposed to his own; Kepler's disciple Rhodius, writing in 1611 and in general sympathy with this scholar, had referred to him only as "a certain noble anatomist" when correcting his statements about image formation.[42] But the horrible public spectacle of Jessen's death, conveyed in newsletters across Europe, has the retroactive effect of rendering Scheiner's silence suspect, and of presenting the unmentionable anatomist, in error about the true mode of vision, and seemingly blind to Catholic supremacy, as the very image of the sort of enemy signaled by the "Hieroglyphic Compendium of the Entire Work."

The judicious choice of iconography in the frontispiece of the *Oculus*, while merely the result of a series of good guesses about the locale and outcome of the eventual conflict, likewise presents the author as commendably farsighted in political matters, a characteristic seemingly connected to his expertise in optics, and popularly associated with his order. What seems worth emphasizing is the possible relevance of the dedication and frontispiece to Henry Wotton, whose impression of the *camera obscura* as an idealized mechanism for the rapid and accurate transmission of news appears to have been motivated, at least in part, by his own tardy and inaccurate reports of the disaster at White Mountain. Scheiner's iconography, in this reading, involves that same association of the device with the military encounter, the crucial difference being that what Wotton imagined as mere reportage after the fact figures as a kind of augury in the *Oculus*.

The Private Man

The *camera obscura* reappears in another work conceived in just these historical circumstances, René Descartes' *Discourse on Method*, and in the eventual treatise that took up the issue of interpretation of those reversed and inverted retinal images, the *Dioptrique*, both first published in 1637. As is known from an early biography, from his letters to the Dutch natural philosopher Isaac Beeckman, and from a few remarks in the *Discourse*, Descartes came from the United Provinces to Germany in 1619 as a detached observer of the drama of confessional conflict. Neither his preparation at the Jesuit college of La Flèche in France from 1606 to 1614, nor his course in law in Poitiers from

1614 to 1616, nor his study of mathematics and military architecture in the service of Maurice of Nassau from 1618 to 1619 fully explain his decision.

> The news that they brought to Breda about the great unrest in Germany awoke in him that curious desire to be a spectator of everything of note in Europe. People were talking about a new Emperor, they were talking about the revolt of the Bohemian Estates against their king, and about a war that had broken out between Catholics and Protestants over this issue.[43]

Descartes proceeded in leisurely fashion, maintaining a certain distance from "wars and rumors of wars": as he wrote to Beeckman in April 1619, "the tumults of battle surely do not yet summon me to Germany, and I suspect that though many men may be in arms, of wars there are yet none."[44] He witnessed Ferdinand's imperial coronation in Frankfort late in the summer of 1619, and as he related nearly two decades later, "the onset of winter stopped me in a locale, where having no social contacts to distract me, and by good fortune being without cares or preoccupations that might trouble me, I remained shut in alone all day in a heated chamber, where I was entirely at leisure to entertain my thoughts."[45] Here he began the reflections which would culminate in the *Discourse on Method*.

Though the Thirty Years' War serves as a backdrop for both the *Oculus* and the *Discourse*, very different principles animate these works. Scheiner's treatise, as the prefatory matter makes clear, is martial in impulse, and the Jesuit scientist appears utterly loyal to Ferdinand during this low point in the latter's career; the young Descartes, by contrast, enrolled as a volunteer in the army of Archduke Maximilian of Bavaria, cousin to the emperor Ferdinand, with the wish to serve under this Catholic ruler "just as he had under the [Protestant] prince" Maurice of Nassau, with the firm intention of playing the role of spectator, and with no sure knowledge of "against which enemy the troops were being prepared."[46] Even in the most copious of his early biographies, the *Life of Monsieur Descartes* of Adrien Baillet, it is difficult to ascertain what it is that the French philosopher did or desired to do in this unlikely military role, so thoroughgoing is the portrait of his pacifism and neutrality: "As he cared little about entering into the interests of these states and rulers—Providence not having had him born under their domination—he had no intention of carrying a musket in order to advance the cause of one, nor to destroy that of the other. He enrolled therefore in the Bavarian troops [of Maximilian] as a simple volunteer, without taking up a specific task."[47]

Descartes' lack of strong commitment to either side, here presented as a kind of studied nonpartisanship, had a wretched analogue in entries in the wartime diaries of the Jesuits Johann Buslidius and Jeremias Drexel, then respectively the confessor and the court preacher of Archduke Maximilian of Bavaria. Both priests recorded the hanging deaths of a number of French deserters near Linz in August 1620, and remarked particularly on a soldier from the philosopher's own milieu: "Among the criminal men was one who had studied with our Society at La Flèche. He knew Latin so well that it was hardly to be believed, when he was well prepared for death, how he commended himself to God and to the saints, and to the Blessed Ignatius among them, now in Latin, now in French."[48]

The works of both Scheiner and Descartes address, though very differently, the motif of imminent destruction. The *Oculus* does so in the curious fashion mandated by its suture of optical and confessional issues, by recalling in its dedication Christ's promise of coming annihilation, where "there shall not be left here one stone upon another, that shall not be thrown down," and by suggesting in its frontispiece some connection between Ferdinand's Protestant opponents and Scheiner's professional rivals. The *Discourse* begins by contrast with a meditation on the very inadvisability of a thoroughgoing demolition of any sort, and with the observation that "it is not reasonable for a single individual to undertake to reform a state, in changing everything from its very foundations, and in toppling everything in order to build it up again, nor even to reform the sciences, or the order established in schools for the study of them," before proceeding to the private task of assenting solely to those things that could be demonstrated by reason.[49] This is not to say that Scheiner was without methodological concerns—his treatise is, in fact, devoted to rigorous optical experiments—but rather that his starting point as a Jesuit scientist was a scholastic and Aristotelian notion of the value of those individual *experimenta*.[50] For his part, Descartes appeared later in life to regard his own training with the Jesuits with particular despair:

> I had been schooled in letters since childhood, and because they told me that, in this way, one could acquire clear and certain knowledge about everything that was useful to life, I had a great desire to master them. But no sooner had I finished this course of study—at the end of which one is customarily admitted to the ranks of the learned—

than I completely changed my outlook. I found myself encumbered
by so many doubts and errors, that it seemed to me that I had gained
nothing, in trying to educate myself, than a growing sense of my
own ignorance. And yet I had been enrolled in one of the best
schools of Europe, [the Jesuit college of La Flèche], where I thought
that there would be learned men, if any were to be found on the
face of the earth.[51]

But Scheiner's project would not have been wholly alien to Descartes'
interests. For one thing, the Jesuit's treatise, like Descartes' *Dioptrique*, was
clearly an elaboration of Kepler's contributions in the *Paralipomena ad Vitellionem* and the *Dioptrice*; for another, the young Frenchman had been interested in optical issues, if not since his days at La Flèche when he read those
books "treating those sciences that are regarded as the rarest and most recondite" that may or may not have involved the deployment of lenses and
mirrors, at least since his recent sojourn in the Netherlands, where he contemplated the various effects of the dark room.[52] In his discussions of the
camera obscura Scheiner had avoided the jocose spectacles typical of Giambattista della Porta's best-selling *Natural Magick*, but his occasional allusions to
what the uninformed viewer might perceive—"the presence of diabolical
magic, when all was done most naturally," or, in the case of multiple images,
"rather than one house, a whole village, and rather than one man, a whole
army"—recall some of the feats of that work.[53] Such interests were more
pronounced in the youthful Descartes: in notes made by Gottfried Leibniz
in 1675 from a Cartesian manuscript dating to 1619–1621, meditations on the
camera obscura include references to "tongues of fire and chariots of fire and
other shapes [appearing] in the air" of the darkened room.[54]
    Several letters from 1629–1635 and the *Dioptrique* make it plain that Des-
cartes knew the *Oculus*, that he considered Scheiner—along with Galileo—the
best student of optics after Kepler, and that he admired and sought other
works of the Jesuit.[55] Descartes' perception of Scheiner as a covert Copernican in 1634 reveals the French philosopher as a close and sympathetic reader
of the Jesuit's diverse publications, and complements his own reaction to
Galileo's condemnation in this period.[56] And perhaps most significantly, Descartes' desire to adopt the pose of the ancient artist Apelles, in whose guise
the astronomer had appeared during the sunspot controversy with Galileo,
when contemplating the anonymous publication of a work partly inspired by
the Jesuit suggests both an analogue to the early posture of the spectator, and

a certain and relatively uncommon identification with Scheiner: "I ask you not to talk to a soul about this [argument about parhelia and other meteorological phenomena]," he wrote to Marin Mersenne in 1629, "for I have resolved to expose it to the public, as a sample of my philosophy, and to remain hidden behind the painting in order to hear what one might say of it."[57]

However congenial Scheiner's work on the eye and the *camera obscura* would have been to him in this period, it is impossible to state with certainty that Descartes encountered the work in late 1619. But the treatise had only recently emerged from the press in Tyrolean Innsbruck—relatively near to the place, now securely identified as Ulm, and variously described as "on the border of Bavaria" and "on the Danube," where Descartes sequestered himself for the winter—and it is conceivable that he saw the volume or heard reports of it, if not in his immediate environs, at the annual fair in Frankfurt, whose opening in early September coincided that year with the imperial coronation.[58] What most interests me here are, first, certain homologies between the dark room experiments of the *Oculus* and the setting in which the philosopher conceived the *Discourse on Method*, and secondly, his approach in the *Dioptrique*, published with the *Discourse* in 1637, to the problem of the retinal image posed by Kepler. If these likenesses are not explained by Descartes' early knowledge of the *Oculus*, as is perhaps the case, then we must not judge them as purely coincidental, but rather as indicative of a more pervasive cultural reaction to the *camera obscura*.

The Newsfree Zone

Those few details which the mature philosopher offered in the mid- to late 1630s about his stay in Germany "on the occasion of those wars which have yet to finish there" insist on his social and physical isolation as a precondition for his reflections.[59] Descartes' description of his physical surroundings is extraordinarily meager, and consists of the statement that he "remained shut in alone all day in a heated chamber," *tout le jour enfermé dans un poësle*.[60] The *poësle* was mentioned in early modern French literature as a structure associated with colder climates. Pierre de Ronsard, for example, described such locales as places of hard drinking and freewheeling scriptural interpretation, Philippe Desportes made it clear in his *Farewell to Poland* that he would not miss those enclosures crammed with "boys, girls, calves, and bulls all

together," nor a "barbarous nation shut day and night in their *poësles* with
nothing to amuse them but the bottle," and Agrippa d'Aubigné asked the
degenerate Swiss, "You greedy spineless drunks / what divine sign / would
make you emerge / from those *poësles* awash with wine?"[61] These rooms
evidently had some place (among believers) in notions of health; one of
Kepler's correspondents mentioned it as an appropriate setting for a dying
man, while Giovanni Antonio Magini, in enumerating the reasons he could
not move from Italy to a Germanic country, pointed to the unaccustomed
rigors of a heavy diet, beer, and a warm *poësle*.[62]

Such structures, known by their Latin name of *hypocaustum*, were also
the site of many miserable nights in the months leading up to the Battle of
White Mountain, as the Jesuits Buslidius and Drexel make clear in their
wartime journals; an entry made on October 31, 1620, by the irritated and
exhausted Drexel suggests that the *hypocaustum* served a variety of barely
compatible functions.

> Last night was indeed very turbulent. For we had at last found a
> small place where we were to sleep, but because [imperial general]
> Count Buquoy was coming, it had to be yielded up, and we needed
> to move far away to the loveliest *hypocaustum*, one in which our
> horsemen had stabled their horses, which they then led out with
> some indignation. Along with our baggage train, we brought with
> us bundles of straw gathered together in our previous lodgings. The
> night held a triple threat for us, from the enemy, from fire, from the
> structure itself. If we had seen that building by day, we would never
> have entered it, so dilapidated and tumbledown and ready to col-
> lapse as it was. . . . Yet in the morning I celebrated mass in that
> ruinous pagan shack.[63]

This sort of room was also one of the two places named by Scheiner as
appropriate for optical experimentation: "a bedroom or heated chamber or
other site so occluded."[64] The fact that both Scheiner and Descartes insisted
on this particular location for their undertakings has its logic, for the success
of the optical experiments and of the meditations depended upon on the
physical isolation that such rooms might provide. The implied parallel
between the two projects—one which renders Descartes' featureless *poësle* a
model of the mind just as Scheiner's *hypocaustum* is one of the eye—is not
surprising either: in 1690 John Locke would describe Human Understanding
as "not much unlike a Closet wholly shut from light, with only some little

openings left, to let in external visible Resemblances, or Ideas of things without."[65] But given Descartes' puzzling suggestion that very specific personal, historical, and geographic conditions produced the dull dark room in which he worked out in total isolation a much more general method for all readers capable of the use of reason, it seems appropriate to investigate what the *poësle* was meant to exclude before the philosopher emerged with such an inclusive mandate.

Descartes had insisted, in his journey to Germany and his adoption of the role of a volunteer soldier in no one's employ and without particular tasks, on his status as a private man, that is, as an autonomous individual without a public function, but intent upon and capable of following his own desires, a spectator, at most, of battles and contested coronations. To some degree, all that this stint as a *privatus* did not imply—no allegiance to any cause, no debts to others, no real contacts abroad—translated into the requisite privations of the *poësle*, such that all that remained was the residue of intellectual curiosity that had made him undertake the sojourn in Germany and the thought experiment in the first place. But while Descartes presented the *poësle* as a neutral venue, one without the various and disconcerting connotations of Wotton's dark room on the field of defeat, or William Watts's curious "house of pleasure," or Scheiner's sectarian *hypocaustum*, it would be inaccurate to see this featureless structure as any less related than they are to the question of reportage and interpretation, just as it would be imprecise to view the few personal traits the philosopher offered about himself as an observant, tolerably well-off, and independent man focused upon the proper use of reason as anything other than the sort of individual who, in a slightly later period, would traditionally be associated with the public sphere, and with the rational discussion and interpretation of news in particular.[66]

Consider in this connection the very odd linkage of his sometime home, Amsterdam, with the *poësle* where he spent the winter of 1619, both settings in which the philosopher depicted himself as somehow untouched by any influx of extraneous information. While Amsterdam was for Wotton and for most of his peers a city whose commercial traffic also made it a news center, or a "magazen of rumours as well as of other commodities," Descartes claimed to see it otherwise. In an effort to persuade his friend Guez de Balzac to move to Amsterdam in mid-May 1631, he observed first of all that there was no risk of unwanted social contact here, for

in this great city where I am, there being no one except for me who is not engaged in commerce, everyone is so attentive to profit, that

I could spend my whole life without ever being seen by anyone. I go out walking each day amidst the hustle and bustle of an immense crowd with as much liberty and peace as you would have in your *allées*, and I view the men that I see as I would the trees one encounters in your forests, or the animals that feed there. The very sound of their struggles does not interrupt my reveries any more than would that of some stream. And if sometimes I reflect upon their actions, I derive the same pleasure that you would in seeing the peasants who cultivate your fields, for I see that their work serves to embellish the place where I am staying, and to ensure that I lack for nothing there. And if there is some pleasure in seeing fruits grow in your orchards, and finding abundance everywhere one looks, don't you imagine that there is at least as much in seeing ships arrive here, laden with everything that the Indies produce, and with the rarest products of Europe? What other place in the world could one choose where all commodities, and all the curiosities one might desire, are so easily acquired? What other country is there where one can enjoy such complete liberty, where one can sleep with less worry, where one has always a standing army just to protect us, where poisonings, treasons, and calumny are less known, and where so much of the innocence of our forefathers still lingers? I do not know how you can be so fond of the air of Italy, where one breathes so much of the plague, where the daytime heat is unbearable, the cool of the evening unhealthy, and the obscurity of the night a cover for robbers and murderers. And if you fear the winters of the North, tell me what shadows, what fans, what fountains can shelter you as well in Rome from the misery of the heat as would a *poësle* and a great fire here from the cold?[67]

Among the remarkable features of this description is its unreal replication of the essential characteristics of Descartes' lonely sojourn in the winter of 1619: he has in Amsterdam as in that distant, nameless spot, neither contacts nor wants nor debts, and he wanders freely through that wilderness of merchandise toward the inevitable *poësle*. What Descartes' Amsterdam does not produce is either news or public discussion of the latest events; its numberless inhabitants, reduced somehow to traffickers in all but information, or to a forest and feeding animals, generate in the "sound of their struggles" a noise so constant and so inconsequential as to disturb no reverie, a stream of murmurings at once as inarticulate and as seemingly natural as a brook. Apart from the regular influx of exotic goods—which were, in fact, often detailed in the weekly corantos, directly after accounts of more distant battles and campaigns—there is no novelty in Descartes' scenario, for an army provides endless peace, and some unnamed civic order militates against assassination, treason, and slander, such that the Dutch Republic seems to the French philosopher the innocent country of one's ancestors.[68]

Descartes' portrait of Amsterdam is in some ways accurate: the city's thriving market, the relatively low incidence of violence, particularly at night, and the general tractability of the standing army of the United Provinces were all the subject of commentary, especially by foreigners.[69] But the description is far more notable for what it suppresses, an emerging news industry and its correlate, a large public of avid consumers and discussants drawn from most sectors of Dutch society.[70] By way of contrast one might consider, among other similar items, Jan Joris van Vliet's engraving of ca. 1630, *De Pamflettenverkooper*. Here a crowd variously composed of wealthy, middling, and poor citizens gathers about a vendor and an assistant who reads aloud, a figure with a market basket proceeds toward the pair as if to acquire this perishable commodity as well, a thuggish little girl accompanied by her mother appears to be withholding a broadsheet from a crippled fellow, a central muddle of unrelated arms, legs, and torso reach for and acquire another sheet, and a couple of pickpockets set upon a rich burgher too intent upon what he is hearing to notice them.[71]

Whereas most contemporary onlookers saw the relatively unhampered exchange of information about and views of current events either as a special sector of the commercial economy of Amsterdam—some, in fact, designated the traffic in news publications as one plied exclusively by those of rapacious sloth, people "too lazy to work, [who] incessantly walk the streets and trade in books and the latest tidings"—or as a consequence of the Dutch Republic's recent liberation from the Spanish yoke, as in Wotton's vexed allusion to "States of this composition, whoe, having bravely purchased their free-dome, are content (it seemes) that some of it shall appear even in their discourses," or some combination of the two, Descartes left the market and the standing army in place, but suppressed their conventional association with news and its public interpretation, insisting instead on how much this newly and rather violently formed republic resembled France of yesteryear.[72]

Among the more startling suggestions is the implicit interchangeability, in this uneventful world, of a crowd of men and the trees of a forest. It is less the alienating tenor of the observation than its artful erasure of recent history and its relationship to current figures of speech that are of concern here. Just

FIGURE 3. Jan Joris van Vliet, *The Pamphlet Seller*, c. 1630.

Reproduced with permission from Atlas Van Stolk, Rotterdam.

eighteen months earlier, Prince Frederick Henry of Orange had captured Den Bosch—formerly 's-Hertogenbosch, or Bois-le-Duc—from the encircling Spanish and imperial forces, a coup whose reportage sometimes relied on word play about oranges and the newly acquired *bosch*, or forest.[73] The practical difficulty of incorporating the largely Catholic population of that city into the social, economic, and civic fabric of Dutch society received much unfavorable attention both in the United Provinces and in the Spanish Netherlands; after decades under the rule of the latter, those from Den Bosch proved decidedly different from the murmuring crowds Descartes might encounter on the streets of Amsterdam.[74] Though the spectacular siege was regarded with great relief throughout the United Provinces, the question of a truce with Spain made the military event the subject of a very bitter pamphlet war throughout 1629–1630, and this situation is the likely backdrop of Van Vliet's engraving.[75] The deadlock in the province of Holland in particular meant that in April 1631, just weeks before Descartes insisted on the safety provided by the standing army, Prince Frederick Henry, on the verge of another siege in Flanders, found himself confronted by Spanish forces but unsupported by the republic's deputies, who argued that he should risk neither the army nor the state in this encounter, and who forced him to order a retreat of his 30,000 men.[76]

Descartes' effort to persuade Balzac to come to Amsterdam also merits comparison with his correspondent's view of the Dutch. In a letter of 1624 Balzac had deplored "Holland"—meaning either the largest and most prosperous of the United Provinces, and the one in which Amsterdam was situated, or the entire Dutch Republic[77]—because it seemed to school its inhabitants in a glib military parlance, the sort of thing also acquired by students of mathematics, and to drive both groups to discuss their views publicly with even the most uninterested of auditors. In this general diatribe against "big talkers," Balzac professed that

> they are really fatal to me when they have just come from Holland,
> or when they are taking up the study of mathematics. From Milan
> to Siena, I had to deal with one of these opportunists, whose com-
> pany I will count for the rest of my days as the worst among my
> misfortunes. He wanted to remodel every fort that we encountered
> en route, he saw no land that he would not have moved, nor moun-
> tain on which he didn't have some design, he attacked each and
> every one of the cities of the Duke of Florence, he only needed so

much time to take those of Modena, Parma, and Urbino, and I had great difficulties in preventing him from touching the Papal Territories. . . . There is another sort of pest, the number of which is currently multiplying so rapidly in France that it has almost reached infinity. These people can't talk with you for a half an hour without telling you a hundred times that the King is building up his army, that so-and-so has lost credit among his faction, that another is struggling in his affairs, and that a third is embroiled in every intrigue in the court. . . . People persecute me even in villages with this style of speaking, and it is why I hate the State, and public affairs.[78]

Balzac's rough equation of the conversational tactics of those freshly returned from "Holland" with that of those recently embarked on the study of mathematics suggests that the sort of information widely available in the United Provinces, or in this particular province, had to do with the rapid evolution of military fortifications, a field in which the Dutch led the rest of Europe, and that such matters, like the political gossip in France, were regarded as news, and worthy of exchange.[79] Given that there was a genuine overlap, especially in Amsterdam, between early news publishers and cartographers, and that battle accounts printed in the United Provinces were sometimes accompanied by quite detailed maps of the territories and fortresses involved, Balzac's impression of the kind of information one might gain and feel obliged to discuss just by being in "Holland" is a hyperbolic account of the same sort of social dynamic Descartes sought to suppress in his letter of 1631.[80]

Descartes' portrait of Amsterdam, so unlike that offered by his peers, is in some ways consonant with his correspondence, which—in contrast to that of his close friends Balzac and Constantijn Huygens, but like that of Galileo in Italy—makes little reference to current events and to the various channels by which information about the latest developments were conveyed. I am not arguing, however, that either Descartes' status as a resident foreigner or his philosophical preoccupations left him so isolated and estranged from his neighbors as to make Amsterdam seem truly very much like wherever it was that he spent the winter of 1619, but rather that this imaginative construction of the Dutch city in 1631 was an extended meditation upon the issues confronted, but not resolved, in the *poësle*, and that both have important connections to the nascent public sphere of the early modern period.

Discourse

A crucial function of the people, in the sense of an amassed populace composed of private men, or those without official positions, was, of course, its spectatorship to royal and ecclesiastical ceremonies, a venue certainly not designed for any sort of political exchange between ruler and ruled, but rather for the display of the public character of the former, that is, its power, authority, and sacral nature.[81] It is significant, therefore, that shortly before his retreat to the *poësle*, Descartes witnessed the coronation of Holy Roman Emperor Ferdinand II in Frankfurt, a ceremony whose immense political and ecclesiastical importance was quite clearly undercut by Frederick of the Palatinate's simultaneous claims to the crown of Bohemia, and by the support of those who doubled as spectators to the latter's majesty and authority.

Apart from a reference to "those wars which have yet to finish" in Germany, Descartes did not insist upon the resultant nullity of the role of spectator, but began the *Discourse* instead with the ironic observation that of all things in the world, good sense or reason must be the best distributed, for even those greedy for other things did not feel that they were lacking in this.[82] While the suggestion is that the common tendency to judge is not in fact good sense or reason but rather, in most cases, something irrational, unexamined, based on popular repute, and closer to prejudice,[83] what interests me here is the fact that Descartes turned away from the expected and restrictive venue of civil conversation, and explicitly associated the ability to discern true from false with the instincts and experiences of the property owner. In recalling his decision to travel to Germany, he explained that

> it seemed to me that I could encounter a lot more truth in the
> reasoning that everyone makes in regard to the affairs that matter to
> him, and whose outcome will shortly punish him in the event that
> he has judged poorly, than in the reasoning that the man of letters
> makes in his study, regarding speculations that are entirely without
> effect, and which have no consequence for him, apart from the fact
> that the further they are from common opinion, the greater the
> measure of glory he might derive from them, because he will have
> had to use that much more cleverness and artifice to render them
> plausible.[84]

The assumption that those individuals who worked constantly to protect their private interests and the status quo would be best qualified to enter, collectively, a public sphere characterized by rational and purposive debate would become a traditional notion, but Descartes' portrait of this segment of society is here, as elsewhere, atomized and strangely remote from the realm of the political. The assiduous exercise of reason, in other words, neither reaches beyond the threshold of the individual economy, nor coalesces into a generalized conversation about how the public good is to be achieved. Descartes' insistence on intellectual freedom, understood as both an avoidance of arbitrary extremes and the liberty to change one's mind, is likewise concerned with the isolated use of critical reason, and likewise predicated on the fixity of the social and economic order, "promises and contracts" offering a measure of permanence to the civil and financial status of those who have taken religious vows or married, and guaranteeing the outcomes of individual business ventures.[85] The naked conversion of religious and marital vows to contractual arrangements differing little from the quotidian ones of the marketplace is all the more remarkable in that this audacious first step leads nowhere: no public emerges from any sector of the population so exposed. And where Descartes acknowledged the existence of something like popular opinion, it was more the stuff of custom and idle prejudices, for *la pluralité des voix* to which the more critical perspective of a lone individual might be opposed indicated not a multiplicity of views, but a consensus without sense, the unreasoned belief of a numberless mass.[86]

Descartes' studied avoidance of a public of listeners and discussants differs appreciably from the reactions of his peers, which consist for the most part in references either to the inaccuracy of the news consumed by and views held by the populace, or to the comic pretentiousness of those who felt that such timely information allowed them access to a political process. While it is true that political power, particularly in those years of the Dutch Republic, was concentrated in the hands of a few remarkably corrupt individuals, and not in the assemblies of each of the provinces,[87] and still less within the grasp of those who wrote and read about and discussed developments within the state, the veil of silence thrown over a conversation typically regarded as even more raucous than inconsequential is worth note.

Though Descartes' tendency to atomize society, his decision to follow his own critical judgments rather than those of authority or custom, and his suggestions that his discoveries were above all useful "for the preservation of health," and that this utility justified their communication to a wide audience, all have their logic, the entire trajectory sketches out a shadowy alternative to the eventual fiction of the public sphere.[88] In the *Discourse* a lone property-owning individual, relying on reason and writing anonymously, makes public for discussion not the well-being of the body politic, but rather the means of maintaining vigor in each and all of its isolated members. This portrait undercuts the emergent image of a community defined and even constituted by its regular consumption of news. Given that France's nascent public of newsreaders was created and maintained, for the most part, by the energetic efforts of Théophraste Renaudot, at once physician to Louis XIII, founder of the centralized weekly *Gazette* in 1631, and advocate of the connection between public welfare and a citizenry with regular but rigidly controlled access to political information, Descartes' model figures as a significant antitype.

Descartes recognized that both the system he proposed to make public and its particulars had a sort of momentum, and in one of the most interesting analogies of the entire treatise he explicitly compared its potential gains to the exponential growth of the merchant's wealth, and its setbacks to the spiraling losses of the military. A philosopher engaged in the slow discovery of truth, as he saw it, resembled

> those who, beginning to get rich, have less difficulty in making great acquisitions than they had when they were poorer in obtaining a good deal less. Or one might compare [philosophers] to the leaders of an army whose forces tend to grow in proportion to their victories, and who need a great deal more skill to maintain themselves after the loss of a battle than they do in taking additional cities and provinces after a victory.[89]

This pair of analogies would have seemed both apposite and awkward in the late 1630s, when the Dutch experienced a media-fueled boom, the most dramatic of which was the great tulip mania, and Frederick Henry a series of military and political misfortunes, beginning with his shift toward the "war party" of the republic, and a disastrous Franco-Dutch plan to invade, to subdue, and to partition the Spanish Netherlands.[90] This is as close as the *Discourse* gets to current events, and the allusion serves not to illustrate anything about the property-owning public's reasoned evaluation of either its mercantile or its military institutions, but rather the lone philosopher's establishment of a system designed to preserve the health of all.

By way of comparison with Descartes' effort, one might consider the contemporaneous prefaces of William Watts's *Swedish Intelligencer*, for they offer a succinct rationale on the virtues of making public information about current events. Such arguments, undoubtedly self-serving, are more notable for their depiction of reportage as a "common good," for their representation of a heterogeneous public ranging from learned late humanists to those of "meane capacity," and for their strong condemnation of any monopolization of information.[91] But this is not to say that Watts went so far as to imagine that access to what he presented as a "common good" somehow produced critical reasoning in those addressed as "favourable and iudicious" readers; on the contrary, he saw such capacities suspended as his audience fell under the sway of the chronicle's hero, Gustavus Adolphus of Sweden.

> Though this *Story* should goe out of my hands, poorely enough and weakely endighted: yet that the *prevailing Fortunes* of the King of *Sweden*, would so potently assist and go along with it; as that by that time it came into your hands, it would so prevaile with your *affections*, and so gently in the reading, *captivate the attentions* of all the favourers of his actions, that their *Iudgements* (for the time) should become deprived of this part of their due liberties; and forbeare the *power*, though, not of *seeing* yet of *censuring*, my *errors*.[92]

In this observation Watts appears to have collapsed that emerging "public opinion" back into the older sense of *opinio*, the popular impression of others based on their reputation, an unreasoning belief of the sort Descartes attacked in the *Discourse*. Here *opinio* is presented not as something banished by publicity, but rather as an indispensable precondition for the enjoyable consumption of news. Enlightenment for Watts necessarily occurred in one of those "houses of pleasure," a chamber of illusions orchestrated by the English king and his licenser, and reserved for and dominated by an elite clientele.

Bits of Ink

I have presented Descartes' view of both Amsterdam and the *poësle* as zones specifically built about the absence of news and a public sphere, and the locale portrayed in the *Discourse*, in particular, as a kind of vacuum in which the individual and anonymous philosopher performs the work of reason in order to preserve the health of an atomized population. But because of the odd relationship of the *poësle* to Amsterdam, the studied avoidance of news and its public in both, and the contrast with Wotton, Kepler, Watts, and Scheiner, where the *camera obscura* always has to do with the contemplation of current events, the problem addressed in the Fourth Discourse of the *Dioptrique* merits particular attention. Here the philosopher presented the problem of retinal image in terms of the effect of engravings on the soul.

Descartes rejected the conventional comparison of the sensory impression with a painting, reserving this analogy for discussion of the soulless and mechanically produced retinal image in "the eye of a man who has just died or in that of a bull or some other large animal."[93] Not surprisingly, his demonstrations of the distortions and defects of this sort of painting, one that represents "in rather crude perspective" the objects in its field of view, is an elaboration of Scheiner's discussion of the images projected in a dark cubicle or *poësle*.[94] But in the crucial case of the living and sentient being, Descartes compared the sensory *impression* to an engraving, the emphasis being in both cases on the interpretation resulting from what is stamped or imprinted:

> There are things other than images that can stimulate thought, as for example, symbols or words, which resemble not at all the things that they signify. And if, in order to stay as close as possible to conventional views, we prefer to state that the objects that we perceive actually send their images into our minds, we must at least notice that there are no images that need resemble in all particulars the objects which they represent. It is quite enough that they resemble them in a few aspects; indeed it is often the case that the success of the images is due to the fact that they are not as much like these objects as they might be.
>
> You see that engravings—composed only of bits of ink placed here and there on some paper—show us forests, and cities, and men, and even battles and storms even though, of the infinite number of various qualities that such images convey to us of these objects, they resemble them in the matter of *figure* [shape] alone. Even in this regard, the resemblance is incomplete, seeing that engravings depict solid bodies that are variously rounded or hollow on a perfectly flat surface; in keeping with the principles of perspective, moreover, they often better represent circles by ovals than by other circles, and squares by rhombuses rather than by other squares, and so forth, such that often those images are the most successful, and best suggest

an object, when they resemble it less. Now we should think just the same of the images that are formed in our mind, and we should observe that it is just a question of knowing how they offer a means to the soul of sensing all the various qualities of the objects to which they correspond, and not at all how in themselves they are like them.[95]

Descartes' argument insists on the shift from perfect mimesis—the false promise of the pictorial metaphor traditionally used to describe the retinal image—to that larger category in which symbols, words, and "bits of ink placed here and there on a paper" all manage to convey meaning by the persuasive arrangement of recognizable signs.[96] Kepler's reference to diagrams [*schemata*] as the "natural letters of Geometers" in the preface of his *Dioptrice* is a typical example of the perception that pictorial representation somehow avoided the artificial conventions and arbitrary nature of other sign systems and thus more closely approximated the depicted object.[97] Descartes' suggestion, by contrast, is that the opposition of *schemata*, in the sense of rhetorical figures, and those *schemata* of Geometers is a false one.

This is generally consistent with Descartes' interest in a universal language, but also more narrowly related to the frequent association of printed texts and engraved images alike in the context of dark room experiments. It comes as no surprise to see that what the philosopher had called *tailles-douces*, meaning copperplate engravings made without the use of acid, was rendered by the clumsy but apt phrase *icones illas quæ à typographis in libris excuduntur*, "those images that are stamped in books by printers," in the Latin translation of the *Dioptrique*.[98] What is crucial here is the focus upon the conventions governing the composition and interpretation of mechanically produced and thus widely available representations in this discussion, for newsbooks and pamphlets would clearly be the best emblem of such items.

In presenting dark room phenomena as the idealized correlate of wartime news, Wotton and Watts suggested that there was a kind of transparency and inevitability to the particular images they sought to convey: foreign reportage was nothing but the transcript of events in a remote landscape. The appeal of the metaphor, as originally deployed, was that the interpretation of such information, like that of the Keplerian retinal image, was restricted to a civil elite or a slightly larger subset of wealthy news consumers. The prophetic cast of Scheiner's depiction of the *camera obscura*, by contrast, treats it not as an instrument that pretends to record events in neutral fashion but rather as one that foreshadows and even generates them; the sharp division between the blindness of heresy and the insight of religious orthodoxy also undercuts the conventional depiction of vision as the analogue of reason. Descartes' airtight *poësle* and his archaic Amsterdam, seemingly structured around the absence of news, evoke an atomized society, vast clusters of individuals whose constant exchanges breed noise but no discourse, commerce but no community, just the sort of self-interested fragmentation the eventual fiction of the public sphere, with its salubrious focus on current issues of public well-being, was meant to exclude. And his shift away from the purely mimetic character of representation to a larger realm characterized by an admixture of different sign systems, finally, recognizes precisely what the related trope of the *camera obscura* sought to suppress, the social and rhetorical dimension of communication in the public sphere.

This News of the World Without

Thus far I have discussed the association of the *camera obscura* with news of the pan-European confessional conflict that would occupy the continent for three decades, but this is not to suggest that the device did not appear, in this same period, as the complement to concerns somewhat more regional, and even domestic, in scope. By way of conclusion to this chapter, I will examine the way the dark room figured in the 1630s in the household economies of Constantijn Huygens and of Nicolas Claude Fabri de Peiresc.

Huygens's appearance here is no accident. As an adolescent he became acquainted with Pierre Jeannin when the latter was in The Hague for the negotiation of the Twelve Years' Truce, he was a neighbor and a friend of Sir Henry Wotton from 1614, and a neighbor and later an aide to François d'Aerssen in 1620–1621, and his first translation to Dutch, made in November 1619, involved one of Ben Jonson's epigrams. His optical explorations with Cornelis Drebbel in the early 1620s cost him a fair amount of money and his father some worry, he had read Scheiner's *Oculus* by 1629, and he continued to pursue such studies in the 1630s with Descartes as the latter finished and published his *Discourse on Method* and *Dioptrique*.[99]

Huygens was preoccupied with access to news as well; in January 1621, for instance, shortly after a diplomatic mission in Venice, he copied into his commonplace book several lines from Jonson's *Volpone*, where Sir Politic Would-Be proposed "to serve the state / Of Venice with red herrings for

three years, / And at a certain rate, from Rotterdam, / Where I have correspondence."[100] He saw Jonson's *Newes from the New World Discover'd in the
Moone*, whose treatment of the imperative to offer "no news" must have
appealed to him, in February 1621.[101] In December of that year, he repeatedly
expressed his frustration over James's mania for ceremonies and his lack of
interest in matters of both domestic and foreign affairs, referred contemptu-
ously to "this parliamentary comedy" and this "farce," and complained about
the lack of information he had to send home, telling his parents that "the
ambassadors feel that they are becoming monks, because from the moment
of entering their lodgings the first evening to leaving them now, no news; it's
etiquette that enchains them."[102]

The nexus of news and optics emerged in his work in the 1630s. In
October 1635 Huygens described to Descartes his plan to have a large convex
lens ground by an artisan in Amsterdam, and he also confided that a professor
in that city was at work on a device capable of making legible the words of a
letter one mile away, a claim quickly dismissed by Descartes.[103] Descartes
finally conceded in November 1638 that the latest effort of the artisan in
Amsterdam was not without merit, and that while the shape of the glass was
not yet perfect, and the surface lacking in polish, he expected eventual success.[104] A lens of this type would have provided an inverted, sharply focused
image, but with some blurring toward its periphery: such was perhaps the one
used by Vermeer in the 1660s.[105] A mutual friend of Huygens and Descartes,
Henricus Renerius, was working on the *camera obscura* in 1637–1638, and a
letter of January 1, 1638, indicates that he expected Huygens to be abreast of
his research:

> I know three ways of observing upright figures in the *camera obscura*.
> And I think that I would be able to pride myself on having made
> for his Highness [Frederick Henry] a painting of my *invenzione*, one
> which will surpass in its perfection everything that has ever been
> seen, if I were permitted to have it done by a painter who was rather
> good. And with this painting I would even be able to make those
> who were not forewarned swear that they were seeing the thing itself,
> and not a picture. But these matters need to be related *viva voce* to
> be believed, and require an actual demonstration so that they will
> be felt and almost touched.[106]

Such developments would have been of particular interest to Huygens,
for one section of his long poem *The Day's Work*, begun in 1630 and

completed in June 1638, is devoted to the workings of the *camera obscura*. Here Huygens addressed his late wife under the poetic name *Sterre* or "Stella," and explicitly compared the projection of exterior images into a dark room to the reception of news:

> I am running over with rumors[107]
> Which I will report with pleasure
> And which you might blamelessly
> Mix with other joys.
> I shall have this news of the world without
> Brought within the walls to you,
> Just as one can catch in glasses
> Lively images from things
> That in the heat of the day
> Are drawn inside upside down.
> They're upside down, Stella,
> As when a lie labors
> On the newborn truth,
> One just born and as clear
> As things are at mid-day . . .[108]

In the notes Huygens wrote to accompany his poem, he described the dark room instrument as *een geslepen Glas*, meaning "a cut glass," a lens, in other words, whose shape was carefully ground, but also in figurative terms, something sharp, cunning, or artful.[109] To judge from the tone of this section of the poem, it appears that it was written before Susanna's death and before Renerius announced his various innovations to right the images, and it seems in any case that those three methods, however appealing otherwise, would not have served Huygens's rhetorical purpose here, which was to temper the current notion of projection as a direct and faithful transcription of current events.

> Go, you scholarly and worldly quills
> Whose flight these thousand years
> Consists in catching the topsy-turvy course
> Of yes- and no-things,[110]
> And in selling them now as treasure
> As if from undamaged goods,

Though all are born of rumor,
And almost all with a birthmark
That they have lost
Between the childbed and the cradle
Where they are bundled in deceit
And wrapped up in lies.[111]

On the whole, Huygens's verse seems less closely related to a specific quarrel at issue in a particular pamphlet than to a more general argument about whether or not certain events have actually taken place. Though this criticism is addressed to historians and chroniclers—whose business it is "to set the record straight," to right the inverted image in the dark room—the more significant implications regard newswriters and the extremely capacious class of newsmongers. The movement from the mild eroticism and autobiographical tone with which this section begins to the moralizing commentary on the dubious birth and infancy of what will in time pass for historical truth is interesting in that it implicates Huygens: he brings home tales circulating in public, *geruchten*, to amuse his wife before the poem turns away to some allegorical moment in which a topsy-turvy Rumor spawns individual stories that may or may not be accurate. As is often the case, the economy of public and private is peculiarly porous: in the first section, rumors already known on the street acquire a new air of secrecy and intimacy when ushered indoors, while in the second, the domestic details of the birth and swaddling of such tales involve their public presentation as historical fact.

Not surprisingly, there is a decorous silence about the sexual matters that underlie both the autobiographical and allegorical narratives—the poem no more insists on what "other joys" Susanna might mix with news than it reveals the identity of promiscuous Rumor's various partners—but these protagonists are, of course, Constantijn and his figurative counterpart, the private man on whom the generation of public opinion depends. Less surprisingly still, the active agency suggested by *Loop ick over van geruchten*, "I am running over with rumors," is mirrored in the passive construction, frequent in Huygens's correspondence, that describes the private man's acquiescent encounter with gossip, conjecture, and reports, *Er loopt een gerucht*, and *Le bruit court*, "A rumor is running about."[112]

The most striking aspect of the entire vignette is the implicit cloistering of Huygens's wife and the monitoring of her news consumption. Degraded forms of news were routinely distinguished from timely and essential political and economic information by an association with the timeless, the inconse-quential, and the feminine: "idle gossip," "old wives' tales," and in a later period, *faits divers* or "human interest stories." But Huygens's insistence that the "news of the world without" was characterized by dubiety from its very inception, and only subsequently disguised as fact by historians, offers a somewhat different impression of the arena in which such information was first generated, analyzed, and consumed. This version of the nascent public sphere suggests that the object of scrutiny was *not* the event itself, that distant landscape of battles, but rather the various representations of things that may or may not have occurred. Thus while writers such as Wotton and Watts imagined the *camera obscura* as a variant on the telescope, as an instrument offering qualified observers a uniform, undistorted, and realistic image of remote incidents, Huygens depicted it as a device focused on the phantasmic "world without," where rumors and opinions were the protagonists.

The Tyranny of the *Gazettier*

That displacement of interest, this shift from an obsession with news of a particular sort to a much more generalized monitoring of news consumption itself, is especially pronounced in the frenzied activity of the scholar and *parlementaire* of Aix-en-Provence, Nicolas Claude Fabri de Peiresc. Here, too, the dark room becomes the operative figure by the 1630s. Peiresc sub-scribed to manuscript news, maintained an extensive personal correspon-dence throughout Europe, received a steady stream of well-informed visitors, and read and collected the retrospective annuals *Mercurius Gallobelgicus* and *Mercure François*.[113] The latter publication, having emerged at irregular inter-vals in 1605, became an annual in 1621, but the great lag time between events and publication, and the fact that it was increasingly under the control of Cardinal Richelieu may have diminished its appeal for some readers.[114] Peiresc, in any case, sought the complete set in 1625, confessed in a letter of 1627 to having parted with certain volumes of it in return for materials for his vast collection of antiquities, and saw to it in 1636 that a history writer from Genoa had it available as a research tool.[115] He valued the *Mercurius Gallobelgicus* above all for its maps.[116]

But in 1633 and 1634, in the tense phase preceding France's declaration of war on Spain, and when fears of invasion ran high in Provence, Peiresc also expended considerable energy in procuring the most prominent weekly paper in France, Théophraste Renaudot's *Gazette*.[117] Like the *Mercure Fran-çois*, the several pages of the weekly *Gazette* were produced under the watchful surveillance and sometime dictation of Cardinal Richelieu.[118] In mid-September 1631 Peiresc had alluded enthusiastically to "those gazettes or sheets of foreign news, which I find very good and convenient, and which I think will be gathered into collections, just as the *Mercure* is. I would be happy to receive two copies of each sort, whenever they are printed, so that I can retain one even if the other escapes from my hands."[119] His reference was probably to the *Gazette* and its short-lived rival and predecessor, the *Nouvelles ordinaires de divers endroits*, whose title, personnel, and readership Renaudot soon acquired, and to which he would also add the *Extraordinaires* and the monthly *Relations*.[120]

In late 1632 Peiresc expressed some impatience with Renaudot's tardiness in releasing the weekly to printers and to individual subscribers outside Paris, and in January 1633 he reported that the *Gazette* was being printed in Aix, as in Lyon, but at irregular intervals, and perhaps in pirated editions.[121] What seems to have inflamed him most of all was the knowledge that others in his environs acquired the *Gazette* a week before he did; throughout this period his otherwise cheerful and scholarly correspondence with his friend Jacques Dupuy in Paris is punctuated by irritated diatribes against the *gazettier*, and several schemes to insure delivery of his copy before those of his neighbors.[122] Peiresc also made clear at this time his scant regard for the manner in which Renaudot's *bureau d'addresse*, a kind of clearinghouse for information of all sorts, articulated and amplified what seemed to him negligible items better left to the ephemeral attentions of the local town crier. "In the large sheet you sent me," he wrote, "I found some pleasant cries and proclamations about lost stuff, cries that can be heard a lot further than those at the cross-roads and during the weekly mass, but I would have greater regard for them if one could use this method to acquire a rare book when one wants it and it's not to be found in common bookshops, or when someone wants to rid oneself of a few rare books."[123]

In early April 1633 Renaudot, having received several complaints from Dupuy, undertook to send Peiresc the *Gazette* in more timely fashion, but with the ambitious proposal that the latter, in addition to the cost of the subscription, send letters containing news from Provence and beyond in exchange for the more expeditious delivery of the printed weekly. Peiresc expressed some reservations about the whole business, loftily suggesting to

Dupuy, who certainly knew otherwise, that the sort of news that really interested him—just notices of new books and specialized information about the
trade in antiquities, he said—would be of appeal neither to Renaudot nor to
his clients.[124] He also showed real alarm over the impression such an arrangement might convey to others.

> I would hate nothing so much as looking like a newsmonger; I'd
> rather never receive any news at all, and just do without it as I have
> done for a long time during my stay in the country. I would even
> like to avoid looking like someone who is overly curious about what
> is going on in the world, and I would pay dearly for that, because it
> is often a great bother when others come to ask for news: they imag-
> ine that one is not only obliged to inform them when one has some,
> but also obliged to have something to report when one has nothing,
> or at least not what they want.[125]

Peiresc appears to have indicated that from time to time, and without
directly concerning himself with the *gazettier*, he would include in his letters
to Dupuy news of the sort that could be "divulged without regret" to Renaudot in the event that the latter took the pains to seek it out.[126] Renaudot,
who may have been ignorant of or indifferent to Peiresc's unwillingness to
enter into a more formal arrangement, sent the *Gazette* directly to the scholar
for three weeks; when it failed to appear the fourth week, Peiresc vowed a
sort of vengeance, telling Dupuy that the news of the Levant and of the
House of Savoy included in his letter was not to be shared with the *gazettier*.[127] As he put it, "I have no intention of opening a *bureau d'addresse* in
this city."[128]

By fall 1633 Peiresc reached a new arrangement with Renaudot, who was
having some difficulty not only in procuring news, but also in preventing
pirated editions outside Paris, in part because his publication was often stolen
or temporarily borrowed en route to printers and subscribers.[129] Renaudot
was to send the *Gazette* directly to Peiresc, camouflaging it with a seal other
than that of his Bureau; Peiresc, for his part, was to engage in a little charade,
because he was not to reveal to anyone outside his household that he had
received the publication in advance of others in Aix, but had instead to
inquire impatiently about the *Gazette* at the shop of a local printer who was
licensed to publish all or part of the weekly from Paris.[130] Renaudot evidently
still expected something more than the occasional second-hand tidbit from

Peiresc, and whenever the *Gazette* failed to reach him, or came more rapidly to the printer and others in the city, the exasperated scholar attributed such events to the "tyranny of the *gazettier*," and pointedly included in his private correspondence news to be withheld from him.[131]

Relations between Peiresc and Renaudot deteriorated further in 1634; as Peiresc wrote their unenviable mediator, Jacques Dupuy, in January of that year,

> This man asks a bit too much, it seems to me, and he will soon be sorry for the trouble he has caused you, because I would prefer never to see the *Gazette* again than to bother you about it again except to pay him what is due for the past, and thank him for the future. Either he will send his *Gazette* to someone else in this city, and in that case, whoever it is, I am certain that he will not fail to let me see it that same day, or he just won't send it at all, and in that case, when no one else receives it by the courier, I won't miss it any more than anyone else, and I will do without it just as I did before it was printed. And there are even merchants in Lyon and friends of mine who are offering me the one published in Amsterdam[132] via courier, so he can keep his *Gazettes* for the coming year, and rest assured that they will be entirely without credibility when he won't even let us see them, being already rather discredited, and they would be even more so if I had not made a show of believing in them in order to maintain his credibility.[133]

Shortly thereafter, when Renaudot withheld from him a *relation* containing the Inquisition's sentence against Galileo, Peiresc, ever more combative, wrote to Dupuy,

> I hear that there are copies in town, ones which we will see, I am certain, whether he likes it or not, and if it is a day later than it would be if he had sent them directly to us, it's not a great loss for us, and in any case preferable to the tyranny which he would have liked to impose on us by forcing us to write to him, a privilege which I find he hardly deserved, wanting to sell too dear the eggs whose value he himself diminishes in profaning them and divulging them in this way, what with his connections to people of low stature,

this small-time printer [here in Aix] who receives them being in a
very bad way.[134]

Two weeks later, Peiresc wrote indignantly that Renaudot was attempting to enlist other less well-informed citizens in Aix as correspondents, and
he continued to complain throughout this period of the irregularity with
which the *Gazette* arrived, seeking as usual to arrange the delivery of the
weekly through better postal arrangements.[135] He conceded in late March
that he found Renaudot's *relation* on the death of Albrecht von Wallenstein,
the commander of the Imperial Army, "well worth seeing," as were certain
other of his publications, but believed that most of the *Gazette* was a waste
of time, and the editor himself far too capricious in his business practices.[136]
Renaudot had apparently disrupted his delivery because he had received from
the rival correspondent in Aix a report—one that was perfectly true and
familiar to him—that Peiresc was receiving the weekly from someone other
than the local printer. Peiresc protested to Dupuy,

> I swear to you that not a living soul saw those *Gazettes*, apart from
> those within my household, and I never failed to send someone to
> inquire after them whenever the courier arrived, so that no one
> might imagine that I obtained them from elsewhere [than from the
> printer]. . . . Monsieur de Bourdelot[137] spoke of Renaudot as a man
> who in making pleasantries and in putting up with the pleasantries
> of those he is obliged to respect, earns 20,000 *livres* at his Bureau,[138]
> so they say; it's a quintessence or a true philosopher's stone, such
> that he can neglect all correspondence except that which is useful to
> him.[139]

By mid-April Peiresc sought once again to escape "the tyranny of the
gazettier" and to see the paper three days earlier by procuring the copy published in Lyon, an arrangement which soon failed; on June 20, he designated
another middleman to send the edition from that city, and after this point
the references to the troublesome *Gazette* miraculously disappear from his
correspondence for eight months.[140]

The energetic unhappiness that characterizes Peiresc's initial relationship
to the *Gazette* must be ascribed in part to his dependence upon his social
inferiors, for he could not easily extract himself from the services promised,
if not actually provided, by the tyrannical *gazettier*, the "small-time printer,"

and the host of inept couriers and middlemen whose numerous failures his letters detail. But it seems very unlikely, given Peiresc's status as a member of the Parliament of Provence, and the fact that he received prominent visitors, numerous personal letters and manuscript newsletters from abroad, as well as printed separates and newssheets from other venues, that when the *Gazette* actually arrived in Aix that it contained much that was wholly unfamiliar to him. His references to it almost never involve its contents, but rather its tardiness, its inferiority, and especially its regular failure to materialize, and when at last it began to arrive on time, Peiresc simply had nothing at all to say of its substance. His insistence that he was better informed than the rival correspondent in Aix was almost certainly true.

It is surely significant, then, that Peiresc's interest seems above all to have been in knowing what was available to other news consumers in his city, for it is as if the weekly expansions and contractions in the horizons of their information formed a crucial perimeter for what he could reasonably take up in polite conversation.[141] It is therefore noteworthy that as Peiresc sought to make alternative arrangements with Jacques Dupuy to obtain the *Gazette* in the spring of 1634, he also described to him in detail his contemporaneous optical experiments with Pierre Gassendi and a few other privileged visitors, the point of which was to establish what appeared on the retina when vision took place. Whether he was trying to get news that could not have been new to him—and some part of which he himself might have furnished—or conducting research in a dark room, Peiresc wanted to know what others were seeing.

I would like to insist upon the disconcerting homologies between Peiresc's furtive consumption of news and his optical investigations both because Gassendi's biography implies such a connection in its chronicle of the year 1634 by suppressing entirely the negotiations with Renaudot and offering instead an elaborate account of this sort of experimentation, and more generally because the trope of the dark room had been deployed by Peiresc himself, some years earlier, in an enthusiastic description of the sort of insider's information sent to him by Dupuy in his private letters.[142] Certainly the fact that the conventional term for newsmongers and for amateur scientists, and one that Peiresc uses in both contexts, is *curieux*, makes such a parallel plausible, as does the fact that both retinal images and dark room projections onto paper were routinely described as "printing."[143]

Several details drawn from Peiresc's own letters supplement these basic similarities. Recall that a rumor—the oral item that his rival correspondent committed to paper and added to his manuscript newsletter for Renaudot, the result of which was, as usual, no newsprint in Aix—had provoked the scholar to remark that "not a living soul [*âme vivante*] saw those *Gazettes*, apart from those within my household," as if somehow to suggest that he had no control over those occasional circumstances in which the *dead* read things in his house. Peiresc's optical research involved precisely this issue: he and his friends observed how it was that animals saw, and they did so only by examining the retinas of numberless dead cats, fish, and birds within the confines of a dark room. In the most arresting of such descriptions Peiresc reported, "and in the eyes of an owl we saw depicted, in the depths of the concave mirror [of its retina], an entire landscape with all the colors of the sky as well as those of the buildings, tours, and pavilions of the whole Palace, just as if it were in a real concave mirror."[144]

Though Gassendi would report with alarming enthusiasm that "it would have been difficult to find among the various species of birds, fish, and quadrupeds, a single one whose eyes had not been dissected by [Peiresc]," this is the sole discussion of what actually appeared on the retinas of these animals.[145] It may well have been the only vista available: the enormous palace, the seat of the Parlement de Provence, dominated the landscape in both a literal and figurative sense, and guaranteed that this remote city enjoyed the attention of other and grander ones, and some degree of independence from Paris.[146] Put differently, it was precisely because the palace where he served as *parlementaire* figured on the retinas of these animals that Peiresc could adopt the posture of the cloistered amateur scientist, openly curious about optics, and studiedly incurious about the world without: news, when he wanted it, came to him. And it is no accident—though both Peiresc and his biographer Gassendi presented it as an agreeable surprise—that the scholar was interrupted and then assisted in these optical investigations by the papal nuncio Giorgio Bolognetti, en route to Paris, to militate for, among other things, even stricter controls of the French press.[147]

Yet there is something undeniably contrived, or *estrange*, as one Parisian critic would have it, about the exercise of observing what a dead animal, and a nocturnal one, would have perceived in life and by day of the seat of the palace, and it may not be readily distinguishable from the feudal splendor of the image of the ancient palace, then threatened, like other autonomous regional parliaments, by the centralizing policies of Richelieu and Louis XIII.[148] Peiresc's attraction to this ephemeral and almost counterfactual vision bears a certain resemblance to his weekly rage to see what others elsewhere would have seen, had the *Gazette* been available, and had they bothered to buy it from the "small-time printer." The scholar described the retina as embellished with a "horizon" on which images were projected, and asserted that apart from the "window-like opening through which these images entered," this region of the eye was covered with something "very dark brown and black, and capable of staining the fingers like oily ink, almost like mud," as if further to insist on the relationship of the two queries.[149]

Peiresc was aware, of course, that what he was examining was the mechanism of sight in others, and that its proper evaluation depended in some measure on the standard established by his own vision. This sort of experimentation would have a banal analogue in his mental scrutiny of a hypothetical public of newsreaders gathered about the weekly at the printer's shop: he knew perfectly well what they would read because he had already examined the identical text at home, where "not a living soul saw those *Gazettes*, apart from those within [his] household." But Peiresc is at his most interesting in his self-conscious and telling effort to evaluate how and what it is that he himself sees, and here his experimentation involves the nature of afterimages. What he sought was the manner in which the retina retained images of phenomena no longer visible to it, and how these imprints of absent stimuli were superimposed upon his perception of present objects. He reported staring at a window whose left side was covered with dark-colored bars and whose right side was composed, as was then common, of square panes of oiled paper; upon closing his eyes, he still saw the unaltered image, but when he looked instead at a dimly lit wall, the afterimage was a negative of the original, the bars appearing white and the panes of oiled paper black. Significantly, he also transferred his gaze to an open book, noting that there the afterimage of the barred window did not interfere at all with the print of one page, but that that with panes rendered the facing page largely illegible.[150]

For our purposes, what is relevant in Peiresc's experiment is the weirdly apposite nature of the images and afterimages. It would be difficult to imagine a better emblem of the "tyranny of the *gazettier*"—those maneuvers that seemed always to ensure the absence of the weekly—than the heavily barred window whose afterimage did not compromise his access to other sorts of information. Nor are we likely to find a better figure for the sometime delivery of those negligible texts in oily ink than those blank panes of oiled paper, the impact of which offered Peiresc nothing really intelligible, but merely a strong impression of the residual smudge left by the medium. This seems to have sufficed, week after week, in 1633 and 1634. To use the sort of distinction

Renaudot drew in 1632, and one that suggests a rare but fundamental agreement between the *gazettier* and the scholar, "History is the story of the things that happened; the *Gazette*, just the rumors that surround them."[151]

Peiresc's evident fascination with what would much later be identified as the chief feature of news—its flow, that is, its course, trajectory, and diffusion, rather than the particulars of its content—allows for one final modification of the metaphor of the *camera obscura*. In the idealized version exploited by Wotton and Watts, remote events constitute the landscape, and reportage the transcript redacted in the dark room and destined for an eventual elite readership. Scheiner in his *Oculus*, Descartes in his *Discourse*, and Huygens in his *Day's Work* variously dismissed the central fiction of the optical device, the seeming equivalence between event and transcription. In his frontispiece and in some of his arguments the Jesuit author portrayed the instrument as a sort of predictive tool, while Descartes sought to banish the connection between the collective scrutiny of dark room images and the mass consumption of news both by depicting atomized publics and newsfree zones, and by insisting on the complex character of communication within this realm. Huygens's treatment of the *camera obscura* decisively shifts the focus from distant landscapes to that phantasmic public just outside the dark room, constituted solely by the true and false rumors it generates. The correspondence of Peiresc, finally, suggests that the real narrative is not the event transcribed by Wotton and Watts, foreseen by Scheiner, and partially suppressed by Descartes and Huygens, but who buys the story, and when, how, where, and why.

CHAPTER 6

Rapid Transport

The preceding chapters have been devoted to the hyperbolic impressions of optics and of astronomy that emerged in the first decades of the seventeenth century and served to evoke, if not to explain, the sudden availability and wide range of news from remote regions. All are predicated on a more or less stationary consumer, variously confronted with rumors of a Jesuit empire so extensive that it included the moon, with tales of news that seems to come before the event, with improbable presentations of Galileo as a *gazzettante*, and with depictions of dark rooms asserting and undermining the elusive ideal of the timely transfer of current information. While the newsreader's telescopic perspective is unavoidable, it would be misleading to suggest that this relatively immobile audience was at once so sedentary and so unimaginative as to overlook the importance of the physical transportation of news.[1] This concluding chapter examines that interest.

The conventional narrative of new events and instruments implies that the story always outstrips the particular incident, object, or person involved, proliferating wildly and producing a tale much more complex than whatever or whoever prompted it. Matters are necessarily otherwise when transport itself constitutes the novelty. In the case of the items under scrutiny here—an amphibious carriage known as the *zeilwagen* or "sailing chariot," and its descendant, the English mail-coach—the story changes from one of the news of the new or newish invention, where tale and vehicle proceed apace, to a fantasy of pure velocity, and one in which news is effectively left far behind the carrier.

The Latest News

The *zeilwagen* turns up in a number of early modern texts and images: it is described in Latin and Dutch poems, a journal, travel narratives involving trips to China and to the moon, a play, and an epic, and it was depicted in several engravings and etchings, and on fabric.[2] One of the most interesting allusions is a one-line reference in a masque of 1619 by Ben Jonson, *News of the New World in the Moon*, itself the embodiment of the conflation of developments in astronomy with the English obsession with journalism. Before turning to the extraordinary claims made for the *zeilwagen*, then, I want to survey first Jonson's interest in the emerging English news industry, and second, what seems to have been his encounter with the failures of transport.

Jonson's longtime preoccupation with journalism ranges from comments on Venetian gazettes in *Volpone* (1604), to his attack on newsmongering statesmen in "The New Cry" (1608?), to the Twelfth Night masques of the early 1620s, to the bitter lines against the newswriter Thomas Gainsford, or "Captain Pamphlet," in "An Execration upon Vulcan" (1623), to his rejection of news-hungry men as friends in "An Epistle Answering to One That Asked to be Sealed of the Tribe of Ben" (1624?) and finally to the masterful *Staple of News* (1626).[3] Apart from the attention he evidently paid to current events in domestic and foreign politics, it is also certain that Jonson found the welter of scientific and technological information that passed for news both intriguing and ridiculous. As late as his *New Inn* of 1629, for instance, he mocked the fevered discussion between a benighted courtier, Sir Glorious Tipto, and the host and the parasite of the inn, about the recent achievements of the Dutch engineer Simon Stevin, who had been dead for close to a decade:

Sir Glorious Tipto: Host peremptory,
. . . Whence had you this [news]?
Host: Upo' the road, a post, that came from thence,
Three days ago, here, left it with the tapster.
Fly: Who is indeed a thoroughfare of news,
Jack Jug with the great belly, a witty fellow!
· · · · ·

Yes, and he told us,
Of one that was the Prince of Orange's fencer.
Sir Glorious Tipto: Stevinus?
Fly: Sir, the same had challenged Euclid
At thirty weapons more than Archimedes
E'er saw, and engines: most of his own invention.
Tipto: This may have credit and chimes reason, this![4]

Jonson was especially interested in the newsmaking activity of another Dutchman, Cornelis Drebbel, who worked as an alchemist, lens grinder, and inventor at the courts of James I and of Rudolph II.[5] He associated Drebbel's perpetual motion device with the vulgar "motions" of a puppeteer and with the upward trajectory of the career of his collaborator and rival, Inigo Jones, in "On the New Motion," mentioned the device again in *Epicœne, or The Silent Woman*, and for a third time in *The Staple of News*. He alluded to Drebbel's attempts to read distant letters in his *News of the New World in the Moon* and to his *camera obscura* in *The Masque of Augurs*, referred to his submarine in *The Staple of News*, and in *The New Inn* described people "poring through a multiplying glass / Upon a captivated crab-louse, or a cheese-mite, / To be dissected as the sports of nature" in reference to his microscope.[6]

While the number, frequency, and range of such allusions suggest that Jonson found much to admire in the Dutchman's various devices, the context generally dictates a humorous treatment of this magus-like figure. Such a perspective was doubtless complemented by Drebbel himself, who, according to the admirers he found in Peiresc and Huygens, "behaved like a simple and ignorant person," "had the appearance of a Dutch farmer," and who provoked Peter Paul Rubens to comment that there was "something in that badly dressed man and in his coarse clothes that fills one with surprise and that would make any other man ridiculous."[7] In 1612, moreover, Drebbel appealed to Prince Henry for permission to establish a lottery, alleging that he had no other means of subsistence; this request suggested that the so-called magus had been unable to make good on his other schemes, and that he then impressed some onlookers as a failure.[8]

Jonson clearly associated optical instruments, some of them developed or used by Drebbel, with the news, but he also appears to have had some independent familiarity with such devices, since April 1609, when he composed *The Entertainment at Britain's Bourse*. This court masque, one of several which Jonson declined to print in the 1616 collection of his works, was only discovered in 1997 in the papers of Sir Edward Conway, an English statesman who had been lieutenant governor of the cautionary town Den Briel in the Netherlands.[9] The contrast between Jonson's dramatic intentions and the actual performance of the masque is worth scrutiny, for the outcome was likely the result of the difference between the hyperbolic rumors concerning the Dutch telescope, and its physical presence at the English court. Here, in other words, the transport of the newsworthy object emerged as a practical problem.

Undertaken as a celebration of the New Exchange in London, the *Entertainment* featured a long list of worthless trifles promoted by a mountebank, and was meant to end with a few costly and worthwhile items, suitable as gifts for the royal family and court members. The shoddy wares, taken from a "China Howse" or emporium of imported goods, included "hour glasses, looking glasses, burning glasses, concave glasses, triangular glasses, convex glasses, crystal globes," and the pitch ended with the crescendo "Spectacles! See what you lack!" The mountebank and his assistant had a typical exchange about the "jewel" of the collection, a "perspective" or pre-telescopic device, which the seller claimed would allow an observer in Covent Garden to "decipher at Highgate the subtlest character," and to identify individuals, and to read their lips, "ten, nay twenty miles off." The combination of chinoiseries and optical devices, the latter likely from the glassworks in Middelburg, anticipates the "gift" of cash from the merchants of the town in Zeeland and of the Dutch East India Company used in part to underwrite another Jonsonian masque in 1618.[10]

Given the emphasis in the *Entertainment* on the extravagance of the mountebank's claims, and on the inferior quality of his goods, the stage seems set for the gift of a bona fide telescope, which had emerged just six months earlier in The Hague, to a member of the royal family or of the court. Edward Conway had been in The Hague at that moment, and like other members of the English diplomatic and military corps in the United Provinces and in the Spanish Netherlands, had almost certainly heard of the device.[11] The expected presentation, however, does not take place. Though correspondence surrounding the masque suggests that there was some difficulty in obtaining all the props, and though the masque itself was subjected to considerable last-minute revisions, we cannot determine if the telescope or instructions for its manufacture did not arrive from the continent in time, or if the Dutch device or its English equivalent looked more like the mountebank's trash than like a valuable gift. The substituted item is an automaton, said to be of better quality than the one created by Jonson's "antagonist at Eltham," Cornelis Drebbel.[12]

The hurried alterations to the masque must have been very trying to Jonson, but these same irritants might have evoked a novel conceit for the treatment of news. If, as the action suggests, the original offering had been a telescope, it seems clear that news of the device, once more, far outstripped the actual instrument. Word of mouth messages, handwritten letters, and printed accounts, each slightly more substantial than the last, inevitably traveled more quickly and more widely than the object they celebrated, and the abyss between the expectations engendered by news of the commodity and the item itself is an obvious source of humor. Put differently, when the latest news is the slow-moving thing prompting the rumors, it is necessarily an occasion for the comedic depiction of disappointment.

But the gag is also shopworn: the mountebank's patter, in the first part of the masque, exploits a not dissimilar gap between the verbal presentation of the "novelties" from the China House and the shabby stuff onstage. The greatest innovation, it seems, would be to remove or radically alter the differential between the news and the material object by making the means of transport the true novelty. This is what happened over time with the literary treatment of the *zeilwagen*.

The Strangest Piece of News

Jonson invoked the nexus of optical instruments and news in his *News from the New World Discovered in the Moon*, a masque he had begun composing during his walking trip to Scotland in the summer of 1618.[13] This work exploits both the strictures lately placed by James I on English newswriters and the particulars of what Henry Wotton had called "the strangest piece of news . . . ever received from any part of the world," Galileo's *Starry Messenger*.[14] It satirizes competing representatives of the news industry—the Chronicler of the troubling events leading to the Thirty Years' War, the Factor, who trafficked in handwritten newsletters from abroad, and that bold latecomer, the Printer of News—all of whom contributed to the growing popular dissatisfaction with James's foreign policy.[15] It has been suggested, moreover, that this masque was inspired at least in part by an unsuccessful proposal for a newsletter network made to Sir Dudley Carleton by the news correspondent John Pory and an unknown associate between 1616 and late 1618.[16]

The general context for the public's avid consumption of astronomical news was undoubtedly the advent of three comets in the fall of 1618: these bright flares were noted by most newswriters of this period. Many commented on the likely political effect of the comets, some reported public interpretations of their significance, and a few related discussions of their position to ongoing discussions of cosmological import. The well-informed John Chamberlain suggested that the comets were widely regarded as either the sole news worth discussing in the fall of 1618, or as a cipher for all other events of the period:

> [Wednesday] was the first day we tooke notice here of the great blasing-star, though yt was observed at Oxford a full weeke before: yt is now the only subject almost of our discourse, and not so much as litle children but as they go to school talke in the streets that yt foreshewes the death of a king or a quene or some great warre towards. Upon which occasion (I thincke) yt was geven out all this town over that the Quene [Anne] was dead on Thursday: but yesterday I heard for certain that she is in a fayre way of amendment.[17]

The newswriter John Pory, for his part, reported that "our astronomers holde that the newe prodigious starre is above the sphere of the moone and that being situate in Lanx Borea of Libra [i.e., in the upper pan of the Scales], it is against some great Monarke of the north (as the [Holy Roman] Emperour or the Moscovites. . . . Some say, it foretelles great warres about Religion."[18] Jonson's friend Richard Corbet mocked the widespread public interest in the comet not just as a political portent but also as an astronomical phenomenon, an obsession which rendered every common laborer an aspiring observer, and which converted ordinary tools and even musical instruments into Jacob staffs, and the hand into a perspective or telescope:

> Physitians, Lawyers, Glovers on the Stall,
> The Shopp-keepers speak *Mathematiques*, all.
> And though wee read no Gospell in the Signes,
> Yet all Professions are turn'd Divines.
> All weapons from the *Bodkin* to the *Pike*,
> The Mason's *Rule*, the Taylors *Yard* alike
> Take *Altitudes*; and th'early Fidling Knaves
> Of Fluites, and Hoe-Boyes [oboes], make them *Iacobs-staves*.
> Lastly, of fingers glasses wee contrive,
> And every Fist is made a Perspective.[19]

Such public participation in the comets' progress plainly irritated James, who allegedly called one flare "nothing else but Venus with a firebrand in her arse," and whose verses "On the blazeing starr" associated the effort to interpret the celestial event with newsmongering.[20] In a comparison that portrayed the comet as a kind of bulletin—but one much more difficult to read than the newly emergent Dutch coranto—he admonished the "men of Brittayne" to "misinterpret not with vayne conceyte / The character which you see on heavens heighte: / Which though it bringe the World some newes from fate, / The letter is such as none can it translate."[21]

The death of James's spouse, Anne of Denmark, in early 1619, however, seemed to the English populace to confirm the validity and ultimate legibility of the celestial portent, and it is possible that when Jonson began work on this masque, he intended to criticize the public's presumed regard for astronomical information as a species of news.[22] The focus of *News from the New World Discovered in the Moon*, of course, is not the comets but the lunar body, and the astronomical work to which it was indebted was not Galileo's treatise on the comets of 1618, which did not emerge until 1623 and offered little in the way of prognostication, but rather the still newsworthy *Starry Messenger*. Thus even as it mocks the marketing and reception of fantastical news about the world in the moon—for this exotic world bears a strong resemblance to contemporary London—the masque preserves at once James's insistence that there was, in fact, no news worth reporting, *and* the great and startling announcement of Galileo's work, that of the overwhelming similarities of the earth and its satellite.

There was no better way of suggesting, in a period in which both domestic and foreign news were unreliable, dangerous, and only with difficulty distinguished, that the safest reports, and in some sense the most newsworthy, concerned the moon and the remote regions beyond it. And given that another slightly later masque, Jonson's *Gypsies Metamorphosed*, circulated in piecemeal fashion far beyond the court, participating in a larger and much less strictly controlled political culture of news exchange, libel, and commentary, the surprising feature here is the apparent docility of the playwright's stance.[23]

Two letters in the epistolary collection of the English recusant Sir Toby Mathews illuminate this masque's treatment of journalistic issues.[24] The first, allegedly written by "One Friend in London, to another, who was then upon the Confines of Scotland," offers an oblique description of some form of news repression in the English city:

The Politique shrug, and the saying, *it's dangerous to write news*, keeps us here in darknesse. The onely light which we have is from a certain merry man, who is resident with you to that purpose; his Characters of you all are communicated to us here, in strong lines;

and his being there keeps us in hope, of knowing somewhat from time to time, if our capacities may be able to reach it. As for us, the Town is empty. . . . *Hide-Park* is full of Widows, and the *Spring-Garden* full of Maids.[25]

It seems likely that the "merry man" then in Scotland was Jonson, whose well-publicized walking trip there in 1618 had been anticipated since early 1617, when Sir Dudley Carleton, English ambassador to the United Provinces, learned that "Ben Jonson is going on foot to Edinburgh and back, for his profit."[26] The second letter, seemingly written by Jonson himself, is from "A Gentleman to a great Lady, from the Borders of Scotland." It comments on the irrelevance of news from the English city to the Scottish countryside:

Madam, I have, with much ado, out-Marcht the Spring, and am at length fallen into a part of the world, where nothing prospers but Rebellion. Roses, and such foolish Emblems of a Summer, are things onely heard of here; and discourses of a Spring-Garden, or Hide-park, have but the same faith and credit with the people of this Country, which the stories of *Cataia* [China] find with us.[27]

Both letters find resonance in the *News from the New World Discovered in the Moon*, where the lunar vehicles are said to be driven by the wind, "like China wagons," and where the meeting places of these coaches are described as "Above all the Hyde-parks in Christendom, far more hidden and private."[28] Two conclusions may be drawn from these faint echoes. First and most obviously, the initial presentation of a stylized China as a double of the moon would seem to suggest that only infinitely remote locales, and ones with very little bearing on English foreign or domestic policy, could safely be regarded as worth journalistic attention. Second, however, the implied similarity of London and the telescopic moon with its chinoiseries erases the various physical and cultural distances separating the English city, a kingdom of the Far East, and our satellite, this radical relativism being both characteristic of Galileo's insistence on the likeness of the terrestrial and lunar globes and visible in the emergence of emporia such as the "China Howse."[29]

The antimasque insists on what Henry Wotton found to be "of all the strangest" announcements in the *Starry Messenger*, the contention that the moon was at the end and beginning of its monthly cycle "illuminated with the solar light by reflection from the body of the earth."[30] Thus the Second

Herald begins a ponderous speech with the remark that "We have all this while (though the muses' heralds) adventured to tell your majesty no news," and then gestures to the masquers "rapt above the moon far in speculation of virtues, have remained there intranced certain hours, with wonder of the piety, wisdom, majesty *reflected by you on them, from the divine light, to which only you are less.*"[31] One of the songs likewise bids the dancing masquers, now descended from chilly lunar globe, to view the English monarch as representative of the entire earth, and the source of reflected solar light and warmth:

> Now look and see in yonder throne,
> How all those beams are cast from one!
> This is that orb so bright,
> Has kept your wonder so awake;
> Whence you *as from a mirror* take
> The sun's reflected light.[32]

Despite his apparent acceptance of the earth's ability to reflect light, Jonson does not go on to promote the concomitant conclusion of that body's movement and status as a planet: James, though a stand-in and sole representative of our globe, is quite clearly an unmoved first mover, rather than a participant in "the dance of the stars."

The motto of the masque—*Nascitur è tenebris; & se sibi vindicat orbis* ["a world is born from darkness and asserts itself"]—would apply indifferently to the earth or to the moon, Galileo having shown in the *Starry Messenger* that the opacity and rough surface of both those dark bodies enhanced their ability to project light.[33] Though the first newswriter's assertion that some mild form of journalistic censorship kept those in the city "in darknesse" is conventional, it is worth considering the contrast between the "onely light" borne to Londoners, the news relayed by the "merry man" in Scotland, and the implications of Jonson's letter and masque, with its curious conflation of cities, countries, and satellites. The masque resolves such tensions by focusing on the only acceptable news of the day, that which insists on the mutual illumination of the earth and the moon, the relative similarity of all land masses on both bodies, and, in effect, the interchangeability of the city of London, the Kingdom of Cataia or China, and the entire lunar globe.

The last lines of the masque, a reference to "Fame, that doth nourish the renown of kings, / And keep that fair which Envy would blot out," effectively paraphrase the opening lines of the *Starry Messenger*, which declare that "a most excellent and kind service has been performed by those who defend from Envy the great deeds of excellent men and have taken it upon themselves to preserve from oblivion and ruin names deserving of immortality."[34] What is most relevant to our purposes is the metamorphosis of *Fama* or Rumor from the petty and mutable stuff of daily news into a respectable and static icon for deserving rulers like James I, or in Galileo's case, Grand Duke Cosimo II de' Medici, for whom Jupiter's newfound satellites were named.

What Jonson's masque offers is a spectrum of news, one which ranges from the high-velocity, low-fidelity oral reports circulating chiefly in the provinces and among the unlettered, many of which concerned the numberless sudden deaths monarchs of that day seemed to undergo, to the somewhat slower and only slightly more reliable accounts of their various political and social vicissitudes in manuscript and printed newsletters, to the spare and unchanging poses struck by those rulers in their quest for posterity.[35] At the lower end of this range, those alarming chimeras that were called both "rumors without an head" and "the reporte of the common people, . . . *belluam multorum capitum*, [a many-headed monster]" sometimes brought their propagators to assize court. The circulation of "lavish and licentious speech of matters of state," more regular in its rhythms and plausible in content, appears to have resulted in the suppression of some early newsbooks and corantos.[36]

The chronicling of the immobile resplendence of James, however, presumably increased the king's considerable regard for his playwright. Jonson's presentation of James as a stupendous and static representative of the globe is not merely a gesture to the expected resistance to Copernicanism, but a typical way of conveying power and authority without other particulars, the newsless news of pageantry. It is clear that such "representative publicness," to use Habermas's term for its early modern manifestation within absolutist cultures, regularly appeared in news publications, though some readers distinguished between it and the sort of information they preferred.[37] "Italy being now in such a profound torpor," Paolo Sarpi complained in summer 1611, "that we are kept far not just from news, but also from schemes and plans, such that even the *gazzettanti* have no material other than a few banquets and festival decorations."[38] Once the emphasis shifts to transport itself, the tension between this newsless news of royal presence and prerogative and that appetite for novelties both past and future will be a persistent feature.

China Wagons and Stevin's *Zeilwagen*

One detail in particular in the *News from the New World Discovered in the Moon* points to the studied effort to displace the epicenter of the news industry, the United Provinces, throughout the masque. As noted, Jonson compared the lunar vehicles to carriages in Hyde Park and to "China wagons," these last being the *zeilwagen*, or wheeled chariots equipped with sails and made for rapid travel on land. References to such vehicles had first emerged in Portuguese and Spanish descriptions of the Far East in the second half of the sixteenth century, some of which soon appeared in translations to Latin and the other European vernaculars.[39] While the Chinese pedigree of the "sailing chariot" or "wind wagon" was well established through these accounts and through several pictorial works, what is more crucial is that the invention quickly became a Dutch specialty in the popular mind, and that like the telescope it was often presented as an instrument that, whatever its origins, was realized by a Dutchman, offered to Maurice of Nassau, demonstrated to foreigners against a backdrop of strife, and associated with rapid communication.[40]

The commentary accompanying a celebrated engraving based on a drawing by Jacques de Gheyn II, published in Latin and Dutch versions in 1603, stated that the great engineer Simon Stevin designed a *zeilwagen* that traveled along the coast from Scheveningen, just beyond The Hague, north northeast to Petten, a distance normally covered in fourteen hours, in only two, in the spring of 1602.[41] The vehicle was steered, somewhat erratically, by Maurice of Nassau, and it carried diplomatic personnel from France and the Holy Roman Empire, prominent visitors from England and Denmark, and a celebrated Spanish prisoner of war.[42] The jurist Hugo Grotius, then a young man, was among the twenty-eight passengers, and he referred to the trip in both his prose, where he noted the Chinese analogue, and his verse, some of which was published as early as 1603.[43]

The engraving and Grotius's poems both circulated long after the novelty of the vehicle itself had worn off, and the advent of the telescope, in particular, seems to have complemented the lyrical depiction of the *zeilwagen*.[44] Grotius had emphasized, for instance, the surreal nature of the venture: once the winds had filled the sails, "the distances of places disappeared, and the swift carriage / Was borne as if by the oars of the rushing air / Along endless new shores / And the mountain peaks were no sooner seen than they vanished."[45] Given that the nearest mountains would have been almost 150

FIGURE 4. Jacques de Gheyn, *The Sailing Chariot*, 1603.
Reproduced with permission from the Rijksmuseum.

miles away, this last detail suggests that the passengers somehow enjoyed the privilege of telescopic vision in addition to high-speed travel. And yet the vehicle was not merely an analogue of the telescope, but an invention so spectacular as to seem celestial, and to diminish, some seven years before their discovery, the Jovian satellites. Thus in a poem appearing in both Latin and Dutch, the chariot itself states, "Jupiter, give me the starry fire in heaven; / I will not be the first, for there are already two wagons there."[46]

The French scholar Peiresc made a point of seeing the *zeilwagen* in 1606, and he went for a ride as well, recalling with amazement years later that "though he was borne by the swiftest of winds, he did not feel it, for the chariot went just as fast, and that they went flying over ditches and dikes, skimming the surface of the water encountered here and there, and that the couriers ahead of them seemed almost to be running backwards, and that things that had seemed most distant were overtaken in an instant."[47] Johannes Walchius offered a typical assessment of the vehicle in a compendium of new inventions in 1609, in the paragraph preceding his description of that other recent Dutch marvel, the telescope. "Though some might assert that it was also driven by the Chinese," Walchius related, "lately on the Dutch coast, amidst the roar of arms, a type of carriage or chariot never before seen by this age has been contemplated by a certain ingenious man, and he completed his thoughts with work, and brought the thing into being."[48] Significantly, yet another observer of the *zeilwagen*—and one with close ties to both Ben Jonson and Sir Robert Mansell, leader of the English glass industry—seems to have fused the two tales of invention. In 1623, in the course of a visit to the United Provinces, James Howell described

> a Wagon, or Ship, or a Monster mix'd of both, like the *Hippocen-taur*, who was half Man and half Horse: This Engine hath Wheels and Sails that will hold above twenty People, and goes with the Wind, being drawn or mov'd by nothing else, and will run, the Wind being good and the Sails hois'd up, above fifteen miles an hour upon the even hard Sands. They say this Invention was found out to entertain *Spinola* when he came hither to treat of the last Truce.[49]

Though Grotius suggested that the Dutch and the Chinese shared this and many other inventions because they lived on the same latitude, the effort to minimize or to improve upon the alleged Eastern origins of the *zeilwagen* proved a greater preoccupation for many writers. In 1606 Tommaso Campanella, writing from the depths of a Neapolitan prison, proposed an ambitious plan of evangelical reform, one element of which involved making "weighted carriages run on land, better perhaps than what they have in China."[50] In 1612 the young poet Constantijn Huygens alluded to no Chinese prototype, but in the midst of a poem celebrating the ultimate domestication of the device—its (quite unfounded) utility to farmers—gestured vaguely to some central Asian outpost, noting only that "the carriage has sails, which when filled with air, let it cut the dry shores more swiftly than the Scythian arrow."[51] What Huygens promoted as available and practical at home, John Milton would return some sixty years later to the Chinese; in *Paradise Lost*, however, the chariot is no sign of someone else's civilization, but rather an isolated detail in a landscape of desolation and endless despair. Satan, in his approach to Eden, is like "a vulture on Imaus bred,"

> Whose snowy ridge the roving Tartar bounds,
> Dislodging from a region scarce of prey
> To gorge the flesh of lambs or yeanling kids
> On hills where flocks are fed, flies toward the springs
> Of Ganges or Hydaspes, Indian streams;
> But in his way lights on the barren plains
> Of Sericana, where Chineses drive
> With sails and wind their cany wagons light:
> So on this windy sea of land, the Fiend
> Walked up and down alone bent on his prey. . . . [52]

It is possible that Jonson heard something of the sailing chariot while he was in Leyden in the spring of 1613, as it was then being readied in nearby Amsterdam for the visit of the newly wed Elizabeth and Frederick of the Palatine.[53] What is distinctly unlikely is that he knew of it only as a Chinese artifact. His depiction of a wagon that "goes with the wind" is overshadowed by his subsequent depiction of the lunar carriages as convenient places, like those of Hyde Park, for amorous trysts. But the suppression of the Dutch analogue of the "China wagon" would certainly have been noticed by his audience, and likely seemed evidence of his dutiful silence about what many Englishmen regarded as that "great Staple of News."[54] The association of the disguised *zeilwagen* with news, or the timely delivery of current information, however, is not limited to the fact that the United Provinces managed, in a

more or less literal sense, to put their stamp on both stories about Chinese
vehicles and other tales from elsewhere in the world. It is connected rather to
the speed with which the two idealized products were associated, a velocity
best expressed, and perhaps *only* expressed, by an insistence on the tardy
motion of more quotidian means of transport and communication.

Recall that Jonson, making his way on foot from London to Scotland in
1618, prefaced his comparison of news from Hyde Park with stories from
"Cataia" by telling his addressee that he had "out-Marcht the Spring," meaning that in heading north, he had encountered weather still wintry enough
to make reports of "Roses, and such foolish Emblems of a Summer" as
implausible as those tales from China. Jonson referred to his Scottish journey
in the course of his masque, exploiting it for a predictable fat joke about how
"one of our greatest poets (I know not how good a one) went to Edinburgh
on foot,"[55] but the importance of the slow trip and the swift vehicle emerges
when both are compared to Peiresc's experience. When the Frenchman went
for a ride in the *zeilwagen*, "the couriers ahead of them seemed almost to be
running backwards," so slowly did even these professional messengers proceed in comparison to the sailing chariot. Thus just as Jonson, after consum-
ing part of the summer in his journey, nonetheless found himself in Scotland
in weather that seemed more like that of the previous winter, so the couriers
entrusted with mail and other messages appeared, from the perspective of the
passengers, to retrograde even as they made their way toward their destina-
tions. According to the fantastic logic of the travelers, if the trip in the *zeilwagen* had lasted longer, the couriers would have not yet set out, and the events
relayed by them would have not yet occurred.

Within the masque, the reference to the wind-driven "China wagon"
would conform to the overall emphasis on no news: it so outstrips the pace
at which the rest of the world moves that all other travel and all other actions
are as if suspended. The figure of this impossible vehicle is useful in that it
makes visible a curious aspect of early modern news consumption. But it is
also in some way expendable, for the fantastic impression that conventional
means of travel were not just sluggish purveyors of stale information, but also
the foundation of an oddly malleable and porous past, was entirely characteristic of the period. The endless influx of tardy reports from distant places
made the present moment seem more capacious, in other words; the piecemeal pace by which political matters were decided, revised, revisited—the
standard fare of the weekly paper by the mid-seventeenth century—differed
significantly from the definitive past of the rare royal spectacles that had
dominated pamphlets one hundred years earlier.[56] Ordinary accounts about

recent events became things as unaccomplished as any future, a form eventually recognizable in the emergent novel.[57]

Thus the peculiar condition produced by travel in the sailing chariot—that it leaves in its wake not the unaltered terrain of the quotidian, but rather tokens of events yet to happen—was the illusion regularly conferred on the newsreader, especially, it seems, the audience of battle accounts. There is some hint of this tendency in 1629 in Jonson's depiction of three-day-old "news" carried by the post and picked up in a roadhouse concerning the putative inventor of the *zeilwagen*; Stevin, who had died in 1620, was portrayed as "the Prince of Orange's fencer" and as one who had challenged Euclid to a duel, an observation which prompts a skirmish between the newsmongering discussants about whether or not the Dutchman had vanquished the French scholar Joseph Justus Scaliger, deceased some twenty years earlier. Two more instances of this recognizable posture will suffice: a Doctor Bevilacqua in Northern Italy was criticized in the late 1630s for his reactions to news of the wars in Flanders, for according to a chronicler of the period, "while he held forth—and this is truly ridiculous—with his gestures and his movements he prepared the field, joined battle, wiped out cities, reported the princes' views, and the king's, and the captains', and he seemed like a secret counselor, one who knew both the battle orders and the deliberations of war and of peace."[58] A later report suggests that the scuffling between newsreaders became more intense later in the century as options increased; all seemed laboring under the illusion that the outcome of past battles somehow still lay in the future:

> Everyone had his own ideas, and when a gazetteer's article took a particular line it became a riotous excuse for each to test his mettle. It was a war about another war; they skirmished more with their wheeling tongues than the soldiers had done with their sharpened swords. . . . They sometimes grabbed each other and determined battles with blows, the number of deaths with injuries, and the campgrounds with contumacious language.[59]

The Glory of Motion

I have been suggesting, then, that these sorts of retrogressions were best embodied by the unreal difference between the fantastic speed of the privileged passenger in the *zeilwagen* and the pace of the quotidian, but what is more important is the availability of the motif to nontravelers, and its persistence even when that remarkable vehicle had fallen into desuetude, or where it had never appeared. One version of this differentiated tempo is the increasing prevalence of news-related publications emphasizing tardiness.

The initial references to the *claudus tabellarius* or *Hinkende Bote* or *Courrier boiteux* or *Corriere zoppo* proposed the limping messenger as a guarantor of truth: the account offered by the latest herald was implicitly more trustworthy than those of his swifter rivals. News bearing the title *Hinkende Bote* and the image of a limping messenger appeared in Brunschweig as early as 1607. The image was evidently familiar in Italy as well: in 1609, the well-placed prelate Bonifacio Vannozzi noted that "the rumor of a distant object is generally bigger than the truth, and thus the wise should not give assent. He who can wait for the lame fellow's news does best."[60] One year after its publication, the *Starry Messenger* was followed by a report from a self-described *claudus tabellarius* at the Collegio Romano whose task it was to provide no new novelties, but merely to comment on those proposed by Galileo. In a utopian text of 1631, Moravian scholar and reformer Johannes Comenius would describe news brought by men on crutches as particularly sought out by prudent consumers.[61]

By the late seventeenth century, the title had broadened to indicate a popular almanac, and the limping news bearer was often depicted as a wounded soldier. As the contents increasingly involved future or seasonal events such as fairs, prognostications, amusing, freakish, or moralizing stories, and exotic accounts of the Far East, the tardiness and implicit distance from the reportage of the recent events, usually reduced to a survey of the crucial incidents of the last year, mattered still less.[62] Critics associated this sort of news publication throughout the eighteenth and nineteenth centuries with ignorant consumers, and more interestingly, with a programmatic censorship; in 1781, William Blakey, an English engineer variously resident in Paris, Amsterdam, and Liège, complaining of the poor circulation of technical and scientific information, wrote to the editor of the *Journal des Savants* that

> In the enlightened country of France, the press and the book trade
> are under the rod of censorship. Thus one can imagine how tardily
> new discoveries emerge in published documents, with the exception
> of announcements of baptisms, marriages, and deaths, tales of some
> bolt of lightning or great blast of wind uprooting a tree, nice songs,
> and all their folderol. Monsieur, if one publishes anything in France,

it's always by this limping courier. In England, it is as difficult to invent anything new as elsewhere, but what has been accomplished in politics or in the sciences is known in just a few hours throughout the Kingdom, right down to the last hamlet.[63]

Though the engineer did not describe that happy English alternative to the limping messenger, the fantastic speed of the *zeilwagen*, and the corollary of a recent past that somehow still lay within the grasp of the future, had already been addressed within the emerging novel. In the 1760s, a strangely modified version of Peiresc's trip to Scheveningen was regarded as a noteworthy event in Laurence Sterne's *Tristram Shandy*.[64] Peiresc's effort is championed, and misrepresented, by the narrator's Uncle Toby, who, like a parody of the quarrelsome newsreader, revises both past history and his own future by reenacting battle campaigns based on stale information found in the *Gazette*. As Uncle Toby presents it, Peiresc's journey appears to be a noble effort to outmarch the spring, as if to suggest that only legendary expeditions of prodigious slowness would serve to commemorate this vehicle once all reports of its immense speed have died away:

> You might have spared your servant the trouble, quoth Dr. *Slop*, (as the fellow is lame) of going for Stevinus' account of [the sailing chariot], because, in my return from *Leyden* thro' the *Hague*, I walked as far as *Schevling*, which is two long miles, on purpose to take a view of it.
>   —That's nothing, replied my uncle *Toby*, to what the learned *Peireskius* did, who walked a matter of five hundred miles, reckoning from *Paris* to *Schevling*, and from *Schevling* to *Paris* back again, in order to see it,—and nothing else.[65]

Stevin's book, borne by Uncle Toby's limping manservant, arrives at last. "In popp'd Corporal *Trim* with *Stevinus*:—But 'twas too late—all the discourse had been exhausted without him, and was running in a new channel."[66] The volume is searched by Uncle Toby, but contains nothing of the *zeilwagen*, for the engineer had never written anything of it. A manuscript sermon falls out instead, based on an actual sermon Sterne himself had preached in 1750 and published separately, and it is read aloud, with infinite interruptions and commentary, over the course of the next few chapters. The lame corporal Trim performs the reading, effectively occupying the role of the limping courier; Uncle Toby is the consumer of accurate, if tardy, news about Stevin's irrelevance to the *zeilwagen*, and of a fresh performance of a text long since available in print.[67]

Tristram Shandy thus continues the artful erasure of the *zeilwagen* begun in *News from the New World Discovered in the Moon*. If we read the gap between the speeding vehicle and the tardy arrival of more ambiguous and even contradictory reports as the crucial point, we find that the motif reappears even after the departure of its putative inventor. A late (but surely not the last) variant on the swift amphibious vehicle is Thomas De Quincey's *English Mail-Coach*. First published in serial fashion in October and December 1849, in the aftermath of the failed European revolutions of the previous year, *The English Mail-Coach or The Glory of Motion* takes the form of a Victorian-era meditation on the Romantic sublime; reductively put, it is an analysis of the relationship between transport and transportation. Given the implicit distance between the author and his youthful and rather shrill narrator, it can be read as a disclosure of sorts, focusing on the state's perennial effort to control public discussion of current affairs through pageantry performed at what was then a breakneck speed, about fourteen miles an hour.

This pseudobiographical essay recalls the wide network of carriages bearing post and passengers throughout England during the Napoleonic Wars of 1805–1815. It begins with a comparison of the establishment of a rapid and punctual nationalized system of communications to an astronomical finding that had been newsworthy two centuries earlier, but recently surpassed as a timekeeper:

> Some twenty or more years before I matriculated at Oxford, Mr. Palmer, M.P. for Bath, had accomplished two things, very hard to do on our little planet, the Earth, however cheap they may happen to be held by the eccentric people in comets: he had invented mail-coaches, and he had married the daughter of a duke. He was, therefore, just twice as great a man as Galileo, who certainly invented (or *discovered*) the satellites of Jupiter, those very next things extant to mail-coaches in the two capital points of speed and keeping time, but who did *not* marry the daughter of a duke.[68]

From this startling beginning, presented as if obvious to any but the eccentric extraterrestrials who evidently watch us, the narrator turns to a set of familiar concerns. The mail-coach is "a national organ for publishing [the] mighty events" of Trafalgar, Salamanca, Vitoria, and Waterloo; its exact connection with "the state and the executive government" was "obvious but yet not strictly defined." Those who, like the narrator, rode on the outside of the coach—literal hangers-on—benefitted from this ambiguous official authority: "[it] invested us with seasonable terrors," and "had its own incommunicable advantages."[69]

De Quincey's mail-coach is a reincarnation, at once uncanny and absurd, of the early modern depiction of the *zeilwagen*. As if to recall and to dismiss the dispute over its Dutch or Chinese origins, the narrator observes that prior to the establishment of Mr. Palmer's network, passengers had been divided into two categories, "Delft ware" and "porcelain," and he further alleges that the Chinese were recently given a state-coach, entirely unfamiliar to them, by the ambassadors of George III. Of the Dutch there can be no mention apart from the narrator's single allusion to the "secret word," "Waterloo and Restored Christendom;" under French domination by Napoleon's brother Louis Bonaparte until 1810, the newly reconstituted United Kingdom of the Netherlands was a crucial force in Napoleon's defeat in 1815.[70]

The hybrid quality of the *zeilwagen*—the amphibious nature that persuaded James Howell to class it as a "monster"—reappears in the reptilian driver of one mail-coach, repeatedly described as a crocodile for his "monstrous inaptitude for turning round," and in another coachman, who, no less grotesque, was "a monster in point of size." A dispute among the passengers over the seating arrangements is termed a "mutiny," and the mail-coach's sometime proximity to the shoreline is emphasized: its most harrowing incident, in the precincts of Lancashire, takes place as it is "nearing the sea" and "running on a sandy margin of the road."

The narrator's memory of his brief dalliance with a coachman's granddaughter on the outskirts of Bath is recalled, nearly thirty-five years later, as "a rose in June." Thus the events of a time "about Waterloo," or around 1815, are presented as fragments of the world Ben Jonson had outdistanced in 1618 in his walk to Scotland, "I have, with much ado, out-Marcht the Spring, and am at length fallen into a part of the world, where nothing prospers but Rebellion. Roses, and such foolish Emblems of a Summer, are things onely heard of here."[71]

For his part, De Quincey's narrator speaks as one indifferent to news, and thus comically unaware of the spring and summer of political crises that had engulfed so many European nations in 1848: "Roses, I fear, are degenerating, and without a Red revolution, must come to dust." His nostalgic conviction of the exceptional nature of the Napoleonic venture would have been viewed with some irony in late 1849, for the failure of the French revolution of the prior year had resulted in the election of Louis Napoleon and had raised the specter of Bonapartism. While the narrator claims that even "the meanest [English] peasant" had been able to distinguish the battles of the Napoleonic conflict from "the vulgar conflicts of ordinary warfare," "often but gladiatorial trials of national prowess," those currently observing developments in France attributed Louis Napoleon's triumph at the polls in December 1848 to the inability or unwillingness of the French peasantry to differentiate between the late emperor and his nephew. Whether critical of that group's "ignorance," its "reminiscences of the glory of the Empire," and its susceptibility to the "magic influence" of a name, or in sympathy with the "poetical souvenirs of the peasant and the artisan," onlookers agreed that Louis Napoleon was "not the future, but *history*."[72] A purported letter from a Frenchman, published in London several weeks after this first presidential election, presented France's rural population, newly enfranchised and ill informed, as a phantasm, an oversized projection, and the antitype, at once archaic and perplexingly modern, of the docile English provincials of De Quincey's story:

> The *name* was still more powerful than [the provisional government] had thought; it raised the peasant with irresistible strength. There was something like superstition in the unanimous impulse of the country-people. Those who could not read numbered with their fingers the twenty-two letters of Louis Napoleon Bonaparte printed on their tickets. Some of them had been told that the votes for Bonaparte would be, by some secret chemical process, converted into votes for Cavaignac, [his strongest rival], when once in the ballot-box: they came in whole bodies with their muskets to watch over the boxes, day and night; and as a distinctive sign, they had folded their tickets into a triangular shape, meaning the *petit chapeau* of the Great Emperor. There is something awful and portentous in that entirely new power which has so revealed itself: that savage giant, the Peasant, who appears for the first time on the stage, is as yet a mystery.[73]

Against this backdrop De Quincey's outmoded English mail-coach travels. Rather than a variety of news, it brings a single message to those on its route. "From Trafalgar to Waterloo," or from 1805 to 1815, those tidings appear to concern particular victories of the Napoleonic Wars—"Badajoz for ever!" or "Salamanca for ever!"—but the true message is simply the imperative to make way.[74] Thus De Quincey's narrator chronicles a relentless effort to displace, or rather to occupy, the public sphere, by substituting a high-speed pageantry for that autonomous space where reasonable, informed, and property-owning citizens might discuss, evaluate, and choose to contest or to support the state's current activities, which were at that moment, and had been for a long but hazily defined prior period, mainly focused on its war effort.[75]

Whereas the typical strategy of groups attempting to monopolize the public sphere is to mask their particularity and bias by claiming the disembodied central perspective of a neutral observer, De Quincey's narrator recalls the struggle of actual individual bodies to secure an outside position on the mail-coach and to benefit from its "incommunicable advantages." The "public" seeking this place is figured by the narrator as singular, "a well-known character, particularly disagreeable, though slightly respectable," "generally from 30 to 50 years old," and his jockeying for position is termed "a perfect French revolution." Once en route, the narrator seeks with comic transparency to present himself as part of a large and physically powerful state-sponsored group: "We, on our parts (we, the collective mail, I mean) did our utmost to exalt the idea of our privileges by the insolence with which we wielded them." A commercial carriage attempting to outdistance the mail-coach is judged "jacobinical" and "seditious."[76]

Though the reason for the mail-coach's haste is, in theory, the urgency of the news it is bearing in the form of recent letters and gazettes, the message is uniform, timeless, and wholly embodied in the medium of transportation. Distracting particularities, such as the news of the fate of a particular regiment in the "imperfect" battle of Talavera of 1809, are suppressed; for the mother who celebrates that victory without realizing that it probably entailed the loss of her son, his death will remain "imperfect," in the grammatical sense of "unaccomplished," until the arrival of something other than this official news. Redundancies also reduce the scope of discussion. As the coach picks up speed on the open road outside of London, the narrator believes that all in the gathering crowd "understand the language of our victorious symbols," and imagines a young girl informing her companions that the laurels on the fast-moving vehicle indicate a great victory in Spain. He bids the guard to toss a *Courier* emblazoned with "some such legend as GLORI-OUS VICTORY" into an oncoming carriage, conceding that simply to see the paper emerging from the racing triumphal mail-coach, rather than to read and to assess its claims, "explained everything."[77]

When the contents of the mailbags are threatened by fire from the pipe of "an obstinate sailor" riding on the outside of the carriage, the narrator termed the action "treason" and *laesa majestas*, but it is not at all clear that the loss of letters and gazettes constitutes the offense. For though the flame, fanned by their swift motion, "threatened a revolution in the republic of letters," "even this left the sanctity of the [driver's] box unviolated." Like those early modern humanists who camouflaged their requests for news in displays of classical erudition, the narrator vainly attempts to warn the driver of the fire, and signals his particular disregard for the post, by quoting, first in Latin and then in translation, a line from the *Aeneid* concerning the prog-ress of the flames through Troy.[78]

The mail-coach appears thus literally and figuratively burdened by the letters and gazettes that it carries: this ballast of news slows its progress, and evidently complicates, or even erodes, the simpler message of power and authority the vehicle itself conveys. News from abroad, in particular, is an impediment, and the narrator relates that on the evening of his collision in the outskirts of Lancaster, an initial delay has been caused by "a large extra accumulation of foreign mails." The need to overcome that delay, to outrun the news by going still faster, was one factor in the accident; the other causes were the driver's fatigue and a thoroughfare so entirely empty that the mail-coach was able to fly along, mile after mile, on the wrong side of the road. Both circumstances, the narrator explains, were caused by the summer assizes in "Lilliputian Lancaster": the exhausted citizens of that county, far from congregating for discussion of public affairs, were fully absorbed in protecting their private property in the civil and criminal cases then being heard. And the narrator is unable to rouse the slumbering guard because he is blocked by the cumbersome foreign mails on the roof.[79]

Thus far, then, De Quincey's mail-coach appears a hyperbolic version of the *zeilwagen*, its speed evacuating the news it carries of discursive content. Not surprisingly, the much slower vehicle it strikes evokes the precedents in Milton and Jonson, being a "light gig" or "cany carriage" occupied by a courting couple, the lonely road where they collide recalling that landscape "above all the Hyde-parks in Christendom, far more hidden and private."

More interesting than the fact that the mail-coach literally displaces this cart is their positioning, much like the "China wagon" in Jonson's masque, in a zone where the violence of the effort to banish news is obliquely acknowledged. The scene of the accident is romantic: a moonlit terrain near the Irish Sea, covered with "a slight silvery mist, motionless and dreamy," a long, tree-lined avenue resembling a cathedral aisle. The hints of disturbances either from nearby Ireland, where rebellious factions, aided by revolutionaries in 1798 during the "Year of the French," had been forcefully subjugated by the Act of Union in 1801, or from the still restive continent, are no sooner signaled than dismissed. "Except for the feet of our own horses, which, running on a sandy margin of the road, made little noise, there was no sound abroad." Given the narrator's careful description of "the sea on our left," the drift of the northbound mail-coach onto a coast-like right-hand edge of the road, and the Swiftian inflection of "Lilliputian Lancaster," it is as if the entire region were nothing but a narrow corridor nervously awaiting invasion from either the Irish or the French, the carriage itself moving away from the slightly more remote memory of the Irish Rebellion to attend to the pressing menace of the continent.[80]

The narrator's initial perception of the occupants of the light gig in the path of the unattended mail-coach, made from a distance of around 450 yards, participates in the usual telescopic fantasy of the early modern period of lipreading and character scrutiny: "Ah, young sir! What are you about? If it is necessary that you should whisper your communications to this young lady—though really I see nobody about at this hour, and on this solitary road, likely to overhear your conversation—is it, therefore, necessary that you should carry your lips forward to hers?"[81] Like the landscape, this vision is tinged with a hint of political paranoia: the amorous conversation appears, for the briefest of moments, conspiratorial.

The sustained effort to depict the victims as private—naïve, domestic, removed from all concerns but those of the heart, helpless, hardly differentiated from the horse that draws them, almost invisible—naturally distances them from that better-informed and critical public whose demands for current news of appropriate scope and complexity burden the mail-coach. But it also exposes the logic of Jonson's masque, where the insistence on the "hidden and private" locale of the wagons, used for amorous trysts, only partially disguises the state's preoccupation with the equally hidden and private precincts of the seditious, the rebellious, and what would in time be termed the "jacobinical." Put differently, the elaborate production of sentimentalized, private, and domestic characters is the obverse face of the effort to control public consumption of news. The pairing of the trivial courting couple and the implicit embargo on current information is as natural here as that of the trysts in the "China wagons" and the imperative to report "no news" in Jonson's masque, or that of the "politique shrug" and the unescorted women of Hyde Park in his correspondence.

In keeping with their private status, the fate of the couple, after the oblique blow to their carriage, is not described in explicit fashion. But the narrator himself supplies the familiar retrograde perspective of the slow-moving couriers, depicting first the collision, then the moments preceding it, and then a nightmarish sequence in which the young woman appears first as a lady in a sinking sailboat, then as a girl struggling through deep sands along a shore, and finally as an infant in a fragile carriage, the "ransom for Waterloo."[82] By the time these attempted revisions of the original event emerged, the mail-coach itself had fallen into disuse, having been replaced by locomotives, and "Waterloo" generally referred not to the battle, nor to the news that outdistanced that particular conflict, nor even to the "secret word" that the vehicle communicated, in absence of all specifics, in its headlong flight, but to a new railway station.[83] But *The English Mail-Coach* also looks back, well beyond the years "from Trafalgar to Waterloo," to Jonson's "China wagon," when details of Galileo's *Starry Messenger* substituted for all the news that could not be divulged.

Conclusion

Evening News has sought to explain the convergence of journalism, optics, and astronomy in the early modern period in terms of both particular historical conditions and the reactions of writers and readers to the seeming simultaneity of developments in reportage and natural philosophy. While those circumstances, especially the emergence of the Dutch telescope and of serial news, the avalanche of novel astronomical information, and the looming threat of a pan-European confessional conflict have vanished, remnants of the issues confronted by those early seventeenth-century news consumers are with us still. A brief survey of newspaper names in English and in most European vernaculars—*Globe*, *Sun*, *World*, *Star*, and even *Comet*—shows the trace of the lofty cosmological perspective announced in Galileo's *Starry Messenger*, *Reflector*, *Mirror*, *Observer*, *Examiner*, *Outlook*, and their many French, Italian, German, Spanish, Dutch, and Portuguese analogues emphasize, to varying degrees, the acuity and reasoned analysis some associated with optical projections. Less attuned to the fiction of objectivity, and more reminiscent of the brooding watchfulness so prevalent on the eve of the Thirty Years' War are titles such as *Argus*, *Eagle*, *Intelligencer*, *Sentinel*, and *Guardian*. Names such as *Messenger*, *Courier*, and *Mercury*, all connoting the professional and even preternatural speed of the approaching news bearer, are complemented by the stolid immobility of *Recorder*, its naturalized equivalent *Echo*, and its curious mechanical alternative, *Pantagraph*, this last being both the singular title of a newspaper and a device for copying, reducing, and enlarging images whose inventor was Galileo's archrival Christoph Scheiner.

More striking still are the many newspapers whose names gesture to antiquated technologies of communication: *Herald*, *Crier*, *Voix*, *Post*, *Mail*, *Telegram*, *Telegraph*, and, inevitably, *Press*, *Stampa*, and *Tagblatt*. Such obsolescence, once a leisurely signal of longevity, and in some cases, a bid for an evanescent regional flavor defined by the reach of the voice or of local mail services, now seems the symptom of an imploding media landscape. Such are also the connotations of those few titles that, like the original *gazzetta* and the subsequent penny-papers, feature loose change from long vanished currencies, *Picayune*, *Kronen Zeitung*, and *Il Resto del Carlino*. But it is worth noting that similar hints of a mismatch between the best means of spying out novelties, serial news and the telescope, and the information that newly formed consumers desired, came with extraordinary rapidity; the perception that neither the popular image of astronomical revelation, the costly courtier Galileo, nor the emergent news publishers, bent upon broadening their customer base, were worth the civic unrest they had the potential to engender followed soon thereafter. In Galileo's case, we might date the peak of his activities as celestial messenger, and the beginning of his fall, to around 1613, when the novelties ceased, and the damaging impact of his *Letters on Sunspots* became clear to some, but certainly not all, onlookers. For printed serial newsletters, both the ascent and the decline were slightly later, and involved not the censorship and trial of one celebrated individual, but rather the suppression of some forms of reportage, stronger editorial control, or self-control, of most others.

As *Evening News* has argued, that widespread perception of the debatable character of news, whether celestial or terrestrial, coalesced in a general transfer of interest from the motif of the telescope, now effectively outmoded, to that of the *camera obscura*. Inevitably, the dream of an instrument whose intake of data would be no sooner received than transcribed and exhibited in unambiguous form to an elite and subsequently propagated, more schematically, to a larger but still governable audience, almost immediately devolved into fantasies of optical apparatuses that actively produced rather than passively recorded incidents, or admitted but one observer and excluded current events entirely, or took as their prime focus not news itself, but rather the phantasm of a newsreading public. This shift of attention away from news to its conjuring or creation, to its erasure, or to the spectacle of its reception is with us today, largely in critiques of the media, variously focused upon hype, upon the disappearance of traditional journalism and the resultant atomizing of communities, and upon the impact of social news websites.

Evening News has attributed the insistent pairing of Galileo with his friend Traiano Boccalini, then assumed to have been assassinated by Spanish agents for his satirical *News from Parnassus*, to the simultaneity of their enterprises, to the tenor of their writings as their newslessness emerged, and to their disastrous fates. At stake here is the connection between the doubling and the evident inadequacy of these two "imaginary *gazzettanti*": but why wouldn't one failed journalist do? The pairing of the fraudulent or washed-up newsman with some sort of sidekick—a photographer, blogger, or freelance investigator, typically a figure with technical skills, but deprived of professional credentials and social standing—is a feature of recent novelistic treatments of journalism, a pronounced departure from the comic or tragic isolation of protagonists from Evelyn Waugh in the late 1930s through Antonio Tabucchi in the late 1990s, and an unambiguous gesture to the inevitability of integrated media. It is possible, however, that the bland logic of the newspaper merger itself offers the more revealing analogy: the frequent amalgamation of individual newspapers is predicated upon the expectation that the resultant institution will offer a perspective sufficiently general and persuasive to mask and eventually to mend ideological rifts and provincial narrowness. More pressing, at least initially, than the quotidian business of peddling news is the need to reshape the consumer, to promote an outlook higher in style than in divisive particularities, to purvey as much an evolving posture as volatile and ephemeral information about current events. In short, the tendency that underwrote the aims of both Galileo and Boccalini as they shifted from topical disclosures to long-term perspectives, and from informing their readers by gesturing to recent developments to reforming them as knowing, detached, skeptical humanists, appears a parodic anticipation of the rhetorical efforts associated with the newspaper merger.

The focus of the final chapter, the correlation between rapid delivery and vacuous content, is a familiar one; my concern, however, is less the inevitability of this connection than the vehicles with which such maneuvers were accomplished. The evolution of the fantastic amphibious "sailing chariot" of Ben Jonson's *News of the New World in the Moon* into Thomas De Quincey's English mail-coach clearly involves velocity, but the threat of violence and the imperative to make way before these state-sponsored vehicles are the more crucial characteristics in these two meditations on censorship and national security. The drama of the speedy delivery, in other words, masks other and more troubling features. Timeliness, now distributed somewhat unevenly throughout today's media landscape, will soon cease to be the cardinal trait; what we discuss, and why, will remain an ongoing concern.

INTRODUCTION

All translations are mine unless otherwise indicated.

1. On the role of misinformation, see Dooley, *The Social History of Skepticism*; on Amsterdam, see Smith, "The Function of Commercial Centers."

2. Servolini, "Un tipografo del seicento"; Infelise, *Prima dei giornali*, 85; Callard, *Le prince et la république*, 136–38.

3. Remmert, *Picturing the Scientific Revolution*, 66–67. Pietro Cecconcelli's firm, the earlier establishment, published a work on comets inspired if not written by Galileo in 1619, and issued other title pages bearing the words "alle stelle medicee" ["at the sign of the Medici Stars"]. The press run by Amadore Massi and Lorenzo Landi used the image as a printer's device in 1639 on Paganino Gaudenzi's comparison of Origen and Plato, and in the 1640s on the technical works of Galileo's followers Vincenzo Renieri, Bonaventura Cavalieri, and Evangelista Torricelli.

4. Plutarch, *Moralia*, 6: 473.

5. Ibid., 485.

6. Ibid., 507.

7. Marchesini, "Almanacchi Italiani"; Bollème, *Les almanachs populaires*; Pagani, "A *Lunario*"; Pettegree, *The Book in the Renaissance*, 277–79, 335–38; on allusions to early modern astronomical developments in these works, see Herbst, "Galilei's Astronomical Discoveries," and Kremer, "Mathematical Astronomy and Calendar-Making."

8. Carleton, *Dudley Carleton to John Chamberlain*, 225. The duke of Urbino's collection of manuscript newsletters was said to date to the 1550s; see Infelise, *Prima dei giornali*, 12; "Diario di cose romane"; Orbaan, "La Roma di Sisto V"; Van Houtte, "Un journal manuscrit intéressant"; D'Onofrio, "Gli 'Avvisi' di Roma." On such collections in general see Pettegree, *The Book in the Renaissance*, 149, 334.

9. Franzesi, "Sopra le nuove," 207–8; on Franzesi, see Richardson, *Manuscript Culture*, 4, 19, 33, 45, 89; on what Spaniard diplomats regarded as the Italian craving for *novedades*—regime changes that might result in news—see Levin, *Agents of Empire*.

10. Catholic Church, Pope, *Bullarum*, 7: 969–71. On other bulls directed against the "arte nuova" of the journalists, see Infelise, *Prima dei giornali*, 10, 155–58; Richardson, *Manuscript Culture*, 159–60. For the French context, see Soman, "Press, Pulpit, and Censorship," esp. 441, 447–48.

11. *Bullarum*, 8: 12–13. See Farinacci, *Responsorum criminalium*, 116–17. On composing and copying, and on Farinacci, see Dooley, *The Social History of Skepticism*, 16, 26.

12. *Bullarum*, 8: 646–50.

13. Infelise, *Prima dei giornali*, 23–24.

14. Bellettini, Campioni, and Zanardi, *Una città in piazza*, 170–71.

15. *King Lear* 1.2.26; on the various sorts of letter writers, see Brownlees, *The Language of Periodical News*, 12–24.

16. Thus Cornwall to Gloucester, "Since I came hither, Which I can call but now, I have heard strange news." *King Lear* 2.1.86–87.

17. *King Lear* 1.2.103

18. *King Lear* 1.2.149

19. Galilei, *Sidereus Nuncius*, ix; for the reaction of astrologers at Naples *before* the work was printed, see Galilei, *Opere*, 10: 295.

20. Politi, *Dittionario Toscano*, 359.

21. Garasse, *Le Rabelais Reformé*, 40–41; see also *La Cholere de Mathurine*.

22. Galilei, *Opere*, 14: 103; Dooley, *The Social History of Skepticism*, 31; Mayer, *Roman Inquisition*, 53, 85, 107.

23. See Tatlock, *Seventeenth Century German Prose*, 47–60, especially 49.

24. Galilei and Scheiner, *On Sunspots*, 25–34, 282–83, 282 n. 50.

25. See Vail, " 'The Bright Sun Was Extinguis'd' "; *Gazzetta di Milano* no. 225, 236, 243 (August 12, 23, and 30, 1816), 890, 931, 962; Riccati, *Tableau historique et raisonnée*, 2: 328–39.

26. Frearson, "The Distribution and Readership of London Corantos."

27. On forerunners see Pettegree, *The Book in the Renaissance*, 130–50; on serial news see Morison, *Selected Essays*, 2: 325–45; Sommerville, *The News Revolution in England*, 17–20. On periodicity and other criteria, see Allen, "International Origins"; Raymond, "The Newspaper, Public Opinion, and the Public Sphere," especially 131–33. On timeliness see Woolf, "Genre into Artifact," especially 332–34, and Woolf, "News, History, and the Construction of the Present." On the inconsistencies that make the Flemish newsletter difficult to classify, see Simoni, "Poems, Pictures and the Press."

28. Schröder, "The Origins of the German Press"; Lankhorst, "Newspapers in the Netherlands"; Vittu, "Instruments of Political Information"; Arblaster, "Policy and Publishing"; Weber, "The Early German Newspaper," 69–75; Sommerville, *The News Revolution in England*, 20; Castronovo, "I primi sviluppi," especially 20–23; Infelise, *Prima dei giornali*, 81–85, 195 n 3; Bongi, "Le prime gazzette," especially 336–37. On the tardiness of the Spanish periodical, see Díaz Noci, "Dissemination of News," 411–12; on problems classifying the early English coranto, see Raymond, *Pamphlets and Pamphleteering*, 132–34, 150–52; on the transnational character of early serial news publications, see Raymond, "Newspapers: A National or International Phenomenon?"

29. On the physical characteristics of the Dutch, English, and French corantos and newsbooks, see *Dutch Corantos*, 18–23, and Dahl, "Amsterdam—Cradle of English Newspapers"; Raymond, *Pamphlets and Pamphleteering*, 128–138; Brownlees, *The Language of Periodical News*, 25–96; on early German, Dutch, and English versions, see Morison, "The

Origins of the Newspaper," 332–44; on their Italian counterparts, see Dooley, *The Social History of Skepticism*, 64–65, and Infelise, *Prima dei giornali*, 108–10.

30. On travel times, see Šimeček, "The First Brussels, Antwerp and Amsterdam Newspapers," 1100–1101, 1106–7; Delumeau, *Rome au XVIe siècle*, 12–24; Kintz, "Opinion publique," 86–90; Feyel, *L'Annonce et la nouvelle*, 134–35; Infelise, *Prima dei giornali*, 111–21; Arblaster, "Posts, Newsletters, Newspapers"; Borreguero Beltrán, "Philip of Spain."

31. On the stop press addition, see *Dutch Corantos*, 23–26.

32. Richardson, *Manuscript Culture*, 57–58, 114–26, 157–60.

33. Mayer, *The Roman Inquisition*, 10, 24, 31–34, 44, 58–59, 67, 85, 88, 91, 94–95, 117, 124, 131, 133–35, 140, 149, 212, 249, 253, 285, 316; on protocols involving secrecy, 160–61.

34. Rome, Vatican Library, Fondo Urbinate Latino, Avvisi manoscritti, 1078, part 2: fol. 417r.

35. See Dooley, *The Social History of Skepticism*, 36–44; Infelise, *Prima dei giornali*, 11–78; De Vivo, "Paolo Sarpi and the Uses of Information"; Infelise, "News Networks."

36. See Lambert, "Coranto Printing in England"; Baron, "The Guises of Dissemination"; Atherton, "The Itch Grown a Disease"; Levy, "The Decorum of News"; Raymond, *Pamphlets and Pamphleteering*, 130.

37. Jouhaud, *Les pouvoirs de la littérature*, 97–150; Feyel, *L'annonce et la nouvelle*, 181–84; Duchêne, "Lettres et gazettes."

38. Barbarics-Hermanik, "Handwritten Newsletters."

39. On pamphlets see Watt, "Publisher, Pedlar, Pot-Poet"; Clark, *The Elizabethan Pamphleteers*, 86–120; Harline, *Pamphlets, Printing, and Political Culture*. On the separate, see Cust, "News and Politics," 63–64, and Levy, "The Decorum of News," 27; on *occasionnels* and *canards* in France before 1631, see Vittu, "Instruments of Political Information," 163–64; for their place in the journalistic enterprises of Théophraste Renaudot, see Solomon, *Public Welfare, Science, and Propaganda*, 105–7, 135–39, 141–43, and Feyel, *L'annonce et la nouvelle*, 153–54.

40. See Infelise, *Prima dei giornali*, 94–97; Dooley, *The Social History of Skepticism*, 70–78; Levy, "How Information Spread Among the Gentry," 22–23; Baron, "The Guises of Dissemination," 45–47; Schröder, "The Origins of the German Press," 127–28.

41. Vannozzi, *Della Suppellettile*, 1610, 2: 288–89.

42. See Vittu, "Instruments of Political Information," 166; Levy, "The Decorum of News," 27; Rome, Vatican Library, Fondo Urbinate Latino, Avvisi manoscritti, 1076: 1, fol. 453v. On the fluidity of the categories of elite and popular news, the interrelationship of oral, manuscript, and printed accounts, and the reversibility of the flow from manuscript to print, see De Vivo, *Information and Communication in Venice*, 3–12, 58–61, 67, 71, 96–156, 176–78, 185–87.

43. On the various types of Jesuit letters, their purposes, the systems for their printing and distribution, see Lach, *Asia in the Making of Europe*, 1: 256–61, 314–31, 427–67; on the differences between the public letters destined for publication and private reports from the Far East, see Cooper, *Rodrigues the Interpreter*, 163–81.

44. See Reynolds, "The Classical Continuum"; *Pasquinate romane del cinquecento*; *Pasquino e dintorni*; Barkan, *Unearthing the Past*, 126–27, 208–31, 273–74; Richardson, *Manuscript Culture*, 117–23. On elite writers' simulation of popular language and concerns in writings associated with the Venetian *Gobbo* and elsewhere, see De Vivo, *Information and Communication*, 130, 137–39, 194–95.

45. See, for example, Di Banco, *Bizzarrie politiche*, 9–13; Galindus, *A Further Discovery of the Mystery of Jesuitisme*, 16–17; Bernegger, *Observationes*, 118–19.

46. Naudé, *Le Marfore*, 6. Naudé ends with a line from Horace's *Satire* 2.7.82, *[ducimur] ut nervis alienis mobile lignum*. On the question of a bourgeois appropriation of plebeian complaints and rhetoric in the pasquinades, see Aquilecchia, "Presentazione."

47. Barbarics-Hermanik, "Handwritten Newsletters"; Randall, "Epistolary Rhetoric."

48. Rubens, *Correspondence*, 3: 468–69.

49. Arblaster, "Policy and Publishing," 184–85; on the distinction between public and private manuscript newsletters in Italy, see Infelise, *Prima dei giornali*, 30–32, 79; on Antwerp as a news center, see Arblaster, "Antwerp and Brussels," especially 198–205.

50. On the barbarous Latin of the *Mercurius Gallobelgicus* see Raymond, *Pamphlets and Pamphleteering*, 129; for another such example of cultural compensation in the pamphlet *Late News out of Barbary* (1613), see Brownlees, *The Language of Periodical News*, 16.

51. *Correspondence de Rubens*, 4: 295.

52. Juvenal, *Satire* 3. 34–40. I have used the translation of G. G. Ramsay in Juvenal and Persius, *Works*, 35.

53. *Lettere d'uomini illustri*, 37.

54. Tertullian, *Adversus Valentianos*, III.

55. Fritsch, *Discursus de Novellarum*, I.i–I.v, I–vii.

56. Balzac, *A New Collection*, 99–100.

57. Balzac, *Les entretiens*, 1: 53–54. What the ambassador Giustiniani relates about the castle is drawn from Francesco Berni's *Orlando Innamorato*, III.7.54.

58. Balzac, *Les entretiens*, 1: 54.

59. Alexander Pope's depiction of the Temple of Rumor, an elaboration of a similar structure in the third book of Chaucer's *House of Fame*, is the culmination of such a tradition in English verse.

60. See Infelise, *Prima dei giornali*, 11–12; Schröder, "The Origins of the German Press," 133–34.

61. See Randall, "Joseph Mead, Novellante."

62. *Mercurius Gallobelgicus* (1615): 53; *Mercure François* (2nd ed., 1627): 303–4; *The Swedish Intelligencer, the Third Part* (1633): fol. 2.

63. Rome, Vatican Library, Fondo Urbinate Latino, Avvisi manoscritti, 1077: fol. 123r–123v.

64. Rome, Vatican Library, Fondo Urbinate Latino, Avvisi manoscritti, 1082: fol. 183r.

65. See Bireley, *The Jesuits and the Thirty Years War*, especially 10, 31, 35, 36, 37.

66. Rome, Vatican Library, Fondo Urbinate Latino, Avvisi manoscritti, 1076: fol. 170v.

67. Rome, Vatican Library, Fondo Urbinate Latino, Avvisi manoscritti, 1082: fol. 108v-109r, 142r-142v.

68. Rome, Vatican Library, Fondo Urbinate Latino, Avvisi manoscritti, 1082: fol. 449v, 451v, 470r.

69. Rome, Vatican Library, Fondo Urbinate Latino, Avvisi manoscritti, 1079: fol. 657v. Another account with somewhat different details, but of the same tenor, appears on fol. 683v-684r.

70. Rome, Vatican Library, Fondo Urbinate Latino, Avvisi manoscritti, 1076, part 1: fol. 15v.

71. On the simulation of an oral tenor in written news, see Brownlees, "Spoken Discourse in Early English Newspapers"; on the asymmetry of the exchanges between oral and textual worlds, see Pettegree, *The Book in the Renaissance*, 333–34.

72. Buonarroti il giovane, *Satira prima*, vv. 250–57 in *Opere varie*, 226; for the dating of this satire, see Limentani, *La Satira*, 76–78.

73. Chiabrera, *Canzonette, rime varie, dialoghi*, 462.

74. Testi, *Lettere*, 1: 275.

75. Cited and discussed in Dupré, "Galileo, the Telescope, and the Science of Optics," 247–48. I am very grateful to Professor Dupré for alerting me to this passage. For further discussion of Coccio's letter, see Reeves, *Galileo's Glassworks*, 52–55.

76. Grillo, *Lettere*, 3: 349.

77. Van Helden, *Invention of the Telescope*, 30, 34; Reeves, *Galileo's Glassworks*, 74–75.

78. Allegri, *Seconda parte delle rime piacevoli*, fol. B2v–B3.

79. Hobbes, *Correspondence*, 1: 26.

80. Hobbes, *Correspondence*, 1: 25; see also 1: 18, 19, 21 for his vain efforts to get news, and for his interest in Galileo's *Dialogue Concerning the Two Chief World Systems*.

81. On letter reading see Van Helden, *The Invention of the Telescope*, 28, 30, 34, 35; on Galileo's access to a prototype, see Biagioli, "Did Galileo Copy the Telescope?"

82. Galilei, *Sidereus Nuncius*, 36; Hernández, *Rerum medicarum*, 473.

83. Ghilini, *Teatro d'huomini illustri*, 68.

84. Rome, Vatican Library, Fondo Urbinate Latino, Avvisi manoscritti, 1077: fol. 437. The account, which originated in Venice, bears the date September 5, 1609, and concerns the amount of money Galileo received for the device from the Venetian Senate, the range of the instrument, and the fact that a rival inventor had been turned away in the same period without any reward.

85. See Allen, "International Origins," 314. On the Strasbourg weekly, see Kintz, "Opinion publique," 88–91.

86. Daniele, *Trattato della Divina Providenza*, 151–52. Reading glasses were conventionally measured in decades, or occasionally referred to as "an old man's eyeglass"; by the late sixteenth century, lenses might be assigned a number of "points" or "grades" to indicate their strength. See Ilardi, *Renaissance Vision*, 75–115, 225–26.

87. Rome, Vatican Library, Fondo Urbinate Latino, Avvisi manoscritti, 1083: fols. 271v, 272r,v, 273r.

88. Ibid., 1083: fol. 308r.

89. Both before and after the invention of the Dutch telescope, writers associated telescopic vision with the revelation of clothing and particular colors; see Biringuccio, *De la pirotechnia libri X*, fol. 143; Walchius, *Decas fabularum*, 248; Knowles, "Jonson's *Entertainment*," 137; Giovanetti, "Dello Specchio," 26, 33.

90. Balzac, *Letters*, 152.

91. *Swedish Intelligencer, Ninth Part*, 50.

92. Buonarroti il giovane, *La Fiera*, 35, 140.

93. Bellettini, Campioni, and Zanardi, *Una città in piazza*, 164, 169, 170, 209, 210, 212.

94. See Benzoni, "Caprara;" Hanlon, *The Twilight of a Military Tradition*, 207, 213.

95. See de Broglie, *Catinat*, 84; Hanlon, *The Twilight of a Military Tradition*, 285–87.

96. See Hanlon, *The Twilight of a Military Tradition*, 287; Hardÿ de Périni, *Batailles françaises*, 307. The depredations were also mentioned frequently in the *Gazette d'Amsterdam* in 1691–92.

97. See also Dooley, *The Social History of Skepticism*, 58. For other works by Mitelli, see Novati, "La storia e la stampa."

98. On contemporaneous Italian terrestrial refracting telescopes, but ones larger than and of superior quality to the spyglasses in this print, see Turner, "Three Late-Seventeenth Century Italian Telescopes"; on seventeenth- and eighteenth-century telescopes and spyglasses in the Museum of History of Science in Florence, see Van Helden, *Catalogue of Early Telescopes*; for terrestrial telescopes in particular, see 54–60, 66–75, and for spyglasses, see 76–83. On Italian lens-making in the second half of the seventeenth century, see Bedini and Bennett, "'A Treatise on Optics.'"

99. See Turner, "Three Late-Seventeenth Century Italian Telescopes," 44, 48, 52, 53, 54, 62; Van Helden, *Catalogue*.

100. See Turner, "Three Late-Seventeenth Century Italian Telescopes," 48, 49, 52, 53, 55.

101. Manzini, *L'Occhiale*, 246–47.

102. On scrap paper in telescopes see Turner, "Three Late-Seventeenth Century Italian Telescopes," 48; Van Helden, *Catalogue*, 54–55, 66–70.

103. On pasteboard from scrap paper, see Lalande, *L'Art du cartonnier*, 1–2, 23–26; on scrap paper in early modernity, see Pettegree, *The Book in the Renaissance*, 334.

104. See Infelise, *Prima dei giornali*, 208, 209.

105. Scheiner, *Rosa Ursina*, second unnumbered folio.

106. Halasz, *The Marketolace of Print*, 24–27.

CHAPTER 1. JESUITS ON THE MOON

1. On the Galileian documents that remain to us, see Wootton, *Galileo Watcher of the Skies*, 1–4.

2. Galilei, *Opere*, 3(1): 68.

3. Galilei, *Opere*, 10: 275; on Jupiter's satellites, see Gingerich and Van Helden, "From *Occhiale* to Printed Page"; Gingerich and Van Helden, "How Galileo Constructed the Moons of Jupiter."

4. Reeves, "Kingdoms of Heaven."

5. See the excellent edition prepared by Isabelle Pantin, Kepler, *Discussion avec le messager céleste*, 18, 85–86, 110. I have relied extensively on the introduction, notes, and translation provided by Pantin in my study of Kepler's *Discussion with the Starry Messenger*. On Kepler's *Somnium*, see further Chen-Morris, "Shadows of Instruction"; Aït-Touati, *Fictions of the Cosmos*, 17–44.

6. Kepler, *Gesammelte Werke*, 15: 489. This part of the letter, written to the ailing Johannes Pistorius the Younger, at that time the confessor to Rudolph II, is discussed by Caspar in *Kepler*, 163–64. On Pistorius's importance to the emperor's shift toward Catholicism, see Evans, *Rudolf II and His World*, 90–91.

7. Kepler, *Gesammelte Werke*, 15: 418.

8. Kepler, *Gesammelte Werke*, 16: 123, 290–91.

9. Kepler, *Gesammelte Werke*, 16: 179.

10. Kepler, *Gesammelte Werke*, 16: 195.

11. Kepler, *Gesammelte Werke*, 15: 170.

12. Kepler, *Gesammelte Werke*, 16: 302, 314.

13. Caspar, *Kepler*, 69–70; Kepler, *Gesammelte Werke*, 14:441; on Kepler's effort to provide a philosophical pedigree for Galileo's arguments, see Bucciantini, *Galileo e Keplero*, 181–89.

14. The expression *pro aris et focis* is also used by Paolo Sarpi in a letter of May 1609 to Jacques Gillot; at issue is the fervent ultra-Catholic support for the doctrine of papal infallibility. See Sarpi, *Lettere ai gallicani*, 132.

15. Kepler, *Gesammelte Werke*, 4: 286.

16. Kepler, *Gesammelte Werke*, 13: 184.

17. On the use of this tactic in Kepler's Graz, see Caspar, *Kepler*, 97–98, 111–15.

18. Bireley, *The Jesuits and the Thirty Years War*, 89–93, 105–6, 111–14, 160–61; Palmitessa, "The Prague Uprising of 1611."

19. Sacchi, *La Guerre de Trente Ans*, 1: 395–99; Pánek, "The Nobility in the Czech Lands," 282–83.

20. Bireley, *The Jesuits and the Thirty Years War*, 89–90, 111–12, 161, 179–80; on Clavius, see especially Lattis, *Between Copernicus and Galileo*.

21. Caspar, *Kepler*, 78–82, 96–98, 111–13, 119, 177.

22. Kepler, *Gesammelte Werke*, 16:60; cited in Caspar, *Kepler*, 82. In Matthew 10:16 (KJV) Christ tells His disciples, "Behold, I send you forth as sheep in the midst of wolves: be ye therefore wise as serpents and harmless as doves."

23. Bireley, *Religion and Politics*, 3–21.

24. Evans, *Rudolf II and His World*, 68–69, 90–91, 156–57.

25. On the Carolinum, see Evans, *Rudolf II and His World*, 135–36, 145; on the University and Bachazek, see Caspar, *Kepler*, 148, 166–67; Čornejova, "Education in

Rudolfine Prague," especially 327–29; on Jessenius, see Evans, *Rudolf II and His World*, 136–38.

26. On the Clementium, see Voit, *Pražské Klementinum*, with an English summary on 165–69; Šíma, *Astronomie a Klementinum*; Čornejova, "Education in Rudolfine Prague," 324–27, 330. For an anonymous drawing of the Clementinum made in the first half of the seventeenth century, see Čornejova, "The Religious Situation," 313. For a partial roster of pupils studying with the Jesuits in Prague for most years between 1565 and 1624, their place of origin, and their area of specialization, see *Album academiae pragensis Societatis Iesu 1573–1617*. On the Jesuits' early years in Prague, particularly in regard to the Czech Jiří Pontan S.J., better known as Pontanus, see Evans, *Rudolf II and His World*, 157–61; on the universities at Jena and Altdorf, see Parker, *The Thirty Years' War*, 40–41.

27. Schmidl, *Historiae Societatis Jesu*, I:vi:642; II:i:6, 7; II:iii:207, 351; II:iv:448, 511.

28. Schmidl, *Historiae Societatis Jesu*, II:iv:444, 496. On the Calixtine or old-Utraquists, and their relationship to Catholicism in this period, see Evans, *Rudolf II and His World*, 30, 36–37.

29. Bireley, *Religion and Politics*, 32–36.

30. Reeves, *Galileo's Glassworks*, 116–24.

31. Cambilhom, *A Discoverie*, fol. A4v–B.

32. Cambilhom, *A Discoverie*, fol. A2v–A3v.

33. Cambilhom, *A Discoverie*, fol. Bv.

34. Cambilhom, *A Discoverie*, fol. A4v.

35. Cambilhom, *A Discoverie*, fol. A3v–A4.

36. *Relatio Außführlicher Bericht was sich mit den Passawischen Kriegsvolck*, cited and translated in Palmitessa, "The Prague Uprising of 1611," 307, 309.

37. *Mercurius Gallobelgicus*, cited in *Litteræ Annuæ Societatis Iesu* (1611) 380. Two manuscript *avvisi* in the Vatican Library offer versions of the rumor about weaponry, though these charges are contradicted by others in the same collection. Two different reports sent from Prague on February 28, 1611, allege that various defensive and offensive weapons "for a good number of soldiers" were found in the college, and that "among other things 400 muskets and other weapons to arm a thousand people" had been located. See Rome, Vatican Library, Fondo Urbinate Latino 1079, *Avvisi manoscritti*, fols. 204 and 211r.

38. *Nouvelles d'Allemagne* (1613): 18–19.

39. Sarpi, *Lettere ai protestanti*, I: 178; Schmidl, *Historiae Societatis Jesu*, II:v:533–34.

40. Kepler, *Gesammelte Werke*, 4: 293.

41. Pena, *De Usu Optices*; Scheiner, *Accuratior Disquisitio*, in Galilei, *Opere*, 5:61. Kepler commented upon Pena's text in his *Dioptrice*, written later in 1610 and published in 1611; see *Gesammelte Werke* 4: 336–337.

42. On Galileo's reticence, see Biagioli, *Galileo's Instruments of Credit*, and Wootton, *Galileo Watcher of the Skies*.

43. Kepler, *Gesammelte Werke*, 4: 295–96.

44. Kepler, *Gesammelte Werke*, 4: 92–93; Galilei, *Opere*, 11: 445.

45. In a discussion of dark room tricks in the fourth book of his *On the Use of Eyeglasses* of 1623, Benito Daza de Valdes has one speaker claim that someone projected moving images onto the pages of a blank book, leaving the dupe to search in vain for them after the show was over; see Albertotti, "Manoscritto francese," 109.

46. The *Hipparchus*, an astronomical project in which Kepler had been interested since around 1600, remained incomplete at his death in 1630. See Caspar, *Kepler*, 143, 180, 327.

47. Connor, *Dumbstruck*, 146; Webster, *The displaying of supposed witchcraft*, 68–71.

48. Kepler, *Somnium*, 14–15.

49. Kepler, *Somnium*, 72 n 72; 50 nn 35 and 36.

50. Kepler, *Somnium*, 60, note 50. Rosen believed that Kepler was referring to a "talking machine long before it was actually invented," while I am taking the passage as a description of ventriloquism, and as an allusion to the common supposition that all of those who spoke "from the gut" were really possessed by demons, when in Kepler's view some of them were using their voices as instruments to fool others. I have thus modified Rosen's version in two particulars; I have translated *in hac mechanica* as "in this device," not as "in this mechanism," to suggest a trick where parts of the vocal apparatus are the instruments, and *præstigias magicas ars imitetur* as "the technique resembles magical tricks" rather than as "art is copying magical tricks." It is possible that Kepler had the Elizabethan Edward Kelley in mind as an historical example. Kelley, along with John Dee, was at Rudolf's court in Prague some years before Kepler, and his performances might have involved ventriloquism; see Evans, *Rudolf II and His World*, 225–28.

51. Kepler, *Somnium*, 14, 58–60.

52. Caspar, *Kepler*, 167–68.

53. See Rosen's notes in Kepler, *Somnium*, 58–60. On the claim to Jülich-Clèves and the shifting positions of counts Philip Ludwig and Wolfgang William of Neuburg in this period, see Parker, *The Thirty Years' War*, 21–30.

54. Kepler alluded to such prognostications in the conclusion of his *De solis deliquio, quod hoc anno 1605 mense octobri contigit epistola*; see *Gesammelte Werke*, 4: 53. The *Mercure François* reported that "this year on the first [*sic*] of October, there was an eclipse such that around one hour after midday, though the sky was very calm, suddenly there was a half hour of immense obscurity. The astrologers in different countries prognosticate many things from this, as is their custom." See *Mercure François* (1611): 64a. Given that two popes had already died in 1605—Clement VIII on March 5, and Leo XI on April 27— some element of the popular reaction might have been concern or contempt for the health of Paul V, who had been elected in mid-May that year. Finally, in the *Somnium* and the notes, Kepler alluded to the popular association of solar eclipses with disaster, and tried to offer a physical explanation—demons travel through the cone of the shadow produced by the moon to the earth during total solar eclipses—to account for this belief; see *Somnium*, 16–17, 63, 75–76.

55. Crosse, *Belgiaes Troubles*, 15–16. On such locales as news centers, see De Vivo, *Information and Communication*, 86–119, 187–90, 245, 260, and De Vivo, "Pharmacies as centres of communication."

56. Evans, *Rudolf II and His World*, 154–56.

57. Kepler, *Gesammelte Werke*, 4:299.

58. See Janáček, "Les italiens à Prague"; Krčálová, "La Toscana e l'architettura di Rodolfo II"; Evans, *Rudolf II and His World*, 132, 184–85; Polišenský, "Le relazioni tra la Boemia e l'Italia"; Ledvinka and Pešek, "The Public and Private Lives of Prague's Burghers," 291, 295; Vlček, "Architecture in Prague, 1550–1650"; Kaufmann, "Jesuit Art: Central Europe and the Americas," 289–93. On the importance of Italian and Italianate theories of architecture for Prague in this period, see Fučíková et al., *Rudolf II and Prague*, 587–90; on the use of stucco, terra cotta, and brick touches, see Kybalova, "The Decorative Arts," 376–77, 384, and 574–75; Simons, "Archduke Ferdinand II of Austria," 273–75.

59. Schmidl insists in a discussion of the society's efforts to convert Prague's "heretics" in 1608 on the "sheepfold" and "flock" of the true church. See *Historiae Societatis Jesu*, II:iv:498, 513; on the importance of the *domūs professae* in Jesuit education, see Feldhay, *Galileo and the Church*, 115–16.

60. See the *Mercure François* (1611): 385–85v, which discusses the allegation as a rumor having real efficacy, if not necessarily any basis in fact, among the Protestant leaders of Prague; Janáček, "Les italiens à Prague," 41; Palmitessa, "The Prague Uprising of 1611," 307.

61. Janáček, "Les italiens à Prague," 11, 32.

62. Reeves, "Old Wives' Tales," 335–51, and the bibliography noted there.

63. Kepler, *Somnium*, 15.

64. Kepler, *Somnium*, 15.

65. Kepler, *Somnium*, 64, note 57.

66. Kepler, *Somnium*, 63–64, note 56. I have changed Rosen's "in all directions" for *passim* to the more general "here and there." "In all directions," or "indiscriminately" might, however, be more consistent with Kepler's subsequent intention, in the 1620s, to weaken the strong early connection between the lunar inhabitants and the Jesuits.

67. Kepler, *Somnium*, 64–65, note 59.

68. Scaliger, *Scaligerana*, 51. I owe this reference to Gorman, "The Scientific Counter-Revolution," 28–29.

69. Cited in Bouhours, *Les Entretiens d'Ariste et d'Eugène*, 223. The cardinal's remark is repeated in the course of a discussion about the amazement caused by wit in a German speaker.

70. Schmidl, *Historiae Societatis Jesu*, II:iv:446–47, 545.

71. See Rosen's note in Kepler, *Somnium*, 47–48, and Kepler's own statements in *Somnium*, 47, note 28, and 51–52, note 38.

72. Cambilhom, *Discoverie*, fol. B4v.

73. See Backer, Carayon, and Sommervogel, *Bibliothèque de la Compagnie de Jésus*, 3: cols. 1782, 1784.

74. Sarpi, *Lettere ai protestanti*, 2:4.

75. Gretser, *Furiae praedicantium*; Gretser, *Relatio . . . nunc in gratiam praedicantium Lutheranorum*.

76. Gretser, *Praedicantium Augustanorum . . . repetitae furiae*, 867–68. For other more general Jesuit diatribes against newsmongers, see the Flemish Jesuit Cornelius à Lapide's commentary on the *locus classicus* of Acts 17: 21 in his *Commentaria in scripturam sacram*, 17: 322–23.

77. Gretser, *Relatio Cambilhonica castigata*, 803.

78. Gretser, *Relatio Cambilhonica castigata*, 803.

79. Galilei and Scheiner, *On Sunspots*, 43.

80. Gretser, *Relatio Cambilhonica castigata*, 809–10.

81. Gretser, *Relatio Cambilhonica castigata*, 801.

82. Gretser, *Relatio Cambilhonica castigata*, 814.

83. Gretser, *Relatio Cambilhonica castigata*, 811.

84. Kepler, *Gesammelte Werke*, 16: 326.

85. The work in question was Gretser's *Petrus Cnapheus seu Fullo in Thoma Wegelino Theopaschita redivivus Strenæ*, which was published with his *Hæreticus vespertilio sub Bononiensis epistolæ Italolatinæ velo de perfectione et excellentia Jesuitici Ordinis antea delitescens* in early 1609, with the *Lixivium pro abluendo male sano capite anonymi cujusdam fabulatoris, et, ut vocant, Novellantis, qui cædem christianissimi Galliæ et Navarræ Regis Henrici IV in Jesuitas partim aperte, partim tacite confert* and again with the *Haereticus Vespertilio hæreticus* in 1610, after the assassination of Henri IV in May of that year. See Backer, Carayon, and Sommervogel, *Bibliothèque de la Compagnie de Jésus*, 3: cols. 1781–82, 1785–86.

86. *Exemplar litterarum Bononiæ datarum*, 272.

87. *Exemplar litterarum Bononiæ datarum*, 165 (rectè 265), 272, 274.

88. Gretser, *Hæreticus vespertilio*, 884–85.

89. Gretser, *Hæreticus vespertilio*, 885.

90. Gretser, *Hæreticus vespertilio*, 887.

91. These accusations bear a strong resemblance to a poem written in Latin, which supposedly circulated in Rome in the late 1590s before being recited by a young Huguenot to a Jesuit antagonist near Mount Tabor in 1601. The poem was eventually printed in Latin and in English translation in Biddulph's *Travels of certain Englishmen*, 113–14, and appears in English only in Parker, *Early Modern Tales of Orient*, 248–49. Its rhyme scheme and word choice suggest that it was originally written in Italian: see, for instance, *Bonum panem, melius vinum / Non recipiunt peregrinum, / Neque surgunt ad matutinum*, and *Indii Galli atque pavones, / Quorum cibus sunt macherones, / Horum patrum sunt buccones*. On Biddulph, chaplain to English merchants in Aleppo from 1599 to 1607, see MacLean, *The Rise of Oriental Travel*, 49–114.

92. The source of the story about the Jesuits' access to the gold brought by Spain from the New World was Arnauld's *Arrainment of the whole society of Iesuits in France*, 4, 6.

93. Cozzi, *Paolo Sarpi tra Venezia e l'Europa*, 175–80, 197–98; Lazzerini, "Officina Sarpiana."

94. Brizzi, *La formazione*, especially 71–130; Brizzi, "I Gesuiti e i seminari," especially 147–49.

95. Galilei, *Opere*, 10: 419. On the Horky episode, particularly as it relates to Giovanni Antonio Magini, see Baffetti, "Il 'Sidereus Nuncius' a Bologna"; Pantin, *Discussion avec le messager céleste*, xxviii–lxxviii.

96. Tristan l'Hermite, *Le page disgracié*, 72–73. In this chapter, "How the disgraced page was taken for a magician," the page and his master consider an optical trick from della Porta, *Natural Magick*, 17: 1, where participants take on the faces of various animals, but reject it as costly and impractical. They attempt a related trick from *Natural Magick*, 20: 9, one in which their faces turn frightening colors, but are interrupted by their preceptor, whose sudden appearance and nocturnal garb present a more alarming spectacle.

97. Galilei, *Opere*, 10: 365–66; Pantin, *Discussion avec le messager céleste*, lix–lxvii.

98. Kepler, *Gesammelte Werke*, 16: 320.

99. *Nouvelles des régions de la lune*, 294. This work was routinely published with the *Catholicon*.

100. The migration of the motif from Rudolfine Prague to Jacobean England is perhaps significant; for a comparison of the policies and aesthetic tastes of Rudolf II and James I, see Evans, *Rudolf II and His World*, 80–83.

101. Donne, *Ignatius His Conclave*, 2–3.

102. *Ignatius His Conclave*, 116–18.

103. Camerota, *Galileo Galilei*, 200–218.

104. *Ignatius His Conclave*, 127–28.

105. *Ignatius His Conclave*, 129.

106. Fletcher, *Locustæ, vel Pietas Iesuitica*, 56.

107. Langdale, *Phineas Fletcher*, 52–53, 172–73; on Fletcher's neglect of Kepler's *Ad Vitellionem paralipomena* in depicting the eye, see 194–95.

108. See act III, scene ii.

109. Crosse, for instance, refers to the "Engines" of the "Ignatian Crue" in *Belgiaes Trouble, and Triumphs*, 3.

110. On Vitelleschi see Bireley, *The Jesuits and the Thirty Years War*.

111. Pantin, "Galilée, la lune, et les Jésuites."

CHAPTER 2. MEDICI STARS AND THE MEDICI REGENCY

1. Whitaker, "Galileo's Lunar Observations"; Gingerich and Van Helden, "From *Occhiale* to Printed Page."

2. Reeves, *Painting the Heavens*, 23, 231 n. 1.

3. Grillo, *Delle lettere*, 2: 149.

4. Chapin, "The Astronomical Activities of Peiresc," especially 15–18; Lambert, *Catalogue descriptif et raisonné*, 2: 200–201; Buccantini, Camerota, and Giudice, *Il Telescopio*, 165–71.

5. On the missives in the Venetian archive reporting the assassination of Henri IV, see Infelise, "News Networks," 64–65.

6. Heller, *Labour, Science, and Technology*, especially 22, 59–60, 67–69, 99–118, 138–39.

7. Duplessis-Mornay, *Mémoires et Correspondence*, XI: 105.

8. Duplessis-Mornay, *Mémoires et Correspondence*, XI: 106.

9. Richelieu, *Mémoires*, 1: 52.

10. Camden, *Epistolae*, 129–30.

11. Toulmin, *Cosmopolis*, 56–62; Bucciantini, Camerota, and Giudice, *Il Telescopio di Galileo*, 172–82.

12. Kepler, *Gesammelte Werke*, 16: 311; on the Horky affair, see Bucciantini, Camerota, and Giudice, *Il Telescopio di Galileo*, 87–104.

13. Horky, *Brevissima Peregrinatio*, in Galilei, *Opere*, 3: 142.

14. Horky, *Peregrinatio*, in Galilei, *Opere*, 3: 134, 143.

15. Galilei, *Opere*, 10: 342–43.

16. Horky, *Peregrinatio*, in Galilei, *Opere*, 3: 137.

17. Galilei, *Opere*, 10: 342–43. On the likelihood of this diagnosis, see Heilbron, *Galileo*, 161–62.

18. *Mercure François* (1613): 339 B.

19. Reeves, "Representing Invention: The Telescope as News"; Jeannin, *Les négociations*, in Van Helden, *The Invention of the Telescope*, 43.

20. Lacordaire, "Brevets de logements," 196–200; Lacordaire, "Brevets accordés"; Guiffrey, "Logements d'artistes," 6–9, 19–26; Huard, "Les logements des artisans"; Ballon, *The Paris of Henri IV*, 48–50, 307–8; Fuhring and Bimbenet-Privat, "Le Style 'Cosses de Pois.'"

21. Guiffrey, "Logements," 19; *Négociations diplomatiques*, 5: 605.

22. *Négociations diplomatiques*, 5: 82. The king's letter of January 9, 1609, was said to have gone posthaste to Jeannin; see Great Britain, Historical Manuscripts Commission, *Report on the Manuscripts of Lord de l'Isle and Dudley*, 4: 95.

23. Buisseret, *Sully and the Growth of Centralized Government*, 40, 120–39.

24. Van Helden, *Invention of the Telescope*, 44, 46–47.

25. Van Buchel, *Description de Paris*, 97; on Thierry Badovere, whose surname was also written "Badoire," "Baudoyre," and "Baudoire" see Cimber and Danjou, *Archives curieuses de l'histoire de France*, Ser. 1, vol. 7: 55, 151; L'Estoile, *Mémoires-Journaux* 10: 126; "Les victimes de la Saint-Barthélemy à Paris," 43–44; Haag and Haag, *La France protestante*, I: col. 702.

26. D'Aubigné, *Confession catholique*, 359–61; L'Estoile, *Mémoires-Journaux*, 7: 77–78.

27. L'Estoile, *Mémoires-Journaux* 9: 190; Malherbe, *Œuvres*, 3: 79.

28. Dulaure, *Histoire physique, civile et morale de Paris*, 5: 178–80; Duplomb, *Histoire générale des ponts de Paris*, 59–71, 153–57; Blanchet, "Note sur les projets de reconstruction du pont Marchant"; Ballon, *The Paris of Henri IV*, 109, 299–301; Barbiche and Dainville-Barbiche, *Sully*, 281–82.

29. Malherbe, *Œuvres*, 3: 109.

30. Malherbe, *Œuvres*, 3: 114.

31. Galilei, *Opere*, 10: 392.

32. Galilei, *Opere*, 10: 433.

33. Galilei, *Opere*, 11: 173, 174.

34. Brahe, *Carteggio inedito*, 448–49.

35. Matthieu, *Histoire de la mort déplorable de Henry IV*, 33. On the numerous Venetian mirrors in the inventory of 1589, see Bonnaffé, *Inventaire des meubles de Catherine de Médicis*, 156. Friedrich Risner's edition of Alhazen's *Optics*, published in Basel in 1572, begins with a preface dedicated to Catherine, and states that knowledge of optics will dissuade the public from attributing certain illusions to the tricks of the devil.

36. The incident, supposedly predicted by Nostradamus, forms the historical backdrop of Madame de Lafayette's *La princesse de Clèves*.

37. *Ronsard, Oeuvres*, 2: 1279–82.

38. Tassoni, *Lettere*, 1: 81–82; Favaro, *La Libreria,* 70.

39. De la Broderie, *Ambassades*, 4: 228; Jeannin, *Négociations*, 5: 207–8.

40. Anderson, *On the Verge of War*, 74–108; on Henri's military aims and motivations, see 81–82 and 95–98.

41. Matthieu, *Histoire de la mort déplorable*, 23.

42. Matthieu, *Histoire de la mort déplorable*, 29.

43. Great Britain, Historical Manuscripts Commission, *Papers of William Trumbull the Elder*, 2: 291.

44. Malherbe, *Œuvres*, 3: 167, 181; *Mercure François* (1613): fol. 435a; Richelieu, *Mémoires*, 1: 52.

45. Matthieu, *Histoire de la mort déplorable*, 30–34. On the pont Neuf as a news center, see "Les estranges tromperies de quelques Charlatans nouvellement arrivez à Paris"; "Nouveau Règlement générale pour les Nouvellistes," 264–65 and n. 1.

46. Matthieu, *Histoire de la mort déplorable*, 34–39, 45–52.

47. Great Britain, Historical Manuscripts Commission, *Papers of William Trumbull the Elder*, 2: 281.

48. On Pierre Fougeu, seigneur d'Escures, see especially Buisseret, *Henry IV*, 95–96, and Barbiche and Dainville-Barbiche, *Sully*, 612.

49. *Négociations diplomatiques*, 5: 614; Winwood, *Memorials of Affairs of State*, 3: 156; on the madness of the plan, see 154, 155–56.

50. Matthieu, *Histoire de la mort déplorable*, 47–48.

51. Matthieu, *Histoire de la mort déplorable*, 46–47.

52. In his discussion of this episode, Guido Cardinal Bentivoglio insisted on the king's desire to *riavere la principessa*. See *Memorie e lettere*, 284; for Peiresc's discomfort about such candor, see *Lettres de Peiresc aux frères Dupuy*, 2: 136, 246–47.

53. Stewart, *On Longing*, 37–69, especially 54–69.

54. Buisseret, *Henry IV*, especially 151–55, 180–81; Buisseret, "L'atelier cartographique"; Barbiche and Dainville-Barbiche, *Sully*, 206–9.

55. Van Helden, *Invention of the Telescope*, 41–42, 46.

56. Matthieu, *Histoire de la mort déplorable*, 67.

57. Matthieu, *Histoire de la mort déplorable*, 67–68.

58. Henri IV, *Recueil des lettres missives*, 6: 596–602; L'Estoile, *Mémoires-Journaux 1574–1611*, 8: 214.

59. Barbiche and Dainville-Barbiche, *Sully*, 327–36, 340–41.

60. Matthieu, *Histoire de la mort déplorable*, 47.

61. Sully, *Mémoires ou Œconomies royales*, 3: 243.

62. Duplessis-Mornay, *Mémoires et correspondance*, 11: 91.

63. *Anti-Coton*, 47–48.

64. Great Britain, Historical Manuscripts Commission, *Papers of William Trumbull the Elder*, 2: 313.

65. Matthieu, *Histoire de la mort déplorable*, 85–86.

66. Bentivoglio, *Memorie e lettere*, 296–97.

67. Thus Matteo Botti wrote from Paris to the grand duke of Tuscany in mid-June 1610 that the astrologer of the duke of Savoy had predicted Henri's assassination, and the union of the houses of Savoy and Medici in marriage. See *Négociations diplomatiques*, 5: 639.

68. Matthieu, *Histoire de la mort déplorable*, 86.

69. Froissart, *Œuvres*, 11: 189–201.

70. Montaigne, "It is folly," 133; Chateaubriand, *Analyse raisonnée*, 145.

71. Winwood, *Memorials of Affairs of State*, 3: 159, alludes to the collaboration of the sorcerers of the Society of Jesus.

72. See Th. Juste, "François d'Aerssen."

73. Jeannin, *Négociations du Président*, 4: 466–470.

74. See Juste, "Cornelis d'Aerssen."

75. Henri was said to have confronted the Aerssens in spring 1608 about their spying and corruption; see the skeptical account in Great Britain, Historical Manuscripts Commission, *Report on the Manuscripts of Lord de l'Isle and Dudley*, 4: 23.

76. Badovere relied on François d'Aerssen to send letters and books to the poet and diplomat Pieter Corneliszoon Hooft in Amsterdam, and proposed that a sum owed him by Domenicus Baudius be deposited with either Aerssen or Jeannin; see Hooft, *De Briefwisseling*, 1: 96–98, 104–8. In regard to Cornelis d'Aerssen's role as *greffier*, see Van Helden, *Invention of the Telescope*, 39, 40.

77. Israel, *The Dutch Republic*, 452; Harline, *Pamphlets, Printing, and Political Culture*, 270 n. 3, as well as the pamphlet exchanges listed in *Bibliotheek van Nederlandsche Pamfletten*, 1: 183–87, and in *Catalogus van de pamfletten-verzameling berustende*, 1: 506–12. As Van der Mijle was the first ambassador the United Provinces sent to the Venetian Republic, his embassy and the question of the sovereignty of the United Provinces were the subject of some attention; for this see Sarpi, *Lettere ai protestanti*, 1: 101, 109, and 2: 62, 163–64, 207, 229. The anonymous *De Maculis in sole animadversis, & tanquam ab Apelle, in tabulâ spectandum in publicâ luce expositis, Batavi dissertatiuncula*, now attributed to Willebrord Snell, was written at the instigation of Van der Mijle and dedicated to him; see in this connection Vermij, *The Calvinist Copernicans*, 44–45.

78. Harline, *Pamphlets, Printing, and Political Culture*, 27, 112, 157, 188; Great Britain, Historical Manuscripts Commission, *Papers of William Trumbull the Elder*, 6: 479–81.

On Aerssen's political and diplomatic career, see Israel, *The Dutch Republic*, 254, 452, 466–67, 480, 524–25, 527.

79. Richelieu, *Mémoires*, 51. Nicolas Rogier or Roger acted as a courier between Henri IV and Sully in March 1608; see Henri IV, *Lettres missives*, 7: 494. He is also mentioned in two inventories of the queen's jewels; see Bruel, "Deux inventaires de bagues," 186, 198, 205–6.

80. Ilardi, "Renaissance Florence," 530–31; Ilardi, *Renaissance Vision*, 99–100, 113; Hayward, *Virtuoso Goldsmiths*, 326–27; Bimbenet-Privat, *L'orfèverie parisienne*, 102–3, 133, 177–78, 185–86; "Petit dictionnaire." The words *orfèvre* and *lapidaire* appear interchangeable for certain early modern authors; see Mersenne, *L'Impiété des déistes*, 21, 220; Peiresc, *Lettres*, 5: 156. For statutes regarding the different responsibilities of *lapidaires* and *orfèvres-joaillers* in early modern France, see Jacob, *Le livre d'or des métiers*, 106, 199, 210–11.

81. Sirtori, *Telescopium*, 50–51, 54.

82. Houzeau, "Le Téléscope à Bruxelles," 28.

83. Van Helden, *Invention of the Telescope*, 48, 50; Sirtori, *Telescopium*, 24–25.

84. Reeves and Van Helden, "Verifying Galileo's Discoveries: Telescope-Making at the Collegio Romano."

85. Bedini, "Lens Making for Scientific Instrumentation," especially 689–90.

86. L'Estoile, *Mémoires-Journaux*, 8: 192, 193; 9:260, 291–93, 295, 303, 326; 10: 317; 11: 53.

87. Great Britain, Historical Manuscripts Commission, *Papers of William Trumbull the Elder*, 3: 68–69.

88. Galilei, *Opere*, 3 (1): 60.

89. Rubens is the prototypical example of such a figure, but see also Keblusek, "The Business of News," and Noldus, "Dealing in Politics and Art."

90. On a similar move by Matthieu's successor, Charles Bernard, see Ranum, *Artisans of Glory*, 105–8, especially 107 n. 9.

91. Bireley, *The Jesuits and the Thirty Years War*, 67–68; Audiat, *Un fils d'Estienne Pasquier*; Huppert, "A Matter of Quality."

92. Sorel, *La Bibliothèque françoise*, 98.

93. Nicolas Pasquier, "À Monsieur d'Ambleville," cols. 1054, 1061.

94. Pasquier, "À Monsieur d'Ambleville," cols. 1056–57.

95. Pasquier, "À Monsieur d'Ambleville," cols. 1057–58.

96. Pasquier, "À Monsieur d'Ambleville," col. 1054.

97. Pasquier, "À Monsieur d'Ambleville," cols. 1055–56.

98. On the importance of "the people" to Nicolas' father, Étienne Pasquier, see Huppert, "A Matter of Quality," 196–97; on the notion of "the people" in sixteenth-century France, see Davis, *Society and Culture*, especially 190, 193–94, 209; on Estienne Pasquier's attention to popular practices and sayings, see 238–39.

99. Boas, *Vox Populi*; Davis, *Society and Culture*, 232, 251–52.

100. Pasquier, "À Monsieur de Montagnes," col. 1280.

101. Pasquier, *Remonstrances tres-humbles a la Royne*, 9.

102. Boccalini, *Comentarii sopra Cornelio Tacito*, 437; Wilkinson, "'Homicides Royaux,'" especially 132–33.

103. *Gazette* for September 17, 1632.

104. Pasquier, "À Monsieur d'Ambleville," cols. 1059–60.

105. Cited in Schickler, "Hotman de Villiers et son temps," 151.

106. Great Britain, Historical Manuscripts Commission, *Papers of William Trumbull the Elder*, 2: 298.

107. The Venetian ambassador to the French court, Antonio Foscarini, stated that Ravaillac was from Angoulême, that he had a wife in Brussels, that he was a member of the Feuillants, a monastic order associated with the Cistercians, and that various talismanic characters and relics were sewn into his tunic. Foscarini also transformed the name of the killer into *Raguagliace*, which would evoke "bad news," *Ragguagliaccio*. See *Relazioni degli stati europei*, ser. II. 1: 334.

108. Wotton, *The Life and Letters*, 2: 490.

109. Great Britain, Historical Manuscripts Commission, *Papers of William Trumbull the Elder*, 2: 313.

110. Pasquier, "À Monsieur d'Ambleville," col. 1060; *Mercure François*, 1611: fol. 210v–211; *Mercurius Gallobelgicus* (1615): 92–93. The *Mercure François* notes that the story first appeared in a small pamphlet.

111. Estienne Pasquier, *Les Œuvres*, 2: 175; on the business about the bear cubs, see Ziolkowski and Putnam, *The Virgilian Tradition*, 70, 192, 359, 391.

112. McCusker, "The Demise of Distance," especially 299–307.

113. On the *Courant d'Italie & d'Almaigne, &c.*, see *Dutch Corantos*, 53–54, 163–66 v.

114. *Dutch Corantos*, 163. On this feature in the slightly earlier printed weekly of Johann Carolus of Strasbourg, see Weber, "The Early German Newspaper," 73; on the datelines in English corantos of 1620–1621, see Brownlees, *The Language of Periodical News*, 42–46.

115. *Dutch Corantos*, 163.

116. *Dutch Corantos*, 163 v.

117. *Dutch Corantos*, 166–166 v.

118. *Dutch Corantos*, 166 v.

119. *Dutch Corantos*, 166 v.

120. Parker, *Thirty Years' War*, 42, 68, 91–92, 97, 102–3, 130–37; Bireley, *The Jesuits and the Thirty Years War*, 44–56, 63–79, 97–99, 100–105, 114–22; Vittu, "Instruments of Political Information," 164–65.

121. Pasquier, *Le Catéchisme des Jésuites*, 391–412.

122. A more interesting basis for the comparison would be the fact that Pasquier was once portrayed, like the mutilated statue of Pasquino, without hands. On this incident see *La main ou œuvres poétiques*; on the armless Pasquino, and the importance of this bodily mutilation to the statue's expressive function, see Barkan, *Unearthing the Past*, 208–31.

123. Pasquier, *La Chasse du renard Pasquin*, 10.

124. Pasquier, *La Chasse du renard Pasquin*, 11.

125. Pasquier, *La Chasse du renard Pasquin*, 11–12.

126. On Marolles' contacts, see de Boer, "Men's Literary Circles," 730–36, 741–42, 754–56, 761–68, 773–74.

127. Marolles, *Mémoires*, 1–13.

128. Marolles, *Mémoires*, 10. For another newsmongering priest—a prior in Padua in 1620—see Infelise, *Prima dei giornali*, 142.

129. Marolles, *Mémoires*, 13.

130. Marolles, *Mémoires*, 272–77.

131. Miller, *Peiresc's Europe*, 16–48.

132. Gassendi, *Vita Peireskii*, 274.

133. Muñoz Calvo, *Inquisición y ciencia*, 108.

134. Bayard, "Jean Bochart de Champigny." The sojourn in Venice (August 1607–August 1611) was particularly well remunerated; see p. 49 of this study.

135. Sarpi, *Lettere ai protestanti*, 2: 152.

136. Marolles, *Mémoires*, 10.

137. Gassendi, *Vita Peireskii*, 275.

138. Descartes, *Oeuvres*, 6: 81–82.

139. Peiresc, *Lettres aux frères Dupuy*, 80–81; Gassendi, *Vita Peireskii*, 298.

140. Metius, *Institutiones astronomicae et geographicae*, 3–4; cited and translated in Van Helden, *Invention of the Telescope*, 48. In another work of 1614, the *Nieuwe geographische onderwysinghe*, Metius states that his brother has not yet revealed his "invented glasses," the strength of which will allow for the more accurate measure of longitude. See Van Helden, *Invention of the Telescope*, 48.

141. Van Helden, *Invention of the Telescope*, 48–51; Sirtori, *Telescopium*, 23–26; Simón de Guilleuma, "Juan Roget, óptico gerundense"; Simón de Guilleuma, "Notas Bibliográficas"; Settle, "The Invention(s) of the Telescope."

142. On the condemnation of notorious almanac maker Noël Morgart, around 1614, see "Rencontre et naufrage," 212–15; Buisseret, *Sully and the Growth of Centralized Government*, 166–67.

143. Reeves, *Galileo's Glassworks*.

144. Galilei, *Opere*, 10: 255.

145. Galilei, *Opere*, 10: 250.

146. Galilei, *Opere*, 10: 250–51; Biagioli, "Did Galileo Copy the Telescope?" especially 227–28.

147. Galilei, *Opere*, 10: 255.

148. See Galilei, *Opere*, 10: 257, 259–61, 264, 267.

149. Galilei, *Opere*, 10: 267.

150. Galilei, *Opere*, 10: 260–61.

151. Sarpi, "Per l'epistolario di Paolo Sarpi," *Ævum* 10: 23–24, 40, *Ævum* 11: 32, and *Ævum* 11: 276, 281–82, 296–97, 318; Ulianich, "Saggio introduttivo," in Sarpi, *Lettere ai gallicani*, li–liii; Wootton, *Paolo Sarpi*, 102–3.

152. Galilei, *Opere*, 10: 257.

153. Bigourdan, "Joseph Gaultier et la découverte de la visibilité des astres"; Humbert, "Joseph Gaultier de la Valette."

154. Grillo, *Delle lettere*, 2: 149.

CHAPTER 3. GALILEO *GAZZETTANTE*

1. Galilei, *Opere*, 10: 122, 126.

2. On Tedeschi's participation in this Academy in 1609–1610, and again in 1620, see Groote, *Musik in italienischen Akademien*, 36, 40, 48.

3. *Lettere d'uomini illustri*, 163; Pescetti, *Risposta all'Anticrusca*, 31–32, 34, 38, 116.

4. *Lettere d'uomini illustri*, 131.

5. On news in Venice see De Vivo, *Information and Communication*; in the Veneto and beyond, Infelise, *Prima dei giornali*.

6. Bentivoglio, *Lettere del Cardinal*, 26, 29.

7. Grillo, *Delle Lettere*, 3: 225. On Grillo, and on "going out" as a synonym for publishing, see Richardson, *Manuscript Culture*, 3–4, 31–32, 93–94, 98–99.

8. *Lettere d'uomini illustri*, 209, 408–9; Grillo, *Delle Lettere*, 2: 285–89. On the circulation of Machiavelli's works see De Vivo, *Information and Communication*, 235–36; Barbierato, *The Inquisitor in the Hat Shop*, 116–20, 127, 177, 272, 275, 282, 330; Richardson, *Manuscript Culture*, 8, 164–69.

9. Cancellieri, *Dissertazione*, 54–57.

10. Bentivoglio, *Lettere del Cardinal*, 38, 67–68, 88; *Lettere d'uomini illustri*, 408–9, 413–20.

11. Grillo, *Delle Lettere*, 2: 284, 591, 594.

12. Grillo, *Delle Lettere*, 2: 590.

13. Firpo, "In margine al processo"; Vialardi di Sandigliano, "Un cortigiano e letterato"; Richardson, *Manuscript Culture*, 158.

14. Grillo, *Delle Lettere*, 2: 416.

15. Chambers, " 'Bellissimo Ingegno' "; Sluiter, "The Telescope Before Galileo"; Galilei, *Opere*, 10: 367–68, 382, 383, 385, 420.

16. Galilei, *Opere*, 11: 277.

17. Another such figure in a slightly later and more developed media landscape is the artist, cultural broker, and intelligencer Michel le Blon; see Keblusek, "The Business of News."

18. On the move of the newswriter away from clandestine intelligence, see Infelise, "News Networks," 54.

19. On the "invention of facts" for different paradigms, see Stengers, *The Invention of Modern Science*, 49–50.

20. On Galileo's early understanding of Copernicus's work as a powerful philosophical system, see Bucciantini, *Galileo e Keplero*, 53–55; on the refinement of this system as anti-Tychonian, see 129–43, 176–205.

21. On the publication history and translations of the work, see Boccalini, *Ragguagli*, 1: 371; 3: 354–60, 525–31; Firpo, "Traiano Boccalini"; Firpo, *Traduzioni dei Ragguagli*; Dooley, *The Social History of Skepticism*, 117–18, 125–27; Infelise, *Prima dei giornali*, 170–73.

22. Ilardi, *Renaissance Vision*, 82–102, 136–38.

23. On the earlier version of these glasses, see Reeves, *Galileo's Glassworks*, 16, 20, 52, 55–57, 64, 74, 76–77.

24. Boccalini, *Ragguagli*, 1: 10–11.

25. On Caporali's influence in general see Cappelli, "Parnaso bipartito."

26. Caporali, *Opere poetiche*, 98.

27. Vannozzi, *Delle Lettere Miscellanee*, 179, 181, 405–6; Vannozzi, *Della Suppellettile* (1613) 521–22. On Vannozzi's receipt of della Porta's *De refractione*, published in Naples in 1593, see the undated letter in his *Teatro di segretaria copioso*, 177–78.

28. Lancellotti, *L'Hoggidi*, 294.

29. Grillo, *Delle Lettere*, 1: 909; Boccalini, *Ragguagli*, 2: 54, 56, 61, and 332–33; Magagnati, *Meditazione poetica*, fol. A3, unnumbered page 8.

30. Boccalini, *Ragguagli*, 3: 341, 515–16, where Firpo dates the letter to around 1597; for a later, revised estimate closer to 1600, see Firpo, "Lettere di Traiano Boccalini," 22–23.

31. On Boccalini's influence in Spain, see García Santo-Tomás, "Fortunes of the *Occhiali Politici*."

32. Boccalini, *Comentarii sopra Cornelio Tacito*, 360; *Considerationi sopra la vita di Giulio Agricola*, 3.

33. Boccalini, *Ragguagli*, 2: 247–49.

34. Boccalini, *Ragguagli*, 2: 296–97.

35. Boccalini, *Ragguagli*, 1: 327.

36. On the durable connection of observation with both political astuteness and scientific progress in various early modern deployments of Tacitus and Machiavelli, see Soll, *Publishing the* Prince, 36–38, 48–50, 61–68; on the more general association of an interest in natural philosophy, particularly in atomism and materialism, with political and religious restiveness, often among newsmongers, in early modern Padua and Venice, see Barbierato, *The Inquisitor*, 4, 33, 36, 46, 71, 85–86, 99–101, 126, 250–54, 265–67, 271, 293–95, 299, 301–7, 313–15, 330–33.

37. Boccalini, *Ragguagli*, 1: 327.

38. Boccalini, *Ragguagli*, 1: 328.

39. On Sarpi's familiarity with Tacitus and Machiavelli, see Wootton, *Paolo Sarpi*, 9, 42, 48, 69–73; on optical investigations, see Reeves, *Galileo's Glassworks*, 57, 71–78, 81–98.

40. Guerrini, *Ricerche su Galileo*, 14–21.

41. Politi, "Al Signor Giulio Pannocchieschi," 527.

42. Galilei, *Sidereus Nuncius*, 58–63; Caporali, *Opere Poetiche*, 68; on the *stellete* typically prepared for banquets, see Messisbugo, *Libro nuovo*, fol. 47–48, 68v–69v.

43. Van Helden, *The Invention of the Telescope*, 15–20; Reeves, *Galileo's Glassworks*, 8–9, 70–78, 87–89; Galilei, *Opere*, 12: 101.

44. *Lettere d'uomini illustri*, 121. The expression Pignoria used for telling tall tales, *piantar carotte*, is roughly equivalent to *vender fumo*, and occurs at least once in explicit association with the origins of early modern journalism, in verses by Giovanni Maria Cecchi; see Bernardini, *Guida della stampa periodica Italiana*, 31 n. 2.

45. Boccalini, *Ragguagli*, 3: 365.

46. Reeves, "Faking It."

47. Grillo, *Delle Lettere*, 2: 52–53, 284–85, 295–97, 492, 625, 630; Reeves, "Faking It," 71–72; on the delay caused by Gianfrancesco Sagredo's effort to have manuscript copies made of Galileo's first and second letters on the sunspots in Venice, see Galilei, *Opere*, 5: 184–85.

48. Boccalini, *Ragguagli*, 3: 254.

49. Apelles and Tacitus are also compared in Boccalini's *Considerationi sopra la vita di Giulio Agricola*, 2.

50. Boccalini, *Ragguagli*, 3: 208.

51. Boccalini, *Ragguagli*, 2: 10, 330.

52. Galilei, *Opere*, 10: 418; 12: 46.

53. Galilei, *Opere*, 3(1): 185; 11: 11; 19: 229, and 20: 611.

54. Galilei, *Opere*, 5: 121, 136–37; Galilei and Scheiner, *On Sunspots*, 316–30.

55. Galilei, *Opere*, 5: 138, 140; 11: 230–31; Galilei and Scheiner, *On Sunspots*, 237–49.

56. Mayer, "The Censoring of Galileo's *Sunspot Letters*"; Gingerich and Van Helden, "From *Occhiale* to Printed Page"; Wootton, "New Light," 65–66, 74–78; Van Helden, "Galileo and the Telescope."

57. The English coranto likewise moved from an initial emphasis on timely reports to an effort to produce a more uniform and sustained outlook; see Brownlees, *The Language of Periodical News*, 60–65.

58. Mayer, "The Censoring of Galileo's *Sunspot Letters*."

59. Cozzi, "Traiano Boccalini, Il Cardinal Borghese e la Spagna," 240–44.

60. Soll, *Publishing the* Prince, 31–34, 72–88, 95–114.

61. Burke, "Tacitism"; Gajda, "Tacitus and political thought," 264–66.

62. Burke, "Tacitism," 162; Alamo Varienti, *Discorso del Signor Alamo*, 35th and 36th unnumbered folios.

63. Cited in Burke, "Tacitism," 162. The representative of this approach, for Milton, was Virgilio Malvezzi; on his *Discourses on Tacitus*, published in 1635, see Gajda, "Tacitus and political thought," 263–64.

64. Politi, *Dittionario Toscano*, 359.

65. Politi, "Al Lettore," 25th unnumbered folio. Politi's address to the reader dates to 1604, but this remark appeared only in 1618. The substitution of *gazzette di Roma* for *acta* occurs in Book XVI: 22 of the *Annales*; in the Venetian edition of 1618, see 324.

66. Politi, "Al Lettore," 29th unnumbered folio. On Guicciardini's place in manuscript circulation, see Richardson, *Manuscript Culture*, 88, 161, 163, 166.

67. Boccalini, *Ragguagli*, 1: 327.

68. Politi, "Al Lettore," 29th unnumbered folio; see also Politi, "All'illustre . . . signor Francesco Visdomini," 16th and 17th unnumbered folios.

69. On Galileo's lecture notes, heavily indebted to Christopher Clavius's commentary on the *Sphere*, see Wallace, *Prelude to Galileo*, 136–38, 199–201, 216–17.

70. On Bentivoglio's role in the Roman Inquisition, see Mayer, *The Roman Inquisition*, 71–72, 78, 92, 107, 222, 224, 278.

71. Bentivoglio, *Memorie e lettere*, 97.

72. De Vivo, *Information and Communication*, 235–36; Quaglio, "Indicazioni sulla fortuna editoriale"; Connell, "New Light on Machiavelli's Letter," especially 95, 106.

73. A satire of the 1730s, inspired by Boccalini's portraits of Tacitus and Machiavelli, presents the latter as the printer of a *Faithful and comic story of the religion of the cannibals of Rome*; see Artaud de Montor, *Machiavel*, 388–99; Barbierato, *The Inquisitor*, 330.

74. The motif of "dissection" of the body politic, in this case the decadent Venetian Republic, reappears in the work of the Tacitist Abraham-Nicolas Amelot de La Houssaye, translator of Machiavelli; see Soll, *Publishing the* Prince, 63–64.

75. Grillo, *Delle Lettere*, 2: 288, 289.

76. Grillo, *Delle Lettere*, 2: 484.

77. Grillo, *Delle Lettere*, 3: 51.

78. Grillo, *Delle Lettere*, 1: 909; 2: 648–49.

79. Grillo, *Delle Lettere*, 3: 196.

80. Grillo, *Delle Lettere*, 3: 171.

81. Grillo, *Delle Lettere*, 3: 196.

82. Bernardo Davanzati's commentary on the *Annals* of Tacitus, written before 1606 but not published in full until 1637, included a gratuitous reference to Galileo's early explanation of the aurora borealis. See Tacitus, *Opere di Cornelio Tacito*, 1: 66.

83. Brown, *The Return of Lucretius*, 68–87; Favaro, *La Libreria*, 60; Richardson, *Manuscript Culture*, 252–53; Galilei, *Opere*, 11: 290.

84. Galilei, *Opere*, 15: 230.

85. Galilei, *Opere*, 18: 415; 11: 502–3.

86. Tassoni, *Lettere*, 1: 24, 26, 30, 34, 38, 54, 62; Mestica, *Trajano Boccalini*, 125–26.

87. Galilei, *Opere*, 10: 121; 19: 628.

88. *Documenti per la storia della Accademia dei Lincei*, 24; Gabrieli, "Per la storia," 88; Tassoni, *Pensieri e Scritti Preparatori*, 850. On the Academy in this period, see Freedberg, *The Eye of the Lynx*.

89. Galilei and Scheiner, *On Sunspots*, 128, 257, 291, 293–94; Tassoni, *Lettere*, 2: 165–67.

90. Tassoni, *Lettere*, 2: 167.

91. Vianello, "Le postille al Petrarca," 239.

92. Muratori, "Appendice," 55–56.

93. Tassoni, *Lettere*, 2: 77, 78–79, 80, 81, 82–83.

94. Tassoni, *Lettere*, 2: 144.

95. Tassoni, *Pensieri e Scritti Preparatori*, 511–12, 961; Galilei, *Opere*, 5: 282.

96. Tassoni, *Lettere*, 2: 26–27, 31, 41, 43–44, 46; *Pensieri e Scritti Preparatori*, 512–18, 977, 992.

97. Tassoni, *Lettere*, 2: 90.

98. Tassoni, *Pensieri e Scritti Preparatori*, 923.

99. Tassoni, *Pensieri e Scritti Preparatori*, 850, 913, 923.

100. Galilei, *Opere*, 5: 80–81.

101. Galilei, *Opere*, 9: 278–79; 20: 585. On references to the pseudo-Homeric *Batracomiomachia* in the *Secchia Rapita*, see Longhi, "Il vestito sconveniente," 106–7.

102. Tassoni, *La Secchia Rapita* (canto II, 35) 2: 55.

103. *La Secchia Rapita* (canto II, 40–41) 2: 58–59. Curiously, the "two vials," *due pitali*, could be paraphrased as *due boccalini*.

104. Galilei, Opere, 3(1): 322; 10: 396.

105. Mercury's recapitulation of observations drawn from the *Starry Messenger* is offered in canto 10, stanza 35–47 of Marino's *Adone*.

106. For an early version of the argument about the "coiffure" or "headgear" of luminous bodies, see Galilei, *Sidereus Nuncius*, 57–58; for particular references to Jupiter, see Galilei, *Opere*, 11: 194–95; 6: 360–66.

107. Casanova, *The Story of My Life*, 715.

108. Tassoni, *Secchia Rapita* (canto VIII: 46–63) 2: 245–52. On the subsequent and more troubling tale, see Scalabrini, "Gli amori ridicoli."

109. Galilei, *Opere*, 5: 98.

110. Firpo, "La satira politica," 5; Gabotti, "Per la storia della letteratura civile," 327–30.

111. Lagalla, *De phoenomenis*, 54–55.

112. The observation of the Scythian ambassador appears in Quintus Curtius Rufus, *Historiae Alexandri Magni* VII: viii.

113. Besomi and Camerota, *Galileo e il Parnaso Tychonico*, 233.

114. Brahe, *Carteggio inedito*, 245, 484; Bucciantini, *Galileo e Keplero*, 39–48, 56–63, 84–92 123–38, 176–93, 230, 260–74, 279–87; Besomi and Camerota, *Galileo e il Parnaso Tychonico*; Tassoni, *Pensieri e scritti preparatori*, 415, 850, 913, 923.

115. Besomi and Camerota, *Galileo e il Parnaso Tychonico*, 203, 216.

116. Besomi and Camerota, *Galileo e il Parnaso Tychonico*, 135–58, 237–49.

117. Grillo, *Delle Lettere*, 2: 116, 223–24.

118. Galilei, *Opere*, 10: 291–96.

119. Diffley, "Tassoni's Linguistic Views."

120. Galilei and Scheiner, *On Sunspots*, 289–90.

121. Politi, "All'illustre . . . signor Francesco Visdomini," 9th unnumbered folio; Galileo and Scheiner, *On Sunspots*, 120.

122. *Lettere d'uomini illustri*, 131, 137, 153–54, 164, 170, 173–74, 177.

123. Reeves, "From Dante's Moonspots," 206–7.

124. Marcus Welser of Augsburg's mastery of Tuscan, for instance, was signaled as evidence of its portability; see Pescetti, *Risposta*, 16, 112. Isabelle Stengers discusses the need of the innovative scientist to transform history so that others "after them, [who] will write history, are constrained to speak of their invention as a 'discovery' that others could have made." See *The Invention of Modern Science*, 39.

125. Bruni, *Armonia Astronomica*, 52–53.

126. Bruni, *Armonia Astronomica*, fol. A4.

127. Wilding, *Galileo's Idol*, forthcoming, and Florence, National Central Library, Fondo Galileiano, ms. 70, fol. 4r. ,

128. *Paradise Lost* I: 287–88, and III: 588–90.

129. *Paradise Lost* V: 261–63, and 275–77.

130. On optical instruments as prostheses fit for postlapsarian man, see Picciotto, *Labors of Innocence*, 189–93, 205–11; on the association of optical instruments with angelic powers, see Wilding, "Galileian Angels."

131. *John Milton*, 259; on this work within the context of the seventeenth-century pamphlet, see Raymond, *Pamphlets and Pamphleteering*, 262–75; on the encounter of Galileo and Milton in 1638, see Butler, "Milton's Meeting with Galileo."

132. Louthan, "Mediating Confessions"; Galilei, *Opere*, 16: 400, 405. The identification of the author is the work of the editors.

133. Collins, *The Sociology of Philosophers*, 523.

CHAPTER 4. CAMERAS THAT DON'T LIE

1. Dupré, "Inside the *Camera Obscura*"; Lindberg, *Theories of Vision*, 178–208; Alpers, *The Art of Describing*, 33–38, 49–51; Crombie, "Expectation, Modeling and Assent"; on the dark room as a model of the mind see Kofman, *Camera Obscura of Ideology*; Bredekamp, *The Lure of Antiquity*, 37–45; Crary, *Techniques of the Observer*, 25–66; Reiss, *Against Autonomy*, 219–57; Mitchell, *Iconology*, 16–17, 168–90.

2. The distinction is drawn in 1615 in Father Christoph Scheiner's lecture notes; see Daxecker, *The Physicist and Astronomer*, 100.

3. Gestrich, "The Public Sphere"; Randall, "Epistolary Rhetoric, the Newspaper, and the Public Sphere"; Shohet, *Reading Masques*, 10–16, 106, 149–88; Infelise, "News Networks," 66–67.

4. Mah, "Phantasies of the Public Sphere."

5. The classic text is Pena, *De usu optices*, fol. C2. This preface first appeared in the edition of Euclid's *Optics* and *Catoptrics* published in Paris in 1557; see also Hallyn, "Jean Pena et l'éloge de l'optique," especially 219–21.

6. Wotton, *Life and Letters*, 1: 486–87; see also Wootton, *Galileo Watcher of the Skies*, 5–6, 59, 103.

7. Galilei, *Opere*, 3: 147–78; Durkan, "A Post-Reformation Miscellany," 114–17.

8. Great Britain, Public Record Office, *Calendar of State Papers . . . Venice*, 11: 157. On the circulation of news of various sorts in Venice and other major European cities, see Infelise, "News Networks Between Italy and Europe."

9. Great Britain, Historical Manuscripts Commission, *Lord de L'Isle*, 5: 238.

10. Great Britain, Historical Manuscripts Commission, *Papers of William Trumbull the Elder*, 5: 240.

11. Stubbs, "John Beale," 467–71; Boyle, *Letters and Papers*, 7: 13; *The Hartlib Papers*, fol. 8 / 56 / 1a, 2a; 67 / 22 / 5a, 7a. On the polemoscope adapted for domestic use, see "Singuliers effets de catoptrique."

12. Walton, *Lives*, 116. Networks Between Italy and Europe." Great Britain, Historical Manuscripts Commission, *Lord de L'Isle*, 5: 238. Great Britain, Historical Manuscripts Commission, *Papers of William Trumbull the Elder*, 5: 240. Stubbs, "John Beale," 467–71; Boyle, *Letters and Papers*, 7: 13; *The Hartlib Papers*, fol. 8 / 56 / 1a, 2a; 67 / 22 / 5a, 7a. On the polemoscope adapted for domestic use, see "Singuliers effets de catoptrique." Walton,

13. Walton, *Lives*, 76–77.

14. Wotton, *Life and Letters*, 1: 5–6 n. 4.

15. Pantin, "*Simulachrum, species, forma, imago.*"

16. Vesalius, *De humanis corporis fabrica*, 646.

17. Maurolico, *Diaphanorum libri*, 116–17, 130–31; Colombo, *De re anatomica*, 219.

18. Platter, *De corporis humani structura*, 186–87; Lindberg, *Theories of Vision*, 176–77; Crombie, "Expectation, Modeling and Assent: Part 1," 628–29.

19. Guillemeau, *A Worthy Treatise*, thirty-fourth unnumbered page.

20. Du Laurens, *Discourse*, 34; Crooke, *ΜΙΚΡΟΚΟΣΜΟΓΡΑΦΙΑ*, 571.

21. Walton, *Lives*, 77.

22. Van Helden, *Invention of the Telescope*, 12, 15, 19, 28–30, 35.

23. Lindberg, *Theories of Vision*.

24. Colombo, *De re anatomica*, 216; Banister, *Histoire*, 102; Du Laurens, *Discourse*, 37–46; Crooke, *ΜΙΚΡΟΚΟΣΜΟΓΡΑΦΙΑ*, 666–70.

25. Della Porta, *Natural Magick*, XVII: 6, 365; Kepler, *Gesammelte Werke*, 2: 187–88.

26. Van Helden, *Invention of the Telescope*, 12–16, 28–36; Reeves, *Galileo's Glassworks*, 36–46, 49–114.

27. Gentili, *De Legationibus*, 2: 53, 65–68; on Spanish ambassadors as spies, see Levin, *Agents of Empire*, 154–82.

28. Gentili, *De Legationibus*, 2: 170.

29. Gentili, *De Legationibus*, 2: 175–76; Bazzoli, "Ragion di stato," 303–10.

30. Borrelli, "Techniche di simulazione," 108–18.

31. Borrelli, "Techniche di simulazione," 114.

32. Weiss, "Henry Wotton and Orazio Lombardelli."

33. Katritzky, "Was *commedia dell'arte* performed by mountebanks?"; Katritzky, "Mountebanks, mummers, and masqueraders"; Nevinson, "Illustrations."

34. Peiresc, *Lettres*, 1: 52, 6: 238.

35. Rosenheim, "The *Album Amicorum*"; Nickson, *Early Autograph Albums*; Nickson, "Some sixteenth-century albums"; Rudt de Collenberg, "Un *Liber amicorum*"; Nickson, "Some Early English, French, and Spanish contributions."

36. Rosenheim, "The *Album Amicorum*," 308.

37. Jantz, "Goethe and an Elizabethan Poem."

38. Walton, *Lives*, 87.

39. Walton, *Lives*, 93–94; Schleiner, "Scioppius' Pen."

40. Walton, *Lives*, 94.

41. *Reliquiæ Wottonianæ*, fols. e–f 2.

42. *Reliquiæ Wottonianæ*, fol. e.

43. Walton, *Lives*, 94; Winwood, *Memorials*, 3: 407; Chamberlain, *Letters*, 1: 385, 397.

44. Walton, *Lives*, 93.

45. Fuller, *Worthies*, 178.

46. Gilbert, *De mundo sublunari*, 250; Magrini, "Il 'De magnete,'" 28; Abromitis, "William Gilbert."

47. Great Britain, Historical Manuscripts Commission, *Lord de L'Isle*, 5: 225.

48. Venuti, *The Translator's Invisibility*, 43–98.

49. Townshend, *Poems and Masks*, 43–44; Suckling, *Fragmenta Aurea*, 19.

50. D'Addio, *Il pensiero politico*, 609–70; Gabrieli, "La 'Philoteca Scioppiana.'"

51. D'Addio, *Il pensiero politico*, 620.

52. D'Addio, *Il pensiero politico*, 621; Sosio, "Il manoscritto"; Reeves, *Galileo's Glassworks*, 115–38; Duplessis-Mornay, *Mémoires*, 11:260–61, 431.

53. D'Addio, *Il pensiero politico*, 611, 612, 616, 617, 634–35; Gabrieli, "La 'Philoteca Scioppiana,'" 231; Rosen, *The Naming*, 30–37; on the Lincei, see Gabrieli, *Contribuiti*, and Freedberg, *The Eye of the Lynx*.

54. Galilei, *Opere*, 10: 288, 460, 477.

55. D'Addio, *Il pensiero politico*, 609; Rosenheim, "The *Album Amicorum*," 260–61, 301; Katritzky, "Mountebanks, mummers, and masqueraders," 15; for the bibliography on Welser, see Galilei and Scheiner, *On Sunspots*, 107.

56. Wotton, *Life and Letters*, 1: 120–22; Galilei and Scheiner, *On Sunspots*; Galilei, *Opere*, 5: 62; [Borromeo], "Uno scritto inedito," 310–11, 323, 325, 326–27, 334, 336, 338.

57. On such observations, see Heilbron, *The Sun in the Church*, 68; Bartolini, *I fori gnomici*; Settle, "Egnazio Danti."

58. Galilei and Scheiner, *On Sunspots*, 127–28.

59. Lindberg, "The Theory of Pinhole Images" and "A Reconsideration"; Straker, "Kepler, Tycho, and the 'Optical Part of Astronomy'"; Brown, "Stained Glass."

60. Galilei and Scheiner, *On Sunspots*, 11–13, 19–20, 128.

61. Kepler, *Gesammelte Werke*, 4: 92.

62. Kepler, *Gesammelte Werke*, 4: 93.

63. Kepler, *Gesammelte Werke*, 4: 93.

64. Kepler, *Gesammelte Werke*, 16: 288.

65. Galilei and Scheiner, *On Sunspots*, 238.

66. Polišenský, *The Thirty Years War*, 98–115; Gui, *I Gesuiti*, 30–47; *Letters and Other Documents*, 1: 1–3, 17–21.

67. Rabb, "English Readers."

68. Cogswell, *The Blessed Revolution*, 12–20; *Letters and Other Documents*, 1: 37; 2: 14–15, 23, 26, 28, 37, 61, 90.

69. Meroschwa, *Epistola* and *Copia*; in *I Gesuiti*, 20–21 n 13, Gui notes the presence of a French translation of the letter in the Vatican Library, Chigi IV.2228. For the French and Dutch translations, see also *Catalogus van de pamfletten*, 1: 581.

70. Bireley, *The Jesuits*, 44–56, 62, 66–67.

71. Harline, *Pamphlets, Printing, and Political Culture*, 46, 52; Parker, *The Thirty Years' War*, 13, 17, 20, 88–89; Bireley, *The Jesuits and the Thirty Years War*, 33–42.

72. Meroschwa, *Epistola*, 3, and *Copia*, 1.

73. Meroschwa, *Epistola*, 4, and *Copia*, 2.

74. Meroschwa, *Epistola*, 16, and *Copia*, 26.

75. Meroschwa, *Epistola*, 3–4, and *Copia*, 2.

76. Silvestre, "Miroir ou Mica?" The *Seyndt-brief van Wencelaus Meroschvva Bohemer, aen Joannem Traut Norenbergher* relates that "nieuwe sterre-kijckers door hunne buysen in 't firmament des hemels nieuwe sterre / ende in de Sonne nieuwe bleken ghevonden hebben," while the *Lettre de Wenselaus Meroschovva Bohemois* observes that "les modernes Mathematiciens ont trouvé ces tuiaux prospectifs, par lesquels ils voient estoilles dans le Ciel, & taches au Soleil." *Augenglaser* was listed as the equivalent of both *brillen* ("lenses") and *occhiale* ("telescope") in Hulsius, *Dictionarium Teutsch Italienisch*, 64, 267.

77. See also Richard Brathwaite, who alleged that those who ignore the inner eye of faith "doe delight / To glaze their windowes with / *Perspectiue* glasse, / Presenting sundry objects to the sight, / As *Hils, Dales, Seas*, and / Whatsoere shall passe / Within an equall distance . . . ." *A New Spring*, fol. B4v.

78. Grimmelshausen, *Simplicissimus*, book III: i, (164). In book III: vii (180), the hero uses a perspective glass, but to no avail. The motif of the acoustic devices is taken up in one of the continuations of *Simplicissimus*; see p. 376 on this edition. For some cautionary words about Grimmelshausen's usefulness to the historian, see Parker, *The Thirty Years' War*, 269.

79. Wotton, *Life and Letters*, 1: 167–75.

80. Chamberlain, *Letters of John Chamberlain*, 2: 310; Great Britain, Public Record Office, *Calendar of State Papers . . . Venice*, 16: 264, 284.

81. Wotton, *Life and Letters*, 2: 204–5.

82. Bacon, *Novum Organum*, 107.

83. *Stuart Royal Proclamations*, 1: 495.

84. Wotton, *Life and Letters*, 2: 197.

85. Wotton, *Life and Letters*, 2: 198; Sarpi, *Lettere ai protestanti*, 1: 58.

86. Wotton, *Life and Letters*, 2: 199. The original letter about the lying intelligencers appears to have been conveyed by a relative of the man in whose *album amicorum* Wotton had long ago defined the task of the ambassador, John Christoph Fleckamer or Flechammer.

87. Wotton, *Life and Letters*, 2: 199; Great Britain, Public Record Office, *Calendar of State Papers . . . Venice*, 16: 496, 499–500. For the suspicion of Catholic sorcery, see Randall, "Joseph Mead, Novellante," 307; for similar difficulties in obtaining news of the defeat of the Spanish Armada, see Dooley, "Making It Present," 97–103.

88. Wotton, *Life and Letters*, 2: 205–6.

89. Kepler, *Gesammelte Werke*, 18: 63.

90. Kepler, *Gesammelte Werke*, 4: 92, 93; Galilei, *Opere*, 5: 137; Peiresc, *Lettres*, 4: 437; Van Helden, "Telescopes and Authority," 14.

91. Davids, "Successful and Failed Transitions," 227–32.

92. Bacon, *Physical and Metaphysical Works*, 74.

93. Bacon, *Physical and Metaphysical Works*, 194; Kepler, *Gesammelte Werke*, 2: 33–34, 58–59.

94. Watts, *Swedish Intelligencer*, fol. 4; Wood, *Fasti* 1: cols. 383–84; Smith, "Watts, William."

95. Thompson, "Licensing the Press," 671–72.

96. Galilei and Scheiner, *On Sunspots*, 47–51.

97. Thompson, "Licensing the Press," 671–72; Levy, "How Information Spread," 20–25; Cust, "News and Politics"; Frearson, "The distribution and readership of London corantos."

98. Watts, *Mortification*, fol. A2.

99. Great Britain, Public Record Office, *Acts of the Privy Council*, 329; Great Britain, Public Record Office, *Calendar of State Papers, Domestic Series, (1619–1623)*, 198, 201; Great Britain, Public Record Office, *Calendar of State Papers . . . Venice*, 16: 500, 508, 536, 550, 570, 583, 614, 616; Wotton, *Life and Letters*, 2: 474–76; Walton, *Lives*, 104–6. For the court's reaction to Wotton's mission, see Great Britain, Public Record Office, *Calendar of State Papers . . . Venice*, 417, 428, 509; for Wotton's impression of the futility of the undertaking, see *Life and Letters*, 1: 174–75; 2: 207, 208.

100. Walton, *Lives*, 93, and *Reliquiæ Wottonianæ*, n.p.

101. Kepler, *Gesammelte Werke*, 2: 187–89; 4: 372.

102. Lindberg, *Theories*, 182–85, 202–8; Crombie, "Expectation, Modeling, and Assent," I: 632; II: 89–99.

103. Kepler, *Gesammelte Werke*, 2: 153.

104. Kepler, *Gesammelte Werke*, 2: 151–52; Lindberg, *Theories*, 202–8; Crombie, "Expectation, Modeling, and Assent," I: 605–8; II: 96–103.

105. *Letters and Other Documents*, 1: 207, 163–64.

CHAPTER 5. CAMERAS THAT DO

1. Rhodius, *Optica*, fol. A4v, 190, 373, and *De Crepusculis*, 53–57; Galilei and Scheiner, *On Sunspots*, 30–34.

2. Rhodius, *Optica*, 160, 164, 171–72.

3. Rhodius, *Optica*, unnumbered page 20.

4. Daxecker, "Christoph Scheiner's Eye Studies"; "Further Studies by Christoph Scheiner"; *The Physicist*, 119–32; Dear, *Discipline & Experience*, 53–62, 66–67, 102–3. On Wotton's habit of intercepting Jesuit letters, see Wotton, *Life and Letters*, 1: 65–66, 345, 351, 352, 357; 2: 147–48.

5. Wotton, *Life and Letters of Sir Henry Wotton*, 1: 410. Sir Dudley Carleton described Ferdinand as "a stirring Prince and ruled by the Jesuits"; Doncaster called him "a silly Jesuited soule"; see *Letters and Other Documents*, 1: 93, 106. On the influence of his Jesuit confessor on Ferdinand from 1624 to 1637, see Bireley, *Religion and Politics*.

6. Scheiner, *Briefe*; Gorman, "A Matter of Faith?" 307–13; Daxecker, *The Physicist*, 65–68.

7. Scheiner, *Briefe*, 59, 63.

8. Daxecker, *The Physicist*, 19–21, 64–65, 74–80.

9. Daxecker, *The Physicist*, 74–91.

10. On Hans Hauer's landscape, now lost, and the discussion of his *camera obscura* by Daniel Schwenter, professor at the University of Altdorf, in his *Deliciae Physico-mathematicae* of 1636, see Delsaute, "The *Camera Obscura*."

11. Kepler, *Gesammelte Werke*, 17: 149; Alpers, *The Art of Describing*, 50, fig.23, and 246 n 51; Daxecker, "Christoph Scheiner's Eye Studies," 29, and *The Physicist*, 13, 15.

12. Scheiner, *Oculus*, 137. For the "Apellean paintbrush" or helioscope, see Scheiner, *Rosa Ursina*, 9–11, 14, 23, 35, 40–42, 53–54, 59, 61, 65.

13. Scheiner, *Oculus*, 162; *Rosa Ursina*, 106–23.

14. Daxecker, *The Physicist*, 17; Scheiner, *Oculus*, 254.

15. Parker, *Thirty Years' War*, 42–55; Bireley, *The Jesuits and the Thirty Years War*, 33–35.

16. Scheiner, *Oculus*, front matter.

17. Matthew 24: 2, 4–8, KJV.

18. Scheiner, *Oculus*, 125, 18.

19. Ashworth, "Divine Reflections," 184.

20. Ashworth, "Divine Reflections," 184, 192 n. 30.

21. Galilei, *Sidereus Nuncius*, 43; Isabelle Pantin, "Galilée, la lune, et les Jésuites."

22. Van Helden, "The Telescope in the Seventeenth Century," 40–43.

23. Polišenský, *The Thirty Years War*, 105–7.

24. Daxecker, *The Physicist*, 76.

25. Palmitessa, "The Prague Uprising of 1611."

26. Proverbs 1: 10–19, KJV.

27. *Letters and Other Documents*, 2: 1.

28. *Oculus*, 120; Lindberg, *Theories*, 190–93; Crombie, "Expectation, Modeling and Assent"; Pantin, "*Simulachrum, species, forma, imago*," 263–67.

29. Kepler, *Gesammelte Werke*, 2: 150–53, 183–87.

30. Lindberg, *Theories*, 174, 192; Crombie, "Expectation, Modeling and Assent," II: 95; Smolka, "The Scientific Revolution in Bohemia," 219–20; Caspar, *Kepler*, 105–7, 121, 166, 257.

31. *Oculus*, 194.

32. *Oculus*, fol. A.

33. *Oculus*, 18; Smolka, "The Scientific Revolution in Bohemia," 220. The *Anatomiæ Pragæ anno MDC abs se solenniter administratæ historia* was published in Wittenberg by Samuel Selfisch in 1601. On autopsy see Turner, "The Anatomy Lesson"; Cunningham, *The Anatomical Renaissance*, 212–68; Park, "The Life of the Corpse."

34. Gui, *I Gesuiti*, 206; Polišenský, *The Thirty Years War*, 105.

35. Polišenský, *The Thirty Years War*, 98–100.

36. Polišenský, "Tolerance and Intolerance," 10; Gui, *I Gesuiti*, 345 n. 81, following Polišenský, *The Thirty Years War*, 107.

37. Gui, *I Gesuiti*, 343–45.

38. *Courant out of Italy, Germany, &c.* Jessen's name is not on this original list of thirty imprisoned men.

39. *Courant Newes out of Italy, Germany, Bohemia, Poland, &c.*

40. *Corante, or, Nevves from Italy.* For an etching of the execution made five years after the event and published in Frankfurt, see Fučíková, *Rudolf II and Prague*, 653.

41. Gui, *I Gesuiti e la rivoluzione boema*, 344.

42. Rhodius, *Optica*, 162.

43. Baillet, *La Vie*, 53.

44. Descartes, *Œuvres*, 10: 162.

45. Descartes, *Œuvres*, 6: 11.

46. Baillet, *La Vie*, 58, 59.

47. Baillet, *La Vie*, 58.

48. Riezler, "Kriegstagebücher," 115, 151.

49. Descartes, *Œuvres*, 6: 13.

50. Dear, *Discipline*, 53–62, 66–67, 102–3.

51. Descartes, *Œuvres*, 6: 4–5.

52. Descartes, *Œuvres*, 6: 5.

53. Scheiner, *Oculus*, 126, 137, 178.

54. Descartes, *Œuvres*, 10: 216. See Rodis-Lewis, "Machineries," 464–65. For related meditations, see Beeckman, *Journal*, 2: 12.

55. Descartes, *Œuvres*, 6: 106, 361–62, 730, 732; 10: 541–42; Descartes, *Œuvres*, 1: 23, 245, 248, 250, 331; Beeckman, *Journal*, 3: 150–53, 157–58; Peiresc, *Lettres*, 4: 274–75, 353–54, 425; 5: 419–21.

56. Descartes, *Œuvres*, 1: 270–71, 281–83.

57. Descartes, *Œuvres*, 1: 22; see further Schuster and Brody, "Descartes and Sunspots," 31–36.

58. Baillet, *La Vie*, 64, 78. His association in this period with the mathematician and hermeticist Johann Faulhaber, who lived near Ulm, led an early biographer, Daniel Lipstorp, to identify his winter quarters with this area; see Hawlitschek, "Die Deutschlandreise." On this sojourn and the alleged Rosicrucian phase, see Shea, "Descartes and the Rosicrucians," 43.

59. Descartes, *Œuvres*, 6: 11.

60. Descartes, *Œuvres*, 6: 11.

61. Pierre de Ronsard, *Discours à Loys des Masures*, V: 363; Philippe Desportes, *Adieux à la Poloigne*; Agrippa d'Aubigné, *Pièces epigrammatiques*, 22, "Aux dégénérés Suisses."

62. Kepler, *Gesammelte Werke*, 16: 98, 304.

63. For references to the *hypocaustum*, see Riezler, "Kriegstagebücher," 116, 124, 127, 156, 157, 162, 171, 177, 178, 181, 182–83. The passage cited here is from pages 182–83.

64. Scheiner, *Oculus*, 125.

65. Locke, *An Essay*, 163.

66. Habermas, *The Structural Transformation*, 6, 27–28, 46, 55, 74–81, 87–88.

67. Descartes, *Œuvres*, 1: 203–4.

68. See, for example, *Dutch Corantos*, 36v; this list features parcels of various types of indigo, linen, cotton, aloe, coral and other stones, spikenard, borax, saltoeter, ambergris, and sailcloth.

69. Gorski, *The Disciplinary Revolution*, 39–77, especially 51, 63–64, 72–74.

70. On the importance of the Dutch economy—goods and information—to the development of the early modern natural philosophy, see Cook, *Matters of Exchange*, 42–81; on the *gemeente* or "common folk," see Harline, *Pamphlets, Printing, and Political Culture*, 18–22.

71. On children in Dutch art and society, see Schama, *The Embarrassment of Riches*, 480–516, especially 509–12.

72. Harline, *Pamphlets, Printing, and Political Culture*, 87–89.

73. Harline, *Pamphlets, Printing, and Political Culture*, 44.

74. Geyl, *History of the Dutch-Speaking Peoples*, 379–82.

75. Geyl, *History of the Dutch-Speaking Peoples*, 373–79; Israel, *The Dutch Republic*, 506–13.

76. Israel, *The Dutch Republic*, 513–14.

77. Te Brake, *Shaping History*, 173.

78. Balzac, *Lettres Diverses*, I: 147–48.

79. Israel, *The Dutch Republic*, 272–74, 483–84, 568; Parker, *The Thirty Years' War*, 183–86; Gorski, *The Disciplinary Revolution*, 72–74.

80. Thus during the siege of Breda in 1637, Constantijn Huygens asked that an account of Ambrogio Spìnola's siege of that city in 1629, Herman Hugo S.J.'s *Obsidio Bredana*, be retrieved from Frederick Henry's library, that the Dutch might study the maps and plans of the fortress they intended to take back. See Huygens, *Briefwisseling*, 2: 251.

81. Habermas, *The Structural Transformation*, 3–11, 21–22, 31–32, 38–39. On the apathy of both the *popolo* and the elite for public spectacles organized by rulers, see Cashman, "The problem of audience."

82. Descartes, *Œuvres*, 6: 1–2.

83. Habermas, *The Structural Transformation*, 91–102.

84. Descartes, *Œuvres*, 6: 9–10.

85. Descartes, *Œuvres*, 6: 24.

86. Descartes, *Œuvres*, 6: 16.

87. Israel, *The Dutch Republic*, 526–27.

88. Descartes, *Œuvres*, 6: 61, 62, 63, 65; see also Cook, *Matters of Exchange*, 171–74, 226–66.

89. Descartes, *Œuvres*, 6: 66–67.

90. Israel, *The Dutch Republic*, 527–34; Schama, *The Embarrassment of Riches*, 252–53, 350–66.

91. *The German history continued*, fol. 1.

92. *The Swedish Intelligencer*, fol. 3v.

93. Descartes, *Œuvres*, 6: 112, 115.

94. Descartes, *Œuvres*, 6: 115–17, 121–29; Scheiner, *Oculus*, 125–28.

95. Descartes, *Œuvres*, 6: 112–13; see also Crombie, "Expectation, Modeling and Assent" (II) 109–15.

96. Mitchell, *Iconology*, 53–74; Elkins, *The Domain of Images*, 68–91.

97. Kepler, *Gesammelte Werke*, 4: 332.

98. Descartes, *Œuvres*, 6: 599.

99. Huygens, *Fragment*, 24, 35, 57; *A Selection of the Poems*, 11–16, 196–200, 201–15; *Briefwisseling*, 1: 260, 262, 271, 277; Bachrach, "Sir Constantyn Huygens"; *Sir Constantine Huygens and Britain*; Sellin, "John Donne and the Huygens Family."

100. Bachrach, "Sir Constantyn Huygens," 122–23.

101. Bachrach, "Sir Constantyn Huygens," 123; Huygens, *Briefwisseling*, 1: 62, 78. Huygens also had a manuscript copy of Jonson's *Gypsies Metamorphos'd*; see Bachrach, "Sir Constantyn Huygens," 126–27, and Huygens, *A Selection*, 214–15.

102. Huygens, *Briefwisseling*, 1: 72, 73, 74.

103. Huygens, *Briefwisseling*, 2: 119–20, 122–23; on Hortensius, see Van Berkel, *A History of Science*, 36, 52, 562.

104. Huygens, *Briefwisseling*, 2: 418.

105. See Steadman, *Vermeer's Camera*.

106. Huygens, *Briefwisseling*, 2: 339.

107. *Voor-looper*, "forerunner," was a sometime title for popular pamphlets; see *Bibliotheek*, 1: 94; *Catalogus*, 1: 344.

108. Huygens, *Gedichten*, 3: 64; *A Selection*, 96–99.

109. Huygens, *Gedichten*, 3: 64 n. 48.

110. For the *Pasquil van Ja ende Neen* of 1610, see *Bibliotheek*, 1: 112; *Catalogus*, 1: 344.

111. Huygens, *Gedichten*, 3: 64–65; *A Selection*, 98–99.

112. Huygens, *Briefwisseling*, 2: 205, 207; on 249, see the expression *on faisoit courir le bruit en Anvers*.

113. Moureau, *Répertoire des nouvelles*, 6–16; Miller, *Peiresc's Europe*.

114. Solomon, *Public Welfare*, 106–7.

115. Peiresc, *Lettres*, 6: 183; 1: 302–3; 4: 166; Gassendi, *Vie*, 274–75.

116. Peiresc, *Lettres*, 1: 302–3.

117. Parker, *The Thirty Years' War*, 129–37; Kettering, *Judicial Politics*, 51–68. On Renaudot see especially Solomon, *Public Welfare*; Blome, "Offices of Intelligence," 212–14.

118. Solomon, *Public Welfare*, 100–161; Feyel, *L'annonce et la nouvelle*, 132–37, 172–78, 202–3; Moureau, *Répertoire des nouvelles*, xi.

119. Peiresc, *Lettres*, 2: 284.

120. Dahl, Pettibon, and Boulet, *Les débuts*, 19–64; Solomon, *Public Welfare*, 112–18; Feyel, *L'annonce et la nouvelle*, 45–46, 137–49.

121. Peiresc, *Lettres*, 2: 369–70, 371, 394, 400; Feyel, *L'annonce et la nouvelle*, 241–46.

122. Peiresc, *Lettres*, 2: 402, 408–9, 416–17, 425, 432–33, 450–51, 452, 458–59, 465–66, 467–68, 480–81, 490, 493–94. Peiresc was especially aware of the fact that two important outsiders, the marquis de Vitry, the unpopular and ineffective governor of Provence from 1631 until his recall in 1637, and Charles le Roy de la Poterie, a royally appointed *intendant* in Provence from 1630 until 1633, received the *Gazette* before him. On these men, see Kettering, *Judicial Politics*, 81–82, 84–85, 111–12, 118–27.

123. Peiresc, *Lettres*, 2: 468.

124. Peiresc, *Lettres*, 2: 498.

125. Peiresc, *Lettres*, 2: 498–99.

126. Peiresc, *Lettres*, 2: 596.

127. Peiresc, *Lettres*, 2: 517.

128. Peiresc, *Lettres*, 2: 595–96.

129. Peiresc, *Lettres*, 2: 534, 596, 634–35.

130. Peiresc, *Lettres*, 2: 640, 648–49, 650.

131. Peiresc, *Lettres*, 2: 670–71.

132. Though there were no extant French-language corantos emerging from Amsterdam in this period, Peiresc might have been referring to either Broer Janszoon's *Nouvelles de divers quartiers*, or Jan van Hilten's *Courant d'Italie et d'Almaigne*; see Dahl, Pettibon, and Boulet, *Les Débuts de la presse française*, 14–15. On Renaudot's dependence upon other foreign newsletters, see Feyel, *L'annonce et la nouvelle*, 187–88.

133. Peiresc, *Lettres*, 3: 13.

134. Peiresc, *Lettres*, 3: 15; see also Tolbert, "Censorship and Retraction."

135. Peiresc, *Lettres*, 3: 24, 37, 58–59.

136. Peiresc, *Lettres*, 3: 66.

137. Feyel, *L'annonce et la nouvelle*, 179–81, 250–52.

138. Peiresc's annual income is estimated to have been in this range; see Kettering, *Judicial Politics*, 231–32, 235–36; Feyel, *L'annonce et la nouvelle*, 250–53.

139. Peiresc, *Lettres*, 3: 66, 72–73.

140. Peiresc, *Lettres*, 3: 80–81, 101, 113, 120, 130–31, 274.

141. Miller, *Peiresc's Europe*, 48–75.

142. Gassendi, *Vie*, 225–30; Peiresc, *Lettres*, 1: 79–80.

143. Peiresc, *Lettres*, 2: 498; 3: 81.

144. Peiresc, *Lettres*, 3: 165.

145. Gassendi, *Vie*, 224–25.

146. Ambard, *Aix Romaine*, 193–228; Kettering, *Judicial Politics*, 22, 26–29, 33–37, 40–41, 220–21; Bohanan, *Old and New Nobility*.

147. Peiresc, *Lettres*, 3: 92, 93, 95–96; Gassendi, *Vie*, 224; De Caro, "Giorgio Bolognetti"; *Recueil des instructions*, 34–35, 95–96, 180–81; Headley, *Tommaso Campanella*, 126–27.

148. Peiresc, *Lettres*, 3: 157–58; Miller, *Peiresc's Europe*, 90–96.

149. Peiresc, *Lettres*, 3: 81. These were probably chromophanes, or pigments in oil droplets produced in the cones of the eye. The droplets vary in color: they may be clear, purplish red, orange yellow, or greenish yellow. In no circumstances are they brown or

black, but because adjacent colors could easily have been mixed on the hand of the experimenter, and observed only outside the dark room, it is conceivable that they would have taken on an ink-like appearance. See Kühne, "Chemical Processes," 1294–96, 1301.

150. Gassendi, *Vie*, 228–29.

151. Feyel, *L'annonce et la nouvelle*, 152.

CHAPTER 6. RAPID TRANSPORT

1. For helpful overviews, see Arblaster, "Posts, Newsletters, Newspapers"; McCusker, "The Demise of Distance."

2. Devreese and Vanden Berghe, *"Magic Is No Magic."*

3. On Jonson's attention to the news, particularly within the context of his plays and masques, see especially Shohet, *Reading Masques*, 111–15, 178–88; Chartier, *Inscription & Erasure*, 46–62; Nevitt, "Ben Jonson and the Serial Publication"; Sanders, *Ben Jonson's Theatrical Republics,* 123–143; Muggli, "Ben Jonson and the Business"; Sellin, "The Politics of Ben Jonson's *Newes*"; Lanier, "The Prison-House"; Pearl, "Sounding to Present Occasions."

4. *The New Inn*, II: v: 97–113, in Ben Jonson, *Works*, 6: 433–34; on Stevin, see Dijksterhuis, *Simon Stevin*; Devreese and Vanden Berghe, *'Magic Is No Magic.'*

5. On Drebbel, see Tierie, *Cornelis Drebbel*; Colie, "Cornelis Drebbel and Salomon de Caus"; Harris, *The Two Netherlanders*; Grudin, "Rudolf II"; Van Berkel, "Cornelis Jacobusz Drebbel." On the automatic clavichord see Bacon, *Philosophical Studies*, xxvii, 60–61; on a similar figure operating in the Dutch context mid-seventeenth century, see Keblusek, "Keeping It Secret."

6. See "The New Motion," 1–2, in *Works*, 8: 62–63; *The Silent Woman*, V:iii: 62–63 in *Works*, 5: 258; *The Staple of Newes* III: ii: 105–7, in *Works*, 6: 331; *Newes from the New World Discover'd in the Moone*, 97, in *Works*, 7: 516; *The Staple of Newes*, III: ii: 59–62, in *Works*, 6: 326; *The New Inn* I: i: 29–31, in *Works*, 6: 408.

7. See Tierie, *Cornelis Drebbel*, 18, 27, 96 nn. 3, 4; Huygens, *Fragment*, 116, and *Gedichten*, 8: 203; Rubens, *Correspondence*, 5: 153. See further Alpers, *The Art of Describing*, 1–25.

8. Great Britain, Public Record Office, *Calendar of State Papers Domestic Series*, *1611–1618*, 130.

9. See Knowles, "Jonson's *Entertainment*"; Baker, "'The Allegory of a China Shop'"; Ioppolo, *Dramatists*, 159–69; Reeves, "Complete Inventions," 179–82; Shohet, *Reading Masques*, 94.

10. Knowles, "Jonson's *Entertainment*," 134, 137; Shohet, *Reading Masques*, 175.

11. Great Britain, Historical Manuscripts Commission, *Report on the Manuscripts of Lord de l'Isle and Dudley*, 4: 71.

12. Knowles, "Jonson's *Entertainment*," 116; Ioppolo, *Dramatists*, 161–69.

13. For background see Orgel, *The Jonsonian Masque*; Lindley, "Introduction"; Palmer, "Court and Country."

14. On the importance of the news industry to this masque, see Pearl, "Sounding to Present Occasions"; Sellin, "The Politics of Ben Jonson's *Newes*"; Butler, "Jonson's *News*." On the extent and nature of censorship, see Lambert, "State Control," and "Coranto Printing"; Leth, "A Protestant Public Sphere"; Frearson, "London Corantos"; Fox, "Rumour, News"; Thompson, "Licensing the Press." On diffusion, see Cranfield, *The Press and Society*, 1–30; Levy, "How Information Spread"; Clark, *The Elizabethan Pamphleteers*, 86–120; Cust, "News and Politics"; Watt, "Publisher, Pedlar"; Mousley, "Self, State"; Frearson, "The Distribution and Readership."

15. On manuscript newsletters, see Love, *Scribal Publication*, 9–22, 129–33, 191–95; Richardson, *Manuscript Culture*, 57–58, 114–26, 157–60.

16. Lambert, "Coranto Printing," 5–6.

17. Chamberlain, *Letters*, 2: 185.

18. Powell, *John Pory*, microfiche supplement, 51.

19. Corbett, *Poems*, 64.

20. James VI and I, *Poems*, 2: 256.

21. James VI and I, *Poems*, 2: 172–73.

22. James VI and I, *Poems*, 2: 174–75, 256.

23. Knowles, "'Songs of Baser Alloy.'"

24. Seccombe, "Matthew, Sir Tobie."

25. Mathews, *A Collection of Letters*, 195.

26. Bradley and Adams, *The Jonson Allusion-Book*, 100.

27. Mathews, *A Collection of Letters*, 190.

28. Jonson, *Works*, 7: 520–21. In "The Prologue for the Stage" of *The Staple of Newes* a related remark appears when the poet admonishes his audience not to exchange trivial gossip during the performance: "What is it to [the poet's] scene to know / How many coaches in Hyde Park did show / Last spring?" See Jonson, *Works*, 6: 282.

29. On the "China Howse" see Baker, "'The Allegory of a China Shop.'"

30. Wotton, *Life and Letters*, 1: 486; Galilei, *Sidereus Nuncius*, 53–57.

31. Jonson, *Works*, 7: 522. Emphasis added.

32. Jonson, *Works*, 7: 523. Emphasis added.

33. Jonson, *Works*, 7: 513.

34. Galilei, *Sidereus Nuncius*, 29.

35. For background, see Fox, "Rumour, News."

36. Fox, "Rumour, News," 599.

37. Habermas, *The Structural Transformation*, 5–12.

38. Sarpi, *Lettere ai protestanti*, 1: 182.

39. Lach, *Asia in the Making*, 1:2: 741, 770–71, 819; 2:3: 402–3, plates 32–34; Schama, *The Embarrassment of Riches*, 319–20.

40. Forbes, "The Sailing Chariot."

41. Eijffinger, "Zin en beeld."

42. Hegenitius, *Itinerarium*, 83–84.

43. Forbes, "The Sailing Chariot," 5–6; Eijffinger, "Zin en beeld," 259–60.

44. On the connection between vehicular speed and the spectator's view, see Virilio, *Speed and Politics.*

45. Draud, *Bibliotheca Classica*, 2: 965; Grotius, *Poemata*, 385.

46. Grotius, *Poemata*, 388; *Klioos Kraam*, 177. The Dutch version differs slightly: "Jupiter, make a place in Heaven's fire for me / That I see up by the second and third Wagons." The Great and Small Wagons correspond to the Big and Little Dippers.

47. Gassendi, *Vita Peireskii*, in *Opera Omnia*, 5: 265.

48. Walchius, *Decas fabularum*, 247.

49. Howell, *Familiar letters*, 83; on Howell's relationship to Jonson, see *Familiar letters*, 214, 221–22, 258–60, 266–67.

50. Campanella, *Lettere*, 28, 161, 411.

51. Huygens, *Gedichten*, 1: 42.

52. Milton, *Paradise Lost*, 3: 431–41.

53. McPherson, "Ben Jonson Meets Daniel Heinsius"; Teague, "Jonson's Drunken Escapade"; Great Britain, Historical Manuscripts Commission, *Papers of William Trumbull the Elder*, 4: 54, 59, 81; Forbes, "The Sailing Chariot," 6.

54. The phrase is Howell's; see *Familiar Letters*, 9.

55. Jonson, *Works*, 7: 357.

56. Weber, "The Early German Newspaper," 79.

57. On the elasticity of the present, see Woolf, "News, History, and the Construction of the Present."

58. Cited in Infelise, *Prima dei giornali*, 146.

59. Cited and translated, with slight modifications, in Infelise, "News Networks," 66.

60. Vannozzi, *Della Suppellettile* (1609), 361.

61. Galilei, *Opere*, 3(1): 293; Comenius, *Labyrinth of the World*, 209.

62. Lüsebrink, *Les Lectures du peuple.*

63. Blakey, "À l'Éditeur," 459–60.

64. Miller, *Peiresc's Europe*, 149–50; Sherman, *Telling Time*, 246–47; Van Strien, "Holland and the Dutch"; Mayoux, "Variations on the Time Sense."

65. Sterne, *Tristram Shandy*, 83.

66. Sterne, *Tristram Shandy*, 84.

67. Sterne, *Tristram Shandy*, 84–101; Forbes, *The Sailing Chariot*, 3.

68. De Quincey, *The English Mail-Coach*, 191.

69. De Quincey, *Coach*, 193, 198.

70. De Quincey, *Coach*, 193, 194–95, 240.

71. De Quincey, *Coach*, 205, 225, 229, 193, 208, 205.

72. De Quincey, *Coach*, 207, 192; "Revolution in Europe," 842; Wikoff, "Napoleon Louis Bonaparte," 254; "Correspondence."

73. "Original Correspondence from France."

74. De Quincey, *Coach*, 211, 213.

75. On the problem of time in wartime, see Favret, *War at a Distance*, 1–97.

76. De Quincey, *Coach*, 194, 196, 199, 200, 201.

77. De Quincey, *Coach*, 217–19, 214–16.

78. De Quincey, *Coach*, 197–98.

79. De Quincey, *Coach*, 226–27, 228, 231–32.

80. De Quincey, *Coach*, 232, 235, 229–30.

81. De Quincey, *Coach*, 232.

82. De Quincey, *Coach*, 237–43.

83. In a recent discussion of the main epochs of the Republic of Letters, Peter Burke distinguishes between the long era defined by horse-drawn travel, 1500–1800, and the progressively shorter intervals of steamships and locomotives, 1850–1950, jets, 1950–1990, and the Internet, 1990–. *The English Mail-Coach*, on the cusp of the steam era, is at once elegiac in content and prescient in its concerns. See Burke, "The Republic of Letters as a Communication System."

BIBLIOGRAPHY

PRIMARY SOURCES

Alamo Varienti, Baldassar. *Discorso del Signor Alamo.* In Tacitus, *Opere*, tr. Girolamo Canini. Unnumbered fols. 31–36.

Albertotti, Giuseppe, ed. "Manoscritto francese del secolo decimosettimo riguardante l'uso degli occhiali." *Memorie della regia accademia di scienze, lettere, ed arti in Modena* 2nd ser. 9 (1893): 1–124.

Album academiae pragensis Societatis Iesu 1573–1617 (1565–1624). Ed. Miroslav Truc. Prague: Universita Karlova, 1968.

Allegri, Alessandro. *Seconda parte delle rime piacevoli.* Verona: Bartolamio Merlo dalle Donne, 1607.

Anti-Coton. London: Blackfriars, 1611.

Arnauld, Antoine. *Arrainment of the whole society of Iesuits in France.* London: Charles Yetweirt Esquire, 1594.

Bacon, Francis. *Novum Organum.* Ed. and tr. Peter Urbach and John Gibson. Chicago: Open Court, 1994.

———. *Philosophical Studies c. 1611–c. 1619.* Ed. and tr. Graham Rees and Michael Edwards. Oxford: Clarendon, 1996.

———. *The Physical and Metaphysical Works of Lord Bacon.* Ed. Joseph Devey. London: Henry G. Bohn, 1856.

Baillet, Adrien. *La Vie de Monsieur Des-Cartes.* 2 vols. New York: Garland, 1987.

Balzac, Jean Louis Guez de. *Les entretiens.* Ed. B. Beugnot. 2 vols. Paris: Marcel Didier, 1972.

———. *The Letters of Monsieur de Balzac.* London: by I. N. for W. Edmonds, and I. Colby, 1638.

———. *Lettres Diverses.* Paris: Estienne Loyson, 1663.

———. *A New Collection of Epistles of Monsieur de Balzac.* Oxford: Francis Bowman, 1639.

Banister, John. *The Historie of Man.* London: John Day, 1578.

Beeckman, Isaac. *Journal . . . de 1604 à 1634.* Ed. Cornelis de Waard. 4 vols. The Hague: Nijhoff, 1939–1945.

Bentivoglio, Guido, Cardinal. *Lettere del Cardinal Guido Bentivoglio.* Ed. G. Biagioli. Paris: Didot, 1807.

————. *Memorie e lettere*. Ed. Constantino Panigada. Bari: Giuseppe Laterza, 1934.

Bernegger, Matthias. *Observationes historico-politicae xxviii*. Tübingen: J. H. Reiss, 1666.

Biddulph, William. *Travels of certain Englishmen*. London: Th. Haveland for W. Apsley, 1609.

Biringuccio, Vanoccio. *De la pirotechnia libri X*. Venice: Venturino Rossinello, 1540.

Blakey, William. "À l'Éditeur." *Journal des Scavans* 42, 13 (December 1781): 458–70.

Boccalini, Traiano. *Comentarii sopra Cornelio Tacito; Considerationi sopra la vita di Giulio Agricola*. Cosmopoli: Giovann Battista della Piazza, 1677.

————. *Ragguagli di Parnaso*. Ed. Giuseppe Rua and Luigi Firpo. 3 vols. Bari: Giuseppe Laterza e figli, 1934–1948.

[Borromeo, Federico]. Giliola Barbero, Massimo Bucciantini, and Michele Camerota. "Uno scritto inedito di Federico Borromeo: *L'Occhiale Celeste*." *Galilaeana* 4 (2007): 309–41.

Bouhours, Dominique. *Les Entretiens d'Ariste et d'Eugène*. Paris: S. Mabre-Cramoisy, 1671.

Boyle, Robert. *Letters and Papers*. Ed. Michael Hunter, Paul Kesaris, and James P. Hoy. 16 reels. Bethesda, Md.: University Publications of America, 1990.

Brahe, Tycho. *Carteggio inedito di Ticone Brahe . . . con Giovanni Antonio Magini*. Ed. Antonio Favaro. Bologna: Nicola Zanichelli, 1886.

Brathwaite, Richard. *A New Spring*. London: G. Eld, 1619.

Bruni, Teofilo. *Armonia Astronomica & Geometrica*. Venice: Giovanni e Varisco Varischi, 1622.

Buonarroti, Michelangelo il giovane. *La Fiera*. Ed. Uberto Limentani. Florence: Leo S. Olschki, 1984.

————. *Satira prima*. In *Opere varie*. Ed. Pietro Fanfani. Florence: Le Monnier, 1863. 219–27.

Cambilhom, Johannes. *A Discoverie of the most secret and subtile practises of the Iesuites*. London: Robert Boulton, 1610.

Camden, William. *Epistolae*. London: Richard Chiswell, 1691.

Campanella, Tommaso. *Lettere*. Ed. Vincenzo Spampanato. Bari: Giuseppe Laterza & Figli, 1927.

Cancellieri, Francesco Girolamo. *Dissertazione intorno agli uomini dotati di gran memoria*. Rome: Francesco Bovrlie, 1815.

Caporali, Cesare. *Opere Poetiche*. Macerata: Pietro Salvioni, 1614.

Carleton, Dudley. *Dudley Carleton to John Chamberlain, 1603–1624*. In *Jacobean Letters*, ed. Maurice Lee. New Brunswick, N.J.: Rutgers University Press, 1972.

Casanova, Giacomo. *The Story of My Life*. Tr. Willard R. Trask. New York: Everyman's Library, 2006.

Catholic Church. Pope. *Bullarum, diplomatum et privilegiorum sanctorum romanorum pontificum*. Ed. Francesco Cardinal Gaude. 24 vols. Turin: Franco and Enrico Dalmazzo, 1857–1872.

Chamberlain, John. *Letters*. Ed. Norman McClure. 2 vols. Philadelphia: APS, 1939.

Chateaubriand, François René de. *Analyse raisonnée de l'histoire de France*. Paris: Firmin Didot, 1845.

Chiabrera, Gabriello. *Canzonette, rime varie, dialoghi.* Ed. Luigi Neri. Turin: UTET, 1952.

La Cholere de Mathurine, contre les difformez Reformateurs de la France. Paris: Jean Milot, 1616.

Cimber, M. L. and F. Danjou, eds. *Archives curieuses de l'histoire de France.* 27 vols. Paris: Beauvais, 1834–1840.

Colombo, Realdo. *De re anatomica libri XV.* Venice: Nicola Bevilacqua, 1559.

Comenius, Johann Amos. *The Labyrinth of the World and the Paradise of the Heart.* Ed. and tr. Count [Franz] Lützow. London: Swan Sonnenschein, 1901.

Contribuiti alla storia dell'Accademia dei Lincei. Ed. Giuseppe Gabrieli. 2 vols. Rome: Accademia Nazionale dei Lincei, 1989.

Corante, or, Nevves from Italy, Germanie, Hungarie, Poland, Bohemia and France. July 20, 1621. Amsterdam: Broer Ionson, 1621.

Corbett, Richard. *Poems.* Ed. J. A. W. Bennett and H. R. Trevor-Roper. Oxford: Clarendon, 1955.

"Correspondence." *Living Age* 19, 240 (December 1848): 478.

Courant out of Italy, Germany, &c. March 31, 1621. Amsterdam: George Veseler, 1621.

Courant Newes out of Italy, Germany, Bohemia, Poland, &c. July 15, 1621. Amsterdam: George Veseler, 1621.

Crooke, Helkiah. *ΜΙΚΡΟΚΟΣΜΟΓΡΑΦΙΑ: A Description of the Body of Man.* London: William Iaggard, 1615.

Crosse, William. *Belgiaes Troubles, and Triumphs.* London: Augustine Mathewes and John Norton, 1625.

D'Aubigné, Agrippa. *Confession catholique du sieur de Sancy.* Ed. E. Réaume and F. de Caussade. Geneva: Slatkine, 1967.

Daniele, Fidele S.J. *Trattato della Divina Providenza.* Milan: Giovanni Battista Bidelli, 1615.

De Quincey, Thomas. *The English Mail-Coach.* In *Confessions of an English Opium-Eater and Other Writings.* Ed. Barry Milligan. London: Penguin, 2003.

Della Porta, Giambattista. *Natural Magick.* London: Thomas Young and Samuel Speed, 1658.

Descartes, René. *Œuvres.* Ed. Charles Adam and Paul Tannery. 13 vols. Paris: Cerf, 1897–1913.

Di Banco, Lorenzo. *Bizzarrie politiche.* Franeker: Giovanni d'Arcerio, 1658.

"Diario di cose romane degli anni 1614, 1615, 1616." Ed. F. Cerasoli. *Studi e documenti di storia e diritto* 15 (1894): 263–95.

Documenti per la storia della Accademia dei Lincei nei manoscritti galileiani. Ed. Antonio Favaro. Rome: Tipografia delle scienze matematiche e fisiche, 1888.

Donne, John. *Ignatius His Conclave.* London: Nicholas Okes, 1611.

Draud, Georg. *Bibliotheca Classica.* Frankfurt: Nicholas Hoffmann, 1611.

Du Laurens, André. *A Discourse of the Preservation of the Sight.* Tr. Richard Surphlet. London: Felix Kingston, 1599.

Duplessis-Mornay, Philippe. *Mémoires et correspondence.* 12 vols. Paris: Treuttel et Würtz, 1824–1825.

Dutch Corantos 1618–1650. Ed. Folke Dahl. The Hague: Kononklijke Bibliotheek, 1946.

"Les estranges tromperies de quelques Charlatans nouvellement arrivez à Paris." In *Variétés historiques et littéraires*, ed. Edouard Fournier. 3: 273–82.

Exemplar litterarum Bononiæ datarum. In Petrus de Wangen, ed. *Physiognomonis Iesuitica variis opusculis.* Lyon: n.p., 1610. 257–88.

Farinacci, Prospero. *Responsorum criminalium liber primus.* Venice: Giorgio Varisco, 1606.

Fletcher, Phineas. *Locustæ, vel Pietas Iesuitica*, followed by *The Locusts, or the Apollyonists.* Cambridge: Bucke, 1627.

Florence, National Central Library, Fondo Galileiano. Ms. 70.

Franzesi, Matteo. "Sopra le nuove." In *Rime burlesche.* Ed. Pietro Fanfani. Florence: Le Monnier, 1856. 205–8.

Fritsch, Ahasver. *Discursus de Novellarum quas vocant Neue Zeitunge hodierno usu et abusu.* Jena: Bielckian, 1676.

Froissart, Jean. *Œuvres.* Ed. Baron Kervyn de Lettenhove. 25 vols. in 26. Brussels: V. Devaux, 1867–1877.

Fuller, Thomas. *The Worthies of England.* Ed. John Freeman. London: Allen & Unwin, 1952.

Galilei, Galileo. *Opere.* Ed. Antonio Favaro. 20 vols. Florence: Giunti Barbèra, 1969.

———. *Sidereus Nuncius or the Sidereal Messenger.* Tr. Albert Van Helden. Chicago: University of Chicago Press, 1989.

Galilei, Galileo, and Christoph Scheiner S.J. *On Sunspots.* Tr. Eileen Reeves and Albert Van Helden. Chicago: University of Chicago Press, 2010.

[Galindus, Fortunius, a pseudonym for Caspar Scioppius]. "A Discourse of the Reasons Why the Jesuits Are So Generally Hated." In *A Further Discovery of the Mystery of Jesuitisme.* London: T. Dring, 1658.

[Garasse, François S.J.]. *Le Rabelais reformé par les ministres.* Brussels: Christophle Girard, 1619.

Gassendi, Pierre. *Vie de l'illustre Nicolas-Claude Fabri de Peiresc, Conseiller au Parlement d'Aix.* Tr. Roger Lasalle. Belin: Paris, 1992.

———. *Vita Peireskii.* In *Opera Omnia.* 6 vols. Stuttgart: F. Fromann, 1964.

Gazzetta di Milano. Nos. 225, 236, 243. August 12, 23, and 30, 1816.

Gentili, Alberico. *De Legationibus libri tres.* Tr. Gordon J. Laing. 2 vols. Oxford: Oxford University Press, 1924.

The German history continued. The seventh part. London: By Thomas Harper for Nathaniel Butter and Nicholas Bourne, 1635.

Ghilini, Girolamo. *Teatro d'huomini illustri.* Venice: Li Guerigli, 1647.

Gilbert, William. *De mundo sublunari philosophia nova.* Amsterdam: William Boswell, 1651.

Giovanetti, Marcello. "Dello Specchio." In *Marcello Giovanetti (1598–1631): A Poet of the Early Roman Baroque*, tr. William Crelly. Lewiston, N.Y.: Edwin Mellen, 1990.

Great Britain, Historical Manuscripts Commission. *Papers of William Trumbull the Elder, 1605–1610.* Ed. A. W. Downshire et al. 5 vols. in 6. London: HMSO, 1924–1988.

———. *Report on the Manuscripts of Lord de l'Isle and Dudley.* Ed. William A. Shaw and G. Dyfnallt Owen. 6 vols. London: HMSO, 1925–1995.

Great Britain, Public Record Office. *Acts of the Privy Council of England, 1619–1621.* London: HMSO, 1930.

———. *Calendar of State Papers and Manuscripts, Relating to English Affairs Existing in the Archives and Collections of Venice.* Ed. Rawdon Brown. 38 vols. London: Longman Green, 1864–1947.

———. *Calendar of State Papers, Domestic Series, 1611–1618; 1619–1623.* Ed. Mary Anne Everett Green. London: Longman, 1858.

Gretser, Jacob S.J. *Furiae praedicantium Augustae Vindelicorum.* In *Opera Omnia.* 17 vols. Ratisbon: Joannis Conradi Peez, et Felicis Bader, 1734–1741. 11: 823–40.

———. *Praedicantium Augustanorum . . . repetitae furiae.* In *Opera Omnia.* 11: 840–68.

Gretser, *Relatio Cambilhonica castigata.* In *Opera Omnia* 11: 789–815.

———. *Relatio . . . nunc in gratiam praedicantium Lutheranorum.* In *Opera Omnia.* 11: 782–822.

———. *Hæreticus vespertilio.* In *Opera Omnia.* 11: 869–89.

Grillo, Angelo. *Delle Lettere .* Ed. Pietro Petracci. 3 vols. Venice: Deuchino, 1612–1616.

Grimmelshausen, Johann Jacob. *Simplicissimus.* Tr. S. Goodrich. Cambridge: Dedalus, 1995.

Grotius, Hugo. *Klioos Kraam vol verscheiden gedichten.* Leeuwarden: Henrik Rintjus, 1656.

———. *Poemata.* Leyden: Andries Clouck, 1617.

Guillemeau, Jacques. *A Worthy Treatise of the Eyes.* Tr. Anthony Hunton. London: Robert Waldegrave, 1587.

Hartlib, Samuel. *The Hartlib Papers.* Ed. Judith Crawford et al. Ann Arbor: University of Michigan Press, 1995.

Hegenitius, Gotfried. *Itinerarium Frisio-Hollandicum.* Leiden: Henrich Verbiest, 1667.

Henri IV, King of France. *Recueil des lettres missives de Henri IV.* Ed. Jules Berger de Xivrey. 9 vols. Paris: Imprimerie Impériale, 1843–1876.

Hernández, Francisco. *Rerum medicarum Novae Hispaniae Thesaurus.* Rome: Vitale Mascardi, 1649.

Hobbes, Thomas. *Correspondence.* Ed. Noel Malcolm. 2 vols. Oxford: Clarendon Press, 1994.

Hooft, Pieter Cornelisz. *De Briefwisseling.* Ed. H. W. van Tricht. 3 vols. Culemborg: Tjeenk Willink/Noorduijn, 1976.

Horky, Martin. *Brevissima Peregrinatio contra Nuncium Sidereum.* In Galilei. *Opere.* 3: 133–45.

Howell, James. *Familiar letters.* Aberdeen: F. Dougless and W. Murray, 1753.

———. *Lustra Ludovici.* London: Humphrey Moseley, 1646.

Hulsius, Levinus. *Dictionarium Teutsch Italienisch.* Frankfurt: Nicholas Hofmann, 1618.

Huygens, Constantijn. *De Briefwisseling.* Ed. J. A. Worp. 6 vols. The Hague: Nijhoff, 1911–1917.

———. *Fragment eener Autobiographie.* Ed. J. A. Worp. The Hague: Nijhoff, 1897.

———. *Gedichten.* Ed. J. A. Worp. 9 vols. Groningen: Wolters, 1892–1899.

———. *A Selection of the Poems of Sir Constantijn Huygens (1596–1687)*. Tr. Peter David-son and Adriaan van der Weel. Amsterdam: University of Amsterdam Press, 1996.

James VI and I, King of Scotland and England. *The Poems of King James VI of Scotland*. Ed. James Craigie. 2 vols. Edinburgh: Scottish Text Society, 1955–1958.

Jeannin, Pierre. *Négociations*. 5 vols. In Alexandre Petitot, *Collections des mémoires relatifs à l'histoire de France*. Vols. 11–15. Paris: Foucault, 1821–1822.

Jonson, Ben. *Works*. Ed. C. H. Herford, Evelyn Simpson, and Percy Simpson. 11 vols. Oxford: Clarendon, 1925–1952.

Juvenal and Persius. *Works*. Tr. G. G. Ramsay. Cambridge, Mass.: Harvard University Press, 1918.

Kepler, Johannes. *Discussion avec le messager céleste*. Tr. Isabelle Pantin. Paris: Belles Let-tres, 1993.

———. *Gesammelte Werke*. Ed. Max Caspar et al. 22 vols. to date. Munich: C. H. Beck'-sche, 1937–.

———. *Somnium*. Tr. Edward Rosen. Mineola, N.Y.: Dover, 2003.

L'Estoile, Pierre de. *Mémoires-Journaux, 1574–1611*. Ed. Robert Merle. 12 vols. Paris: Tal-landier, 1982.

Lagalla, Giulio Cesare. *De phoenomenis in orbe lunae*. Venice: Tomaso Baglioni, 1612.

Lalande, Joseph Jérôme le Français de. *L'art du cartonnier*. s.l.: s.n., 1762.

Lancellotti, Secondo. *L'Hoggidi*. Venice: Giovanni Guerigli, 1627.

Lapide, Cornelius à, S.J. *Commentaria in scripturam sacram*. 21 vols. Paris: Ludovicus Vives, 1866–1868.

Le Fevre de la Broderie, Antoine. *Ambassades de Monsieur de La Broderie en Angleterre*. 5 vols. N.P., 1750.

Lettere d'uomini illustri. Venice: Stamperia Baglioni, 1744.

Letters and Other Documents Illustrating the Relations Between England and Germany at the Commencement of the Thirty Years' War. Ed. Samuel Rawson Gardiner. 2 vols. Westminster: Nichols and Sons, 1865.

Litteræ Annuæ Societatis Iesu. Dillingen: Ex Typographeo Mayeriana, 1611.

Locke, John. *An Essay concerning Human Understanding*. Ed. Peter H. Nidditch. Oxford: Clarendon, 1975.

Magagnati, Girolamo. *Meditazione poetica sopra i pianeti medicei*. Venice: Heredi di A. Salicato, 1610.

La main ou œuvres poétiques sur la main de Estienne Pasquier. Ed. A. Turnèbe. Paris: Michel Gadoleau, 1584.

Malherbe, François. *Œuvres*. Ed. L. Lalanne. 5 vols. Paris: Hachette, 1862–1869.

Manzini, Carlo Antonio. *L'Occhiale all'Occhio*. Bologna: Hered. Benacci, 1660.

Marolles, Michel de. *Mémoires*. Paris: A. de Sommaville, 1656.

Mathews, Tobie. *A Collection of Letters, Made by Sr Tobie Mathews*. London: Tho. Horne, Tho. Bennett, and Francis Saunders, 1692.

Matthieu, Pierre. *Histoire de la mort déplorable de Henry IV*. In M. L. Cimber and F. Danjou, eds. *Archives curieuses* ser. 1, 15: 11–112. 1837

Maurolico, Francesco. *Diaphanorum libri*. In *Scritti di ottica*, ed. Vasco Ronchi. Milan: Edizioni il Polifilo, 1968: 102–34.

Mercure François. Paris: Jean Richer, 1611; 1613; 2nd ed. 1627.

Mercurius Gallobelgicus. Frankfurt: Sigismundi Latomi, 1615.

Meroschwa, Wenceslas. *Copia Vertrewlichen Schreibens Wentzeln von Meroschvva Behmen / an Johann Trauten Burgern zu Nurnberg*. Augsburg: Sara Mang, 1620.

———. *Epistola ad Iohannem Traut Noribergensem, de statu præsentis belli*. Augsburg: Sara Mang, 1620.

———. *Lettre de Wenselaus Meroschovva Bohemois*. n.p.: 1620.

———. *Seyndt-brief van Wencelaus Meroschvva Bohemer, aen Joannem Traut Norenbergher*. Antwerp: Jan Cnobbaert, 1620.

Mersenne, Marin. *L'Impiété des déistes*. Paris: P. Bilaine, 1624.

Messisbugo, Christoforo di. *Libro nuovo nel qual si insegna il modo d'ordinar banchetti*. Venice: Lucio Spineda, 1610.

Milton, John. *John Milton*. Ed. Stephen Orgel and Jonathan Goldberg. Oxford: Oxford University Press, 1990.

Montaigne, Michel de. "It is folly to measure the true and the false by our own capacity." In *The Complete Essays*, tr. Donald Frame. Stanford, Calif.: Stanford University Press, 1976. 132–35.

Muratori, Lodovico Antonio. "Appendice alla Vita di Alessandro Tassoni." In *Vite di Alcuni Uomini Illustri*. Naples: Gaetano Castellano, 1778. 55–56.

Naudé, Gabriel. *Le Marfore ou discours contre les libelles*. Paris: Louis Boulenger, 1620.

Négociations diplomatiques de la France avec la Toscane. Ed. Abel Desjardins. 6 vols. Paris: Imprimerie Nationale, 1859–1886.

"Nouveau règlement générale pour les nouvellistes." In *Variétés historiques et littéraires*, ed. Edouard Fournier. 8: 261–70.

Nouvelles d'Allemagne. Paris: Jean Richer, 1613.

Nouvelles des régions de la lune: Supplément du Catholicon. In *Satyre Menipée, de la vertu du Catholicon d'Espagne*, ed. M. Charles Labitte. Paris: Charpentier, 1865. 290–95.

"Original Correspondence from France." *The Spectator* 1069 (December 23, 1848): 1231–32.

Pasquier, Étienne. *Le Catéchisme des Jésuites*. Ed. Claude Sutto. Quebec: Sherbrooke, 1982.

———. [Le Sieur Fælix de la Grace]. *La Chasse du renard Pasquin*. Villefranche: Hubert le Pelletier, 1603.

———. *Les Œuvres d'Estienne Pasquier . . . et les lettres de Nicolas Pasquier*. 2 vols. Amsterdam: Compagnie des Libraires Associez, 1723.

Pasquier, Nicolas. "À Monsieur d'Ambleville." In *Les Œuvres d'Estienne Pasquier . . . et les lettres de Nicolas Pasquier*. 2: cols. 1054–68.

———. "À Monsieur de Montagnes." In *Les Œuvres d'Estienne Pasquier . . . et les lettres de Nicolas Pasquier*. 2: cols. 1273–82

———. *Remonstrances tres-humbles a la Royne*. Lyon: Jean Jullieron, 1610.

Pasquinate romane del cinquecento. Ed. Valerio Marcucci, Antonio Marzo, and Angelo Romano. 2 vols. Rome: Salerno, 1983.

Pasquino e dintorni. Testi Pasquinati del Cinquecento. Ed. Antonio Marzo. Rome: Salerno, 1990.

Peiresc, Nicolas Claude Fabri de. *Lettres aux frères Dupuy [et autres].* Ed. Philippe Tamizey de Larroque. 7 vols. Paris: Imprimerie Nationale, 1888–1898.

Pena, Jean. *De usu optices.* In *Opticæ libri quatuor ex voto Petri Rami novissimo per Fridericum Risnerum.* Kassel: W. Wessel, 2nd ed., 1615. fol. C2.

Pescetti, Orlando. *Risposta all'Anticrusca.* Verona: Angelo Tamo, 1613.

Platter, Felix. *De corporis humani structura et usu.* Basel: Officina Frobeniana, 1583.

Plutarch. *On Curiosity.* In *Moralia.* Tr. W. C. Helmbold. 14 vols. Cambridge, Mass.: Harvard University Press, 1939. Vol. 6.

Politi, Adriano. "Al Lettore." In Tacitus, *Opere,* tr. Girolamo Canini. Venice: Giunti, 1618. Unnumbered fols. 23–30.

———. "Al Signor Giulio Pannocchieschi d'Elci." In Tacitus, *Annali e Istorie,* tr. Adriano Politi. Rome: Gio. Angelo Ruffinelli, 1611. 507–28.

———. [Oratio Giannetti, a pseudonym of Adriano Politi]. "All'illustre . . . signor Francesco Visdomini." In Tacitus, *Annali e Istorie,* tr. Adriano Politi. Venice: Roberto Meietti, 1604. Unnumbered fols. 1–19.

———. *Dittionario Toscano.* Rome: Mascardi, 1613.

Recueil des instructions générales aux nonces ordinaires de France de 1624 à 1634. Ed. Auguste Leman. Paris: René Giard, 1919.

Relazioni degli stati europei lette al senato dagli ambasciatori veneti nel secolo decimosettimo. Ed. Nicolò Barozzi and Guglielmo Berchet. Venice: Pietro Naratovich, 1857.

"Rencontre et naufrage de trois astrologues juidiciaires, Mauregard, J. Petit, et P. Larivey, nouvellement arrivez en l'autre monde." In *Variétés historiques et littéraires,* ed. Edouard Fournier. 2: 211–20.

"Revolution in Europe: France." *Tait's Edinbugh Magazine* 15 (December 1848): 840–43.

Rhodius, Ambrosius. *Optica. De Crepusculis.* Wittenberg: Laurentii Seuberlich, 1611.

Riccati, Carlo. *Tableau historique et raisonné des événemens qui ont précedé et suivi le rétablissement des Bourbons en France.* 3 vols. Paris: Delaunay, 1817.

Richelieu, Armand Jean du Plessis, Cardinal and Duke. *Mémoires.* Ed. Alexandre Petitot. 10 vols. Paris: Foucault, 1823.

Rome, Vatican Library, Fondo Urbinate Latino, Avvisi manoscritti. Mss. 1076–83.

Ronsard, Pierre de. *Oeuvres.* Ed. Claude Garnier, Jean and Philippe Galland. 3 vols. Paris: Nicolas Buon, 1623.

Rubens, Peter Paul. *Correspondence de Rubens et documents épistolaires concernant sa vie et ses œuvres.* Ed. and tr. Charles Ruelens. 6 vols. Antwerp: Veuve de Backer, 1887–1909.

Sarpi, Paolo. *Lettere ai gallicani.* Ed. Boris Ulianich. Wiesbaden: Steiner, 1961.

———. *Lettere ai protestanti.* Ed. Manlio Duilio Busnelli. 2 vols. Bari: Giuseppe Laterza, 1931.

———. Pietro Savio. "Per l'epistolario di Paolo Sarpi." *Ævum* 10 (1936): 3–104; 11 (1937): 13–74, 275–322.

———. *Pensieri naturali, metafisici e matematici.* Ed. Luisa Cozzi and Libero Sosio. Milan: Riccardo Ricciardi, 1996.

Scaliger, Joseph. *Scaligerana.* Cologne: Scagen, 1667.

[Scheiner, Christoph, S.J.]. *Accuratior Disquisitio.* In Galileo Galilei, *Opere.* V: 37–70.

————. *Briefe des Naturwissenschaftlers Christoph Scheiner SJ an Erzherzog Leopold V. von Österreich-Tirol 1620–1632*. Ed. Franz Daxecker. Innsbruck: University of Innsbruck Press, 1995.

————. *Rosa Ursina*. Bracciano: 1626–1630.

————. *Oculus, hoc est fundamentum opticum*. Innsbruck: Daniel Agricola, 1619.

Schmidl, Johannes, S.J. *Historiae Societatis Jesu Provinciae Bohemiae*. 4 vols. Prague: Jacob Schweiger, 1749–1759.

"Singuliers effets de catoptrique." *Magasin Pittoresque* 17 (1849): 315–18.

Sirtori, Girolamo. *Telescopium*. Frankfurt: Paulus Jacobus, 1618.

[Snell, Willebrord]. *De Maculis in sole animadversis . . . Batavi dissertatiuncula*. Leiden: Plantin, 1612.

Sorel, Charles. *La Bibliothèque françoise*. Paris: Compagnie des Libraires du Palais, 1664.

Sterne, Laurence. *Tristram Shandy*. Ed. Howard Anderson. New York: Norton, 1980.

Stevin, Simon. *The Principal Works of Simon Stevin*. Ed. Ernst Crone, tr. C. Dikshoorn. 5 vols. in 6. Amsterdam: Swets & Zeitlinger, 1955–1966.

Stieler, Caspar. *Zeitungs Lust und Nutz*. Hamburg: Benjamin Schillers, 1695.

Stuart Royal Proclamations. Ed. J. F. Larkin and P. L. Hughes. 2 vols. Oxford: Oxford University Press, 1973.

Suckling, Sir John. "To His Much Honored, the Lord Lepington, upon his Translation of Malvezzi." In *Fragmenta Aurea*. London: Humphrey Moseley, 1646. 19.

Sully, Maximilian de Béthune, duc de. *Mémoires ou Oeconomies Royales*. 3 vols. Amsterdam: Aletinosgraphe, 1638 and Paris: Augustin Courbé, 1662.

The Swedish Intelligencer. London: Nathaniel Butter and Nicholas Bourne, 1632.

The Swedish Intelligencer, the Third Part. London: Nathaniel Butter and Nicholas Bourne, 1633.

The Swedish Intelligencer, the Ninth Part. London: Nathaniel Butter and Nicholas Bourne, February 6, 1636.

Tacitus. *Annali e Istorie*. Tr. Adriano Politi. Venice: Roberto Meietti, 1604.

————. *Annali e Istorie*. Tr. Adriano Politi. Rome: Gio. Angelo Ruffinelli, 1611.

————. *Opere*. Tr. Girolamo Canini. Venice: Giunti, 1618.

————. *Opere di Cornelio Tacito*. Tr. Bernardo Davanzati. Paris: Vedova Quillau, 1760.

Tassoni, Alessandro. *Le Lettere*. Ed. Giorgio Rossi. 2 vols. Bologna: Romagnoli dall'Acqua, 1901.

————. *Pensieri e scritti preparatori*. Ed. Pietro Puliatti. Modena: Edizioni Panini, 1986.

————. *La Secchia rapita*. Ed. Ottavio Besomi. 2 vols. Padua: Antenore, 1987.

Testi, Fulvio. *Lettere*. Ed. Maria Luisa Doglio. 3 vols. Bari: Laterza, 1967.

Townshend, Aurelian. "To the Right Honourable, the Lord Cary." In *Poems and Masks*, ed. E. K. Chambers. Oxford: Clarendon, 1912. 43–44.

Tristan l'Hermite, François. *Le page disgracié*. Paris: E. Plon, 1898.

Van Buchel, Armand. *Description de Paris*. In *Mémoires de la Société de l'histoire de Paris et de l'Ile-de-France* 26 (1899): 59–195.

Vannozzi, Bonifacio. *Della Suppellettile*. Bologna: Giovanni Rossi, 1609.

————. *Della Suppellettile*. Bologna: Giovanni Rossi, 1610.

————. *Delle Lettere Miscellanee.* Bologna: Bartolomeo Cochi, 1617.

————. *Della Suppellettile.* Bologna: Giovanni Rossi, 1613.

————. *Teatro di segretaria copioso.* Rome: Giacomo Mascardi, 1614.

Vesalius, Andreas. *De humanis corporis fabrica libri septem.* Basel: Johannes Oporinus, 1543.

Walchius, Johannes. *Decas fabularum humanis generis.* Strasbourg: Lazarus Zetzner, 1609.

Walton, Izaak. *The Life of Sir Henry Wotton.* In *The Lives of Doctor John Donne and Others.* Ed. Charles H. Dick. London: Walter Scott Press, 1899.

Watts, William. *Mortification Apostolicall. Delivered in a Sermon in Saint Pauls Church, . . . May 21, 1637.* London: Iohn Cowper, 1637.

Webster, John. *The displaying of supposed witchcraft.* London: Printed by J.M., 1677.

Wedderburn, John. *Confutatio.* In Galilei, *Opere.* 3: 147–78.

Wikoff, Henry. "Napoleon Louis Bonaparte." *Westminster Review* 51 (July 1849): 249–65.

Winwood, Ralph. *Memorials of Affairs of State.* Ed. Edmund Sawyer. 3 vols. London: T. Ward, 1725.

Wood, Anthony à. *Fasti Oxonienses.* Ed. Philip Bliss. 5 vols. London: Rivington, 1813–1820.

[Wotton, Henry.] *The Life and Letters of Sir Henry Wotton.* Ed. Logan Pearsall Smith. 2 vols. Oxford: Clarendon, 1907.

————. *Reliquiæ Wottonianæ.* London: B. Tooke, 1685.

SECONDARY SOURCES

Abromitis, Lois. "William Gilbert as Scientist." Ph.D. Dissertation, Brown University, 1977.

Aït-Touati, Frédérique. *Fictions of the Cosmos: Science and Literature in the Seventeenth Century.* Tr. Susan Emanuel. Chicago: University of Chicago Press, 2011.

Allen, Eric W. "International Origins of the Newspapers: The Establishment of Periodicity in Print." *Journalism Quarterly* 7 (1930): 307–19.

Alpers, Svetlana. *The Art of Describing: Dutch Art in the Seventeenth Century.* Chicago: University of Chicago Press, 1983.

Ambard, Robert. *Aix Romaine. Nouvelles observations sur la topographie d'Aquae Sextiae.* Aix-en-Provence: Association Archéologique Entremont, 1984.

Anderson, Alison Deborah. *On the Verge of War: International Relations and the Jülich-Kleve Succession Crises (1609–1614).* Boston: Humanities Press, 1999.

Aquilecchia, Giovanni. "Presentazione." In *Pasquinate romane del Cinquecento*, ed. Valerio Marucci et al. 2 vols. Rome: Salerno, 1983. 1: ix–xvi.

Arblaster, Paul. "Antwerp and Brussels as Inter-European Spaces in News Exchange." In *The Dissemination of News*, ed. Brendan Dooley. 193–205.

————. "Policy and Publishing in the Habsburg Netherlands, 1585–1690." In *The Politics of Information*, ed. Brendan Dooley and Sabrina Baron. 179–98.

————. "Posts, Newsletters, Newspapers: England in a European System of Communications." *Media History* 11, 1–2 (2005): 21–36.

Artaud de Montor, Alexis-François. *Machiavel, son génie et ses erreurs*. Paris: F. Didot, 1833.

Ashworth, William B. "Divine Reflections and Profane Refractions: Images of a Scientific Impasse in Seventeenth-Century Italy." In *Gianlorenzo Bernini: New Aspects of His Art and Thought*, ed. Irving Lavin. University Park: Pennsylvania State University Press for College Art Association, 1985. 179–207.

Atherton, Ian. "The Itch Grown a Disease: Manuscript Transmission of News in the Seventeenth Century." *Prose Studies* 21 (1998): 39–65.

Audiat, Louis. *Un fils d'Estienne Pasquier: Nicolas Pasquier*. Paris: Didier, 1876.

Bachrach, A. G. H. "Sir Constantyn Huygens and Ben Jonson." *Neophilologus* 35 (1951): 120–29.

———. *Sir Constantine Huygens and Britain, 1596–1619*. Leiden: Leiden University Press, 1962.

Backer, Augustin de, Aloys de Backer, Auguste Carayon, and Carlos Sommervogel. *Bibliothèque de la Compagnie de Jésus*. 12 vols. Brussels: Oscar Schepens, 1890–1932.

Baffetti, Giovanni. "Il 'Sidereus Nuncius' a Bologna." *Intersezioni* 11, 3 (1991): 477–500.

Baker, David. "'The Allegory of a China Shop': Jonson's *Entertainment at Britain's Burse*." *ELH* 72, 1 (2005): 159–80.

Ballon, Hilary. *The Paris of Henri IV*. Cambridge, Mass.: MIT Press, 1990.

Barbarics-Hermanik, Zsuzsa. "Handwritten Newsletters as Interregional Information Sources in Central and Southeastern Europe." In *The Dissemination of News*, ed. Brendan Dooley. 155–178.

Barbiche, Bernard, and Ségolène Dainville-Barbiche. *Sully*. Paris: Fayard, 1997.

Barbierato, Federico. *The Inquisitor in the Hat Shop: Inquisition, Forbidden Books and Unbelief in Early Modern Venice*. Burlington, Vt.: Ashgate, 2012.

Barkan, Leonard. *Unearthing the Past. Archaeology and Aesthetics in the Making of Renaissance Culture*. New Haven, Conn.: Yale University Press, 1999.

Baron, Sabrina A. "The Guises of Dissemination in Early Seventeenth-Century England. News in Manuscript and Print." In *The Politics of Information*, ed. Brendan Dooley and Sabrina Baron. 41–56.

Bartolini, Simone. *I fori gnomici di Egnazio Danti in Santa Maria Novella*. Florence: Polistampa, 2006.

Bayard, Françoise de. "Jean Bochart de Champigny (1561–1630)." *Revue d'Histoire Moderne et Contemporaine* 46, 1 (1999): 39–52.

Bazzoli, Maurizio. "Ragion di stato e interessi degli stati. La trattatistica sull'ambasciatore dal XV al XVIII secolo." *Nuova Rivista Storica* 86, 2 (2002): 283–328.

Bedini, Silvio. "Lens Making for Scientific Instrumentation in the Seventeenth Century." *Applied Optics* 5, 5 (1966): 687–94.

Bedini, Silvio, and Arthur G. Bennett. "'A Treatise on Optics' by Giovanni Cristoforo Bolantio." *Annals of Science* 52, 2 (1995): 103–26.

Bellettini, Pierangelo, Rosaria Campioni, and Zita Zanardi, eds. *Una città in piazza: Comunicazione e vita quotidiana a Bologna tra cinque e seicento*. Bologna: Compositori, 2000.

Benzoni, Gino. "Enea Silvio Caprara." In *Dizionario biografico degli italiani* 19 (1976): 169–177.

Bernardini, Nicola. *Guida della stampa periodica italiana.* Lecce: Fratelli Spacciante, 1890.

Besomi, Ottavio, and Michele Camerota. *Galileo e il Parnaso Tychonico: Un capitolo inedito del dibattito sulle comete tra finzione letteraria e trattazione scientifica.* Florence: Olschki, 2000.

Biagioli, Mario. "Did Galileo Copy the Telescope? A 'New' Letter by Paolo Sarpi." In *The Origins of the Telescope*, ed. Albert Van Helden, Sven Dupré, Rob van Gent, and Huib Zuidervaart. 203–30.

———. *Galileo's Instruments of Credit: Telescopes, Images, Secrecy.* Chicago: University of Chicago Press, 2006.

Bibliotheek van Nederlandsche Pamfletten. Ed. Frederick Muller. 3 vols. Amsterdam: Frederick Muller, 1858–1861.

Bigourdan, M. G. "Joseph Gaultier et la découverte de la visibilité des astres en plein jour." *Comptes Rendus des Séances de l'Académie des Sciences* 162 (January–June 1916): 809–15.

Bimbenet-Privat, Michèle, ed. *L'orfèverie parisienne de la Renaissance.* Paris: Centre Culturel du Panthéon, 1995.

Bireley, Robert, S.J. *The Jesuits and the Thirty Years War: Kings, Courts, and Confessors.* Cambridge: Cambridge University Press, 2003.

———. *Religion and Politics in the Age of the Counter-Reformation: Emperor Ferdinand II, William Lamormaini, S.J., and the Formation of Imperial Policy.* Chapel Hill: University of North Carolina Press, 1981.

Blanchet, Adrien. "Note sur les projets de reconstruction du pont Marchant." *Bulletin de la Société de l'Histoire de Paris et de l'Île-de-France* 40 (1913): 196–200.

Blome, Astrid. "Offices of Intelligence and Expanding Social Spaces." In *The Dissemination of News*, ed. Brendan Dooley. 207–22.

Boas, George. *Vox Populi: Essays in the History of an Idea.* Baltimore: Johns Hopkins University Press, 1969.

Bohanan, Donna. *Old and New Nobility in Aix-en-Provence, 1600–1695.* Baton Rouge: Louisiana State University Press, 1992.

Bollème, Geneviève. *Les almanachs populaires aux XVIIe et XVIIIe siècles.* Paris: Mouton, 1969.

Bongi, Salvatore. "Le prime gazzette in Italia." *Nuova Antologia* 11 (1869): 311–46.

Bonnaffé, Edmond. *Inventaire des meubles de Catherine de Médicis.* Paris: Auguste Aubry, 1874.

Borreguero Beltrán, Cristina. "Philip of Spain: The Spider's Web of News and Information." In *The Dissemination of News*, ed. Brendan Dooley. 23–49.

Borrelli, Gianfranco. "Techniche di simulazione e conservazione politica in Gerolamo Cardano e Alberico Gentili." *Annali dell'Istituto storico italo-germanico* 12 (1986): 87–124.

Bradley, Jesse Franklin, and Joseph Quincy Adams. *The Jonson Allusion-Book.* New Haven, Conn.: Yale University Press, 1922.

Bredekamp, Horst. *The Lure of Antiquity and the Cult of the Machine.* Tr. Allison Brown. Princeton, N.J.: Marcus Wiener, 1995.

Brizzi, Gian Paolo. *La formazione della classe dirigente nel sei-settecento.* Bologna: Il Mulino, 1976.

———. "I Gesuiti e i seminari per la formazione della classe dirigente." In *Dall'isola alla città: I Gesuiti a Bologna,* ed. Anna Maria Matteucci and Gian Paolo Brizzi. Bologna: Nuova Alfa, 1988. 144–56.

Broglie, Emmanuel de. *Catinat, l'homme et la vie (1637–1712).* Paris: V. Lecoffre, 1902.

Brown, Alison. *The Return of Lucretius to Renaissance Florence.* Cambridge, Mass.: Harvard University Press, 2010.

Brown, Sarah. "Stained Glass." In *The Oxford History of Western Art,* ed. Martin Kemp. Oxford: Oxford University Press, 2000. 108–23.

Brownlees, Nicolas. *The Language of Periodical News in Seventeenth-Century England.* Newcastle upon Tyne: Cambridge Scholars Publishing, 2011.

———. "Spoken Discourse in Early English Newspapers." *Media History* 11, 1–2 (2005): 69–85.

Bruel, François L. "Deux inventaires de bagues, joyaux, pierreries et dorures de la reine Marie de Médicis (1609 ou 1610)." *Archives de l'Art Français* n.s. 2, 1 (1908): 186–215.

Bucciantini, Massimo. *Galileo e Keplero.* Turin: Giulio Einaudi, 2007.

Bucciantini, Massimo, Michele Camerota, and Franco Giudice. *Il Telescopio di Galileo: Una storia europea.* Turin: Giulio Einaudi, 2012.

Buisseret, David. "L'atelier cartographique de Sully à Bontin: L'œuvre de Jacques Fougeu." *Dix-septieme Siècle* 174 (1992): 109–16.

———. *Henri IV.* London: Allen and Unwin, 1984.

———. *Sully and the Growth of Centralized Government.* London: Eyre and Spottiswoode, 1968.

Burke, Peter. "The Republic of Letters as a Communication System: An Essay in Periodization." *Media History* 18, 3–4 (2012): 395–407.

———. "Tacitism." In *Tacitus,* ed. T. A. Dorey. New York: Basic Books, 1969. 149–71.

Butler, G. F. "Milton's Meeting with Galileo. A Reconsideration." *Milton Quarterly* 39, 3 (2005): 132–39.

Butler, Martin. "Jonson's *News from the New World,* the 'Running Masque,' and the Season of 1619–20." *Medieval and Renaissance Drama in England* 6 (1993): 153–78.

Callard, Caroline. *Le prince et la république.* Paris: Presses de l'Université Paris-Sorbonne, 2007.

Camerota, Michele. *Galileo Galilei.* Rome: Salerno, 2004.

Cappelli, Federica. "Parnaso bipartito nella satira italiana del '600 (e due imitazioni spagnole.)" *Cuadernos de Filología Italiana* 8 (2001): 133–51.

Cashman, Anthony B. "The Problem of Audience in Mantua: Understanding Ritual Efficacy in an Italian Renaissance Princely State." *Renaissance Studies* 16, 3 (2002): 355–65.

Caspar, Max. *Kepler.* Tr. C. Doris Hellman. New York: Dover, 1993.

Castronovo, Valerio. "I primi sviluppi della stampa periodica fra cinque e seicento." In *La Stampa Italiana dal Cinquecento all'Ottocento*, ed. Carlo Capra, Valerio Castronovo, and Giuseppe Ricuperati. Rome: Laterza, 1986. 3–65.

Catalogus van de pamfletten-verzameling berustende in de Koninklijke Bibliotheek. Ed. W. P. C. Knuttel. 9 vols. in 10. Utrecht: H & S, 1978.

Chambers, D. S. "The 'Bellissimo Ingegno' of Ferdinando Gonzaga (1587–1626), Cardinal and Duke of Mantua." *Journal of the Warburg and Courtauld Institutes* 50 (1987): 113–47.

Chapin, Seymour L. "The Astronomical Activities of Peiresc." *Isis* 48 (1957): 13–29.

Chartier, Roger. "Handwritten Newsletters, Printed Gazettes: Cymbal and Butter." In *Inscription & Erasure: Literature and Written Culture from the Eleventh to the Eighteenth Century*, tr. Arthur Goldhammer. Philadelphia: University of Pennsylvania Press, 2007. 46–62.

Chen-Morris, Raz. "Shadows of Instruction: Optics and Classical Authorities in Kepler's *Somnium*." *Journal of the History of Ideas* 66, 2 (2005): 223–43.

Clark, Sandra. *The Elizabethan Pamphleteers: Popular Moralistic Pamphlets 1580–1640.* Rutherford, N.J.: Fairleigh Dickinson University Press, 1983. 86–120.

Cogswell, Thomas. *The Blessed Revolution: English Politics and the Coming of War, 1621–1624.* Cambridge: Cambridge University Press, 1989.

Colie, Rosalie. "Cornelis Drebbel and Salomon de Caus: Two Jacobean Models for Salomon's House." *Huntington Library Quarterly* 18 (1955): 245–60.

Collins, Randall. *The Sociology of Philosophers: A Global Theory of Intellectual Change.* Cambridge, Mass.: Harvard University Press, 1998.

Connell, William J. "New Light on Machiavelli's Letter to Vettori, 10 December 1513." In *Europe and Italy: Studies in Honour of Giorgio Chittolini*, ed. P. Guglielmotti, I. Lazzarini, and G. M. Varanini. Florence: Florence University Press, 2011. 93–127.

Connor, Stephen. *Dumbstruck: A Cultural History of Ventriloquism.* Oxford: Oxford University Press, 2000.

Cook, Harold. *Matters of Exchange: Commerce, Medicine and Science in the Dutch Golden Age.* New Haven, Conn.: Yale University Press, 2007.

Cooper, Michael S.J. *Rodrigues the Interpreter: An Early Jesuit in Japan and China.* New York: Weatherhill, 1974.

Čornejova, Ivana. "Education in Rudolfine Prague." In *Rudolf II and Prague: The Court and the City*, ed. Eliša Fučíková et al. 323–31.

———. "The Religious Situation in Rudolfine Prague." In *Rudolf II and Prague: The Court and the City*, ed. Eliša Fučíková et al. 310–22.

Cozzi, Gaetano. *Paolo Sarpi tra Venezia e l'Europa.* Turin: Giulio Einaudi, 1979.

———. "Traiano Boccalini, Il Cardinal Borghese e la Spagna." *Rivista Storica Italiana* 68, 2 (1956): 230–44.

Cranfield, G. A. *The Press and Society from Caxton to Northcliffe.* Longman: London, 1978. 1–30.

Crary, Jonathan. *Techniques of the Observer: On Vision and Modernity in the Nineteenth Century.* Cambridge, Mass.: MIT Press, 1995.

Crombie, A. C. "Expectation, Modeling and Assent in the History of Optics: Part I. Alhazen and the Medieval Tradition." *Studies in History and Philosophy of Science* 21, 4 (1990): 605–32.

———. "Expectation, Modeling and Assent in the History of Optics–II. Kepler and Descartes." *Studies in History and Philosophy of Science* 22, 1 (1991): 89–115.

Cunningham, Andrew. *The Anatomical Renaissance: The Resurrection of the Anatomical Projects of the Ancients*. Hants: Scholar Press, 1997. 212–68.

Cust, Richard. "News and Politics in Early Seventeenth-Century England." *Past & Present* 112 (1986): 60–90.

D'Addio, Mario. *Il pensiero politico di Gaspare Scioppio e il Machiavellismo del seicento*. Milan: Giuffrè, 1962.

Dahl, Folke. "Amsterdam—Cradle of English Newspapers." *The Library* 5th ser. 4 (1949): 166–78.

Dahl, Folke, Fanny Pettibon, and Marguerite Boulet. *Les débuts de la presse française: Nouveaux aperçus*. Göteborg: Wettergren & Kerber, 1951.

Davids, Karel. "Successful and Failed Transitions. A Comparison of Innovations in Windmill-Technology in Britain and the Netherlands in the Early Modern Period." *History and Technology* 14 (1998): 225–47.

Davis, Natalie. *Society and Culture in Early Modern France*. Stanford, Calif.: Stanford University Press, 1975.

Daxecker, Franz. "Christoph Scheiner's Eye Studies." *Documenta Ophthalmologica* 81 (1992): 27–35.

———. "Further Studies by Christoph Scheiner Concerning the Optics of the Eye." *Documenta Ophthalmologica* 86 (1994): 153–61.

———. *The Physicist and Astronomer Christoph Scheiner*. Innsbruck: Leopold Franzens University, 2004.

De Boer, Josephine. "Men's Literary Circles in Paris 1610–1660." *PMLA* 53, 3 (1938): 730–80.

De Caro, G. "Giorgio Bolognetti." In *Dizionario biografico degli italiani*. 11: 323–26.

De Vivo, Filippo. *Information and Communication in Venice: Rethinking Early Modern Politics*. Oxford: Oxford University Press, 2007.

———. "Paolo Sarpi and the Uses of Information in Seventeenth-Century Venice." *Media History* 11, 1–2 (2005): 37–51.

———. "Pharmacies as Centres of Communication in Early Modern Venice." *Renaissance Studies* 21, 4 (2007): 505–21.

Dear, Peter. *Discipline & Experience: The Mathematical Way in the Scientific Revolution*. Chicago: University of Chicago Press, 1995.

Delsaute, Jean-Luc. "The *Camera Obscura* and Painting in the Sixteenth and Seventeenth Centuries." *Studies in the History of Art* 55 (1998): 110–23.

Delumeau, Jean. *Rome au XVIe siècle*. Paris: Hachette, 1975.

Devreese, Jozef, and Guido Vanden Berghe. *"Magic Is No Magic": The Wonderful World of Simon Stevin*. Tr. Lee Preedy. Southampton: WIT Press, 2008.

Díaz Noci, Javier. "Dissemination of News in the Spanish Baroque." *Media History* 18, 3–4 (2012): 409–21.

Diffley, Paul B. "Tassoni's Linguistic Views." *Studi secenteschi* 33 (1992): 67–92.

Dijksterhuis, E. J. *Simon Stevin: Science in the Netherlands Around 1600.* The Hague: Nijhoff, 1970.

Dizionario biografico degli italiani, ed. Alberto M. Ghisalberti. 77 vols. to date. Rome: Istituto della Enciclopedia Italiana, 1960–.

D'Onofrio, Cesare. "Gli 'Avvisi' di Roma dal 1554 al 1605 conservati in biblioteche ed archivi romani." *Studi Romani* 10, 5 (1962): 529–43.

Dooley, Brendan, ed. *The Dissemination of News and the Emergence of Contemporaneity in Early Modern Europe.* Burlington, Vt.: Ashgate, 2010.

———. "Making It Present." In *The Dissemination of News,* ed. Brendan Dooley. 95–115.

———. *The Social History of Skepticism: Experience and Doubt in Early Modern Culture.* Baltimore: Johns Hopkins University Press, 1999.

Dooley, Brendan, and Sabrina Baron, eds. *The Politics of Information.* London: Routledge, 2001.

Duchêne, Roger. "Lettres et gazettes au XVIIeme siècle." *Revue d'Histoire Moderne et Contemporaine* 18 (1971): 489–502.

Dulaure, Jacques Antoine. *Histoire physique, civile et morale de Paris.* 10 vols. 4th ed. Paris: Guillaume, 1829.

Duplomb, Charles. *Histoire générale des ponts de Paris.* Paris: J. Mersch, 1911.

Dupré, Sven. "Galileo, the Telescope, and the Science of Optics in the Sixteenth Century." Dissertation, University of Ghent, 2002.

———. "Inside the *Camera Obscura:* Kepler's Experiment and Theory of Optical Imagery." *Early Science and Medicine* 13 (2008): 219–44.

Durkan, John. "A Post-Reformation Miscellany." *Innes Review* 53, 1 (2002): 108–26.

Eijffinger, Arthur. "Zin en beeld. Enige kanttekeningen bij twee historieprenten." *Oud Holland* 93, 4 (1979): 251–69.

Elkins, James. *The Domain of Images.* Ithaca, N.Y.: Cornell University Press, 1999.

Evans, R. J. W. *Rudolf II and His World: A Study in Intellectual History 1576–1612.* Oxford: Oxford University Press, 1973.

Favaro, Antonio. *La Libreria di Galileo Galilei.* Rome: Tipografia delle Scienze Matematiche e Fisiche, 1887.

Favret, Mary. *War at a Distance: Romanticism and the Making of Modern Warfare.* Princeton, N.J.: Princeton University Press, 2010.

Feldhay, Rivka. *Galileo and the Church: Political Inquisition or Critical Dialogue?* Cambridge: Cambridge University Press, 1995.

Feyel, Gilles. *L'annonce et la nouvelle: La presse d'information sous l'Ancien Régime, 1630–1788.* Oxford: Voltaire Foundation, 2000.

Firpo, Luigi. "In margine al processo di Giordano Bruno." *Rivista Storica Italiana* 68 (1956): 325–64.

———. "Lettere di Traiano Boccalini." *Giornale Storico della Letteratura Italiana* 122 (1944): 11–34.

————. "La satira politica in forma di ragguaglio di Parnaso dal 1614 al 1650." *Atti della Accademia delle Scienze di Torino* 87 (1952–1953): 1–51.

————. *Traduzioni dei Ragguagli di Parnaso.* Florence: Sansoni, 1965.

————. "Traiano Boccalini." In *Dizionario Biografico degli Italiani.* 11: 10–19.

Forbes, R. J. "The Sailing Chariot." In *The Principal Works of Simon Stevin,* ed. Ernst Crone. 5: 3–8.

Fournier, Edouard, ed. *Variétés historiques et littéraires.* 10 vols. Paris: P. Jannet, 1855–1863.

Fox, Adam. "Rumour, News and Popular Political Opinion in Elizabethan and Early Stuart England." *Historical Journal* 40, 3 (1997): 597–620.

Frearson, Michael. "The Distribution and Readership of London Corantos in the 1620s." In *Serials and Their Readers 1620–1914,* ed. Robin Myers and Michael Harris. Winchester: St. Paul's Bibliographies, 1993. 1–25.

————. "London Corantos in the 1620s." *Studies in Newspaper and Periodical History* (1993): 3–17.

Freedberg, David. *The Eye of the Lynx: Galileo, His Friends, and the Beginnings of Modern Natural History.* Chicago: University of Chicago Press, 2002.

Fučíková, Eliša et al., eds. *Rudolf II and Prague: The Court and the City.* London: Thames and Hudson, 1997.

Fuhring, Peter, and Michèle Bimbenet-Privat. "Le Style 'Cosses de Pois.' L'Orfèverie et la gravure à Paris sous Louis XIII." *Gazette des Beaux-Arts* 139 (2002): 1–224.

Gabotti, Ferdinando. "Per la storia della letteratura civile dei tempi di Carlo Emanuele I: La politica antispagnuola." *Rendiconti della Reale Accademia dei Lincei. Classe di scienze morali* ser. 5, 3, 1 (1894): 323–40.

Gabrieli, Giuseppe. "Per la storia della prima romana accademia dei Lincei." *Isis* 24, 1 (1935): 80–89.

————. "La 'Philoteca Scioppiana' in un manoscritto laurenziano." *Atti della Reale Accademia d'Italia* ser. 7, 1 (1939–1940): 228–39.

Gajda, Alexandra. "Tacitus and Political Thought in Early Modern Europe." In *The Cambridge Companion to Tacitus,* ed. A. J. Woodman. Cambridge: Cambridge University Press, 2009. 253–68.

García Santo-Tomás, Enrique. "Fortunes of the *Occhiali Politici* in Early Modern Spain: Optics, Vision, Points of View." *PMLA* 124, 1 (2009): 59–75.

Gestrich, Andreas. "The Public Sphere and the Habermas Debate." *German History* 24, 3 (2006): 413–30.

Geyl, Pieter. *History of the Dutch-Speaking Peoples, 1555–1648.* London: Phoenix, 2001.

Gingerich, Owen, and Albert Van Helden. "From *Occhiale* to Printed Page: The Making of Galileo's *Sidereus Nuncius.*" *Journal for the History of Astronomy* 34, 3 (2003): 251–67.

Gingerich, Owen, and Albert Van Helden. "How Galileo Constructed the Moons of Jupiter." *Journal for the History of Astronomy* 42, 2 (2011): 259–64.

Gorman, Michael John. "A Matter of Faith? Christoph Scheiner, Jesuit Censorship, and the Trial of Galileo." *Perspectives on Science* 4, 3 (1996): 283–321.

———. "The Scientific Counter-Revolution: Mathematics, Natural Philosophy and Experimentalism in Jesuit Culture, 1580–c. 1670." Dissertation, European University Institute, 1998.

Gorski, Philip. *The Disciplinary Revolution: Calvinism and the Rise of the State in Early Modern Europe.* Chicago: University of Chicago Press, 2003.

Groote, Inga Mai. *Musik in italienischen Akademien: Studien zur institutionellen Musikpflege 1543–1666.* Laaber: Laaber Verlag, 2007.

Grudin, Robert. "Rudolf II of Prague and Cornelis Drebbel: Shakespearean Archetypes?" *Huntington Library Quarterly* 54 (1991): 181–205.

Guerrini, Luigi. *Ricerche su Galileo e il Primo Seicento.* Pisa: IEPI, 2004.

Gui, Francesco. *I Gesuiti e la rivoluzione boema: Alle origini della guerra dei trent'anni.* Milan: Franco Angeli, 1989.

Guiffrey, J. J. "Logements d'artistes au Louvre." *Nouvelles Archives de l'Art Français* 2 (1873): 1–221.

Haag, Eugène, and Émile Haag. *La France protestante.* 10 vols. 2nd ed. Paris: Sandoz & Fischbacher, 1877.

Habermas, Jürgen. *The Structural Transformation of the Public Sphere.* Tr. Thomas Burger and Frederick Lawrence. Cambridge, Mass.: MIT Press, 1989.

Halasz, Alexandra. *The Marketplace of Print: Pamphlets and the Public Sphere in Early Modern England.* Cambridge: Cambridge University Press, 1997.

Hanlon, Gregory. *The Twilight of a Military Tradition: Italian Aristocrats and European Conflicts, 1560–1800.* New York: Holmes and Meier, 1998.

Hallyn, Fernand. "Jean Pena et l'éloge de l'optique." In *Autour de Ramus: Texte, théorie, commentaire,* ed. Kees Meerhoff and Jean-Claude Moisan. Québec: Nuit Blanche, 1997. 217–32.

Hardÿ de Périni, Edouard. *Batailles françaises: Louis XIV, 1672 à 1700.* Paris: Flammarion, 1904.

Harline, Craig. *Pamphlets, Printing, and Political Culture in the Early Dutch Republic.* Dordrecht: Nijhoff, 1987.

Harris, L. E. *The Two Netherlanders: Humphrey Bradley and Cornelis Drebbel.* Cambridge: Heffer & Sons, 1961.

Hawlitschek, Kurt. "Die Deutschlandreise des René Descartes." *Berichte zur Wissenschaftsgeschichte* 25, 4 (2002): 235–52.

Hayward, J. F. *Virtuoso Goldsmiths and the Triumph of Mannerism 1540–1620.* New York: Rizzoli, 1976.

Headley, John. *Tommaso Campanella and the Transformation of the World.* Princeton, N.J.: Princeton University Press, 1997.

Heilbron, John. *Galileo.* Oxford: Oxford University Press, 2010.

———. *The Sun in the Church: Cathedrals as Solar Observatories.* Cambridge, Mass.: Harvard University Press, 1999.

Heller, Henry. *Labour, Science, and Technology in France 1500–1620.* Cambridge: Cambridge University Press, 1996.

Herbst, Klaus-Dieter. "Galilei's Astronomical Discoveries Using the Telescope and Their Evaluation Found in a Writing-Calendar from 1611." *Astronomische Nachrichten* 330, 6 (2009): 536–39.

Houzeau, J. C. "Le Téléscope à Bruxelles." *Ciel et Terre* 3 (1882): 25–28.

Huard, Georges. "Les logements des artisans dans la grande galerie du Louvre sous Henri IV et Louis XIII." *Bulletin de la Société de l'art français* (1939): 18–36.

Humbert, Pierre. "Joseph Gaultier de la Valette." *Revue d'Histoire des Sciences et de leurs Applications* 1 (1948): 314–21.

Huppert, George. "A Matter of Quality: The Pasquier Family Between Bourgeoisie and Noblesse." *Historical Reflections* 27, 2 (2001): 183–99.

Ilardi, Vincent. "Renaissance Florence: The Optical Capital of the World." *Journal of European Economic History* 22, 3 (1993): 507–42.

———. *Renaissance Vision from Spectacles to Telescopes.* Philadelphia: American Philosophical Society, 2007.

Indice per Materie della Biblioteca Comunale di Siena. Ed. Lorenzo Ilari. Siena: Ancora, 1844.

Infelise, Mario. "News Networks Between Italy and Europe." In *The Dissemination of News*, ed. Brendan Dooley. 50–67.

———. *Prima dei giornali: Alle origini della pubblica informazione.* Rome: Giuseppe Laterza e figli, 2002.

Israel, Jonathan. *The Dutch Republic. Its Rise, Greatness, and Fall, 1477–1806.* Oxford: Oxford University Press, 1998.

Ioppolo, Grace. *Dramatists and Their Manuscripts in the Age of Shakespeare, Jonson, Middleton and Heywood.* London: Routledge, 2006.

Jacob, P. L. *Le livre d'or des métiers: Histoire de l'orfèvrerie-joaillerie et des anciennes communautés et confréries d'orfèvres-joailliers de la France et de la Belgique.* Paris: Seré, 1850.

Janáček, Josef. "Les italiens à Prague à l'époque précédant la bataille de la Montagne Blanche (1526–1620)." *Historica* 23 (1983): 5–45.

Jantz, Harold. "Goethe and an Elizabethan Poem." *Modern Language Quarterly* 12, 4 (1951): 451–61.

Jouhaud, Christian. *Les pouvoirs de la littérature. Histoire d'un paradoxe.* Paris: Gallimard, 2000.

Juste, Th. "Cornelis d'Aerssen" and "François d'Aerssen." In *Biographie Nationale publiée par l'Académie Royale des sciences, des lettres, et des beaux-arts de Belgique.* 44 vols. Brussels: H. Thirry van Buggenhoudt, 1866–1986. 1: cols. 97–106; 106–21.

Katritzky, M. A. "Mountebanks, Mummers and Masqueraders in Thomas Platter's Diary (1595–1600)." In *The Renaissance Theatre: Texts, Performance, Design*, ed. Christopher Cairns. 2 vols. Hants: Aldershot, 1999. 1: 12–44.

———. "Was *commedia dell'arte* Performed by Mountebanks? *Album amicorum* Illustrations and Thomas Platter's Description of 1598." *Theatre Research International* 23 (1998): 104–25.

Kaufmann, Thomas DaCosta. "Jesuit Art: Central Europe and the Americas." In *The Jesuits: Cultures, Sciences, and the Arts 1540–1773*, ed. John W. O'Malley, S.J. Toronto: University of Toronto Press, 1999. 274–304.

Keblusek, Marika. "The Business of News: Michel le Blon and the Transmission of Political Information in the 1630s." *Scandinavian Journal of History* 28 (2003): 205–13.

———. "Keeping It Secret: The Identity and Status of an Early Modern Inventor." *History of Science* 43 (2005): 37–56.

Kettering, Sharon. *Judicial Politics and Urban Revolt in Seventeenth-Century France.* Princeton, N.J.: Princeton University Press, 1978.

Kintz, Jean-Pierre. "Opinion publique, journalisme, et autorité du magistrat au milieu du XVIIᵉ siècle à Strasbourg." In *Pouvoir, ville, et société en Europe 1650–1750*, ed. Georges Livet and Bernard Vogler. Paris: Ophrys, 1981. 79–96.

Knowles, James. "Jonson's *Entertainment at Britain's Burse.*" In *Re-presenting Ben Jonson: Text, History, Performance*, ed. Martin Butler. New York: St. Martin's, 1999. 114–51.

———. "'Songs of Baser Alloy': Jonson's *Gypsies Metamorphosed* and the Circulation of Manuscript Libels." *Huntington Library Culture* 69, 1 (2006): 153–76.

Kofman, Sarah. *Camera Obscura of Ideology.* Tr. Will Straw. Ithaca, N.Y.: Cornell University Press, 1998.

Krčálová, Jarmila. "La Toscana e l'architettura di Rodolfo II: Giovanni Gargiolli a Praga." In *Firenze e la Toscana dei Medici nell'Europa del '500*. 3 vols. Florence: Leo S. Olschki, 1983. 3: 1029–51.

Kremer, Richard. "Mathematical Astronomy and Calendar-Making in Gdańsk from 1540 to 1700." In *Astronomie-Literatur-Volksaufklärung: Der Schreibkalender der Fruhen Neuzeit mit seinen Text- und Bildbeigaben*, ed. Klaus-Dieter Herbst. Bremen: Lumière, 2012. 477–92.

Kühne, Friedrich Wilhelm. "Chemical Processes in the Retina." *Vision Research* 17, 1–2 (1977): 1269–1316.

Kybalova, Jana. "The Decorative Arts." In *Rudolf II and Prague*, ed. Elisa Fučíkova et al. 376–86.

Lach, Donald F. *Asia in the Making of Europe.* 3 vols. Chicago: University of Chicago Press, 1994–1998.

Lacordaire, A. L. "Brevets accordés par les rois Henri IV, Louis XIII, Louis XIV, et Louis XV à divers artistes, peintres, sculpteurs, graveurs, orfèvres, etc." *Archives de l'Art Français* 5 (1853–1855): 189–286.

———. "Brevets de logements sous la grande galerie du Louvre (1628–1675)." *Archives de l'Art Français* 1 (1851–1852): 193–255.

Lambert, C. G. A. *Catalogue descriptif et raisonné de la Bibliothèque de Carpentras.* 3 vols. Carpentras: E. Rolland, 1862.

Lambert, Sheila. "Coranto Printing in England: The First Newsbooks." *Journal of Newspaper and Periodical History* 8, 1 (1992): 3–19.

———. "State Control of the Press in Theory and Practice: The Role of the Stationers' Company Before 1640." In *Censorship and the Control of Print in England and France*

1600–1910, ed. Robin Myers and Michael Harris. Winchester: St. Paul's Bibliographies, 1992. 1–32.

Langdale, Abram. *Phineas Fletcher: Man of Letters, Science, and Divinity.* New York: Columbia University Press, 1937.

Lanier, Douglas. "The Prison-House of the Canon: Allegorical Form and Posterity in Ben Jonson's *The Staple of Newes.*" *Medieval and Renaissance Drama in England* 2 (1985): 253–67.

Lankhorst, Otto. "Newspapers in the Netherlands in the Seventeenth Century." In *The Politics of Information*, ed. Brendan Dooley and Sabrina Baron. 151–59.

Lattis, James. *Between Copernicus and Galileo: Christoph Clavius and the Collapse of Ptolemaic Cosmology.* Chicago: University of Chicago Press, 1994.

Lazzerini, Luigi. "Officina Sarpiana: Scritture del Sarpi in materia di Gesuiti." *Rivista di Storia della Chiesa in Italia* 58, 1 (2004): 29–80.

"Les victimes de la Saint-Barthélemy à Paris." *Bulletin de l'Histoire du Protestantisme* 9 (1860): 34–44.

Ledvinka, Václav, and Jiří Pešek. "The Public and Private Lives of Prague's Burghers." In *Rudolf II and Prague*, ed. Eliša Fučíková et al. 287–301.

Leth, Göran. "A Protestant Public Sphere: The Early European Newspaper Press." *Studies in Newspaper and Periodical History* (1993): 67–90.

Levin, Michael J. *Agents of Empire: Spanish Ambassadors in Sixteenth-Century Italy.* Ithaca, N.Y.: Cornell University Press, 2005.

Levy, Fritz. "The Decorum of News." *Prose Studies* 21 (1998): 12–38.

———. "How Information Spread Among the Gentry, 1550–1640." *Journal of British Studies* 21, 2 (1982): 11–34.

Limentani, Uberto. *La Satira nel Seicento.* Milan: Riccardo Ricciardi, 1961.

Lindberg, David. "A Reconsideration of Roger Bacon's Theory of Pinhole Images." *Archive for History of Exact Sciences* 6 (1970): 214–23.

———. *Theories of Vision from Al-Kindi to Kepler.* Chicago: University of Chicago Press, 1976.

———. "The Theory of Pinhole Images from Antiquity to the Thirteenth Century." *Archive for History of Exact Sciences* 5 (1968): 154–76.

———. "The Theory of Pinhole Images in the Fourteenth Century." *Archive for History of Exact Sciences* 6 (1970): 229–325.

Lindley, David. "Introduction." In *The Court Masque*, ed. David Lindley. Manchester: Manchester University Press, 1984. 1–15.

Longhi, Silvia. "Il vestito sconveniente. Abiti et armature nella *Secchia Rapita.*" *Italique* 1 (1998): 104–26.

Louthan, Howard. "Mediating Confessions in Central Europe: The Ecumenical Activity of Valerian Magni, 1586–1661." *Journal of Ecclesiastical History* 55, 4 (2004): 681–99.

Love, Harold. *Scribal Publication in Seventeenth-Century England.* Oxford: Clarendon, 1993.

Lüsebrink, Hans-Jürgen, ed. *Les Lectures du peuple en Europe et en Amérique du XVII au XX siècle.* Brussels: Éditions Complexe, 2003.

MacLean, Gerald M. *The Rise of Oriental Travel: English Visitors to the Ottoman Empire, 1580–1720*. Houndsmills: Palgrave Macmillan, 2004.

Magrini, S. "Il 'De magnete' del Gilbert e i primordi della magnetologia in Italia in rapporto alla lotta intorno ai massimi sistemi." *Archeion* 8 (1929): 17–39.

Mah, Harold. "Phantasies of the Public Sphere: Rethinking the Habermas of Historians." *The Journal of Modern History* 72, 1 (March 2000): 153–82.

Marchesini, Cesare G. "Almanacchi Italiani." *Gutenberg Jahrbuch* (1940): 177–88.

Marucci, Valerio et al, eds. *Pasquinate romane del Cinquecento*. 2 vols. Rome: Salerno, 1983.

Mayer, Thomas. "The Censoring of Galileo's *Sunspot Letters* and the First Phase of His Trial." *Studies in History and Philosophy of Science* 42 (2011): 1–10.

———. *The Roman Inquisition: A Papal Bureaucracy and Its Laws in the Age of Galileo*. Philadelphia: University of Pennsylvania Press, 2013.

———. *The Trial of Galileo, 1612–1633*. Toronto: University of Toronto Press, 2012.

Mayoux, Jean-Jacques. "Variations on the Time Sense in *Tristram Shandy*." In Laurence Sterne. *Tristram Shandy*, ed. Howard Anderson. New York: Norton, 1980. 570–84.

McCusker, John J. "The Demise of Distance: The Business Press and the Origins of the Information Revolution in the Early Modern Atlantic World." *American Historical Review* 110, 2 (2005): 295–321.

McPherson, David. "Ben Jonson Meets Daniel Heinsius, 1613." *English Language Notes* 44 (1976): 105–9.

Mestica, Giovanni. *Trajano Boccalini e la letterature critica e politica del seicento*. Florence: Giunti Barbèra, 1878.

Miller, Peter N. *Peiresc's Europe: Learning and Virtue in the Seventeenth Century*. New Haven, Conn.: Yale University Press, 2000.

Mitchell, W. J. T. *Iconology. Image, Text, Ideology*. Chicago: University of Chicago Press, 1986.

Morison, Stanley. "The Origins of the Newspaper." In *Selected Essays on the History of Letter-Forms in Manuscript and Print*, ed. David McKitterick. 2 vols. Cambridge: Cambridge University Press, 1980–1981. 2: 325–57.

Moureau, François. *Répertoire des nouvelles à la main*. Oxford: Alden Press, 1999.

Mousley, Andrew. "Self, State, and Seventeenth-Century News." *Seventeenth Century* 6, 2 (1991): 149–68.

Muggli, Mark. "Ben Jonson and the Business of News." *Studies in English Literature* 32 (1992): 321–40.

Muñoz Calvo, Sagrario. *Inquisición y ciencia en la España moderna*. Madrid: Editora Nacional, 1977.

Nevinson, J. L. "Illustrations of Costume in the *alba amicorum*." *Archaeologia* 106 (1979): 167–76.

Nevitt, Marcus. "Ben Jonson and the Serial Publication of News." *Media History* 11, 1–2 (2005): 53–68.

Nickson, Margaret. *Early Autograph Albums in the British Museum*. London: Oxford University Press, 1970.

————. "Some Early English, French, and Spanish Contributions to Albums." In *Stammbücher des 16. Jahrhunderts*, ed. Wolfgang Klose. Wiesbaden: Harrassowitz, 1989. 63–73.

————. "Some Sixteenth-Century Albums in the British Library." In *Stammbücher als Kulturhistorische Quellen*, ed. Jörg-Ulrich Fechner. Munich: Kraus, 1981. 23–36.

Noldus, Badeloch. "Dealing in Politics and Art: Agents Between Amsterdam, Stockholm and Copenhagen." *Scandinavian Journal of History* 28 (2003): 215–25.

Novati, Francesco. "La storia e la stampa nella produzione popolare italiana." *Emporium* 24 (1906): 181–209.

Orbaan, J. A. F. "La Roma di Sisto V negli *Avvisi*." *Archivio della Reale Società di storia patria* 33 (1910): 277–312.

Orgel, Stephen. *The Jonsonian Masque*. New York: Columbia University Press, 1981.

Pagani, Valeria. "A *Lunario* for the Years 1584–1586 by Francesco da Volterra and Diana Mantovana." *Print Quarterly* 8 (1991): 140–45.

Palmer, Barbara D. "Court and Country: The Masque as Sociopolitical Subtext." *Medieval and Renaissance Drama in England* 7 (1995): 338–54.

Palmitessa, James R. "The Prague Uprising of 1611: Property, Politics, and Catholic Renewal in the Early Years of Habsburg Rule." *Central European History* 31, 4 (1998): 299–328.

Pánek, Jaroslav. "The Nobility in the Czech Lands, 1550–1650." In *Rudolf II and Prague*, ed. Eliša Fučíková et al. 270–86.

Pantin, Isabelle. "Galilée, la lune, et les Jésuites." *Galilaeana* 2 (2005): 19–42.

————. "*Simulachrum, species, forma, imago*: What Was Transported by Light into the *camera obscura?*" *Early Science and Medicine* 13 (2008): 245–69.

Park, Katharine. "The Life of the Corpse: Division and Dissection in Late Medieval Europe." *Journal of the History of Medicine and Allied Sciences* 50 (1995): 111–32.

Parker, Geoffrey, ed. *The Thirty Years' War*. London: Routledge, 2002.

Parker, Kenneth, ed. *Early Modern Tales of Orient*. London: Routledge, 1999.

Pearl, Sara. "Sounding to Present Occasions: Jonson's Masques of 1620–5." In *The Court Masque*, ed. David Lindley. 60–77.

"Petit dictionnaire des arts et métiers avant 1789." *Magasin Pittoresque* 50 (1882): 305–7.

Pettegree, Andrew. *The Book in the Renaissance*. New Haven, Conn.: Yale University Press, 2010.

Picciotto, Joanna. *Labors of Innocence in Early Modern England*. Cambridge, Mass.: Harvard University Press, 2010.

Polišenský, Josef. "Le relazioni tra la Boemia e l'Italia al tempo di Giordano Bruno e Galileo Galilei." *Philologia Pragensia* 24 (1981): 6–12.

————. *The Thirty Years War*. Tr. Robert Evans. London: Batsford, 1971.

————. "Tolerance and Intolerance in Rudolfine Prague." In *Rudolf II, Prague and the World*, ed. Lubomír Konečný, Beket Bukovinská, and Ivan Muchka. Prague: Artefactum, 1998. 9–10.

Powell, William. *John Pory, 1572–1636*. Chapel Hill: University of North Carolina Press, 1977.

Quaglio, Antonio. "Indicazioni sulla fortuna editoriale di Machiavelli nel Veneto." *Lettere italiane* 21 (1969): 399–424.

Rabb, Theodore. "English Readers and the Revolt in Bohemia, 1619–1622." In *Aharon M. K. Rabinowicz Jubilee Volume*, ed. Moshe Aberbach. Jerusalem: Bialik Institute, 1996. 152–75.

Randall, David. "Epistolary Rhetoric, the Newspaper, and the Public Sphere." *Past and Present* 198 (February 2008): 3–32.

———. "Joseph Mead, Novellante: News, Sociability, and Credibility in Early Stuart England." *Journal of British Studies* 45 (2006): 293–312.

Ranum, Orest. *Artisans of Glory: Writers and Historical Thought in Seventeenth-Century France*. Chapel Hill: University of North Carolina Press, 1980.

Raymond, Joad. "The Newspaper, Public Opinion, and the Public Sphere in the Seventeenth Century." *Prose Studies* 21 (1998): 109–40.

———. "Newspapers: A National or International Phenomenon?" *Media History* 18, 3–4 (2012): 249–57.

———. *Pamphlets and Pamphleteering in Early Modern Britain*. Cambridge: Cambridge University Press, 2006.

Reeves, Eileen. "Complete Inventions: The Mirror and the Telescope." In *The Origins of the Telescope*, ed. Albert Van Helden, Sven Dupré, Rob van Gent, and Huib Zuidervaart. 167–182

———. "Faking It: Apelles and Protogenes Among the Astronomers." *Kunsthistorisches Jahrbuch für Bildkritik* 5, 2 (2007): 65–72.

———. "From Dante's Moonspots to Galileo's Sunspots." *MLN* 124 Supplement (2009): 190–209.

———. *Galileo's Glassworks*. Cambridge, Mass.: Harvard University Press, 2008.

———. "Kingdoms of Heaven: Galileo and Sarpi on the Celestial." *Representations* 105, 4 (2009): 61–84.

———. "Old Wives' Tales and the New World System: Gilbert, Galileo, and Kepler." *Configurations* 7, 3 (1999): 301–54.

———. *Painting the Heavens*. Princeton, N.J.: Princeton University Press, 1997.

———. "Representing Invention: The Telescope as News." In *Sparks and Seeds: Medieval Literature and its Afterlife. Essays in Honor of John Freccero*, ed. Dana Stewart and Alison Cornish. Turnhout: Brepols, 2000. 267–90.

———. "Speaking of Sunspots: Oral Culture in an Early Modern Scientific Exchange." *Configurations* 13 (2005): 185–210.

Reeves, Eileen, and Albert Van Helden. "Verifying Galileo's Discoveries: Telescope-Making at the Collegio Romano." *Acta Historica Astronomiae* 33 (2007): 127–41.

Reiss, Timothy J. *Against Autonomy. Global Dialectics of Cultural Exchange*. Stanford, Calif.: Stanford University Press, 2002.

Remmert, Volker. *Picturing the Scientific Revolution: Title Engravings in Early Modern Scientific Publications*. Tr. Ben Kern. Philadelphia: Saint Joseph's University Press, 2011.

Reynolds, Anne. "The Classical Continuum in Roman Humanism: The Festival of Pasquino, the *Robigalia*, and Satire." *Bibliothèque d'Humanisme et Renaissance* 49 (1987): 289–307.

Richardson, Brian. *Manuscript Culture in Renaissance Italy.* Cambridge: Cambridge University Press, 2009.

Riezler, Siegmund. "Kriegstagebücher aus dem ligistichen Hauptquartier 1620." *Abhandlungen der Historischen Klasse der Königlich Bayerischen Akademie der Wissenschaften* 23, 1 (1906): 77–210.

Rodis-Lewis, Geneviève. "Machineries et perspectives curieuses dans leurs rapports avec le Cartésianisme." *Dix-septième Siècle* 32 (1956): 461–74.

Rosen, Edward. *The Naming of the Telescope.* New York: Henry Schuman, 1947.

Rosenheim, Max. "The *Album Amicorum*." *Archaeologia* 62 (1910): 251–308.

Rudt de Collenberg, W. H. "Un *Liber amicorum* della Biblioteca Casanatense di Roma (1608–1621)." *Rassegna degli archivi di stato* 46 (1986): 36–52.

Sacchi, Henri. *La Guerre de trente ans.* 3 vols. Paris: Harmattan, 1991.

Sanders, Julie. *Ben Jonson's Theatrical Republics.* New York: Saint Martin's, 1998.

Scalabrini, Massimo. "Gli amori ridicoli dell'eroicomico: Tassoni e la storia di Lucrezia." *MLN* 120 (2005): 223–38.

Schama, Simon. *The Embarrassment of Riches: An Interpretation of Dutch Culture in the Golden Age.* New York: Vintage, 1987.

Schickler, Fernand. "Hotman de Villiers et son temps." *Bulletin historique et littéraire de la société de l'histoire du protestantisme français* 17 (1868): 98–111, 145–61, 401–13, 464–76, 513–33.

Schleiner, Winfried. "Scioppius' Pen Against the English King's Sword." *Renaissance and Reformation* 26, 4 (1990): 271–84.

Schröder, Thomas. "The Origins of the German Press." In *The Politics of Information*, ed. Brendan Dooley and Sabrina Baron. 123–50.

Schuster, John, and Judit Brody. "Descartes and Sunspots: Matters of Fact and Systematizing Strategies in the *Principia Philosophiae*." *Annals of Science* 70, 1 (2013): 1–45.

Seccombe, Thomas. "Matthew, Sir Tobie." In *Dictionary of National Biography*, ed. Sir Leslie Stephens and Sir Sidney Lee. 22 vols. Oxford: Oxford University Press, 1921–1922. 13: 63–68.

Sellin, Paul R. "John Donne and the Huygens Family, 1619–1621." *Dutch Quarterly Review of Anglo-American Letters* 12 (1982–1983): 192–204.

———. "The Politics of Ben Jonson's *Newes from the New World Discover'd in the Moone*." *Viator* 17 (1986): 321–37.

Servolini, Luigi. "Un tipografo del seicento, Amadore Massi." *Accademie & biblioteche d'Italia* n.s. 19 (1951): 207–26.

Settle, Thomas. "Egnazio Danti: le meridiane in Santa Maria Novella a Firenze e gli strumenti collegati." *Giornale di astronomia* 32 (2006): 91–98.

———. "The Invention(s) of the Telescope." *Actes de la III Jornada sobre Historia de l'Astronomia i de la Meteorologia* (2011): 21–40.

Shea, William R. "Descartes and the Rosicrucians." *Annali dell'Istituto e Museo di Storia della Scienza* 4, 2 (1979): 29–47.

Sherman, Stuart. *Telling Time: Clocks, Diaries, and English Diurnal Form 1680–1785.* Chicago: University of Chicago Press, 1996.

Shohet, Lauren. *Reading Masques: The English Masque and Public Culture in the Seventeenth Century.* Oxford: Oxford University Press, 2010.

Silvestre, Hubert. "Miroir ou Mica? A propos d'une *Vita Ambrosii* carolingienne." *Recherches de théologie ancienne et médiévale* 42 (1975): 243–46.

Šíma, Zdislav. *Astronomie a Klementinum. Astronomy and Clementinum.* Prague: Národní knihovna České republiky, 2001.

Šimeček, Zdeněk. "The First Brussels, Antwerp and Amsterdam Newspapers: Additional Information." *Revue Belge de Philologie et d'Histoire* 50, 4 (1972): 1098–115.

Simón de Guilleuma, José M. "Juan Roget, óptico gerundense, inventor del telescopio, y los Roget de Barcelona, constructors del mismo." *Boletín del Instituto Municipal Histórico de Barcelona* 775 (1958): 413–14.

———. "Notas Bibliográficas." *Annals del Institu d'Estudies Gironins* (1958): 413–414.

Simoni, Anna E. C. "Poems, Pictures and the Press. Observations on Some Abraham Verhoeven Newsletters (1620–1621)." In *Liber Amicorum Leon Voet,* ed. Francine de Nave. Antwerp: Nederlandsche Boekhandel, 1985. 353–73.

Simons, Madelon. "Archduke Ferdinand II of Austria, Governor in Bohemia, and the Theater of Representation." In *Rudolf II, Prague and the World,* ed. Lubomír Konečný, Beket Bukovinská, and Ivan Muchka. Prague: Artefactum, 1998. 270–77.

Sluiter, Engel. "The Telescope Before Galileo." *Journal for the History of Astronomy* 28 (1997): 223–34.

Smith, Charlotte Fell. "Watts, William." In *Dictionary of National Biography,* ed. Sir Leslie Stephens and Sir Sidney Lee. 20: 986.

Smith, Woodruff D. "The Function of Commercial Centers in the Modernization of European Capitalism: Amsterdam as an Information Exchange in the Seventeenth Century." *Journal of Economic History* 44, 4 (1984): 985–1005.

Smolka, Josef. "The Scientific Revolution in Bohemia." In *The Scientific Revolution in National Context,* ed. Roy Porter and Mikuláš Teich. Cambridge: Cambridge University Press, 1992. 210–39.

Soll, Jacob. *Publishing the* Prince: *History, Reading, and the Birth of Political Criticism.* Ann Arbor: University of Michigan Press, 2008.

Solomon, Howard M. *Public Welfare, Science, and Propaganda in Seventeenth-Century France. The Innovations of Théophraste Renaudot.* Princeton, N.J.: Princeton University Press, 1972.

Soman, Alfred. "Press, Pulpit, and Censorship in France Pefore Richelieu." *Proceedings of the American Philosophical Society* 120, 6 (1976): 439–63.

Sommerville, C. John. *The News Revolution in England. Cultural Dynamics of Daily Information.* Oxford: Oxford University Press, 1996.

Sosio, Libero. "Il manoscritto dell'Iride e del Calore." In Paolo Sarpi, *Pensieri naturali, metafisici e matematici*, ed. Luisa Cozzi and Libero Sosio. Milan: Riccardo Ricciardi, 1996. 519–46.

Steadman, Philip. *Vermeer's Camera*. Oxford: Oxford University Press, 2002.

Stengers, Isabelle. *The Invention of Modern Science*. Tr. Daniel W. Smith. Minneapolis: University of Minnesota Press, 2000.

Stewart, Susan. *On Longing*. Durham, N.C.: Duke University Press, 1993.

Straker, Stephen. "Kepler, Tycho, and the 'Optical Part of Astronomy': The Genesis of Kepler's Theory of Pinhole Images." *Archive for History of Exact Sciences* 24 (1981): 267–93.

Stubbs, Mayling. "John Beale, Philosophical Gardener of Herefordshire. Part I. Prelude to the Royal Society (1608–1663)." *Annals of Science* 39 (1982): 463–89.

Tatlock, Lynne, ed. and tr. *Seventeenth Century German Prose*. New York: Continuum, 1993.

Te Brake, Wayne. *Shaping History: Ordinary People in European Politics, 1500–1700*. Berkeley: University of California Press, 1998.

Teague, Frances. "Jonson's Drunken Escapade." *Medieval and Renaissance Drama in England* 6 (1993): 129–37.

Thompson, Anthony B. "Licensing the Press: The Career of G. R. Weckherlin During the Personal Rule of Charles I." *Historical Journal* 41, 3 (1998): 653–78.

Tierie, Gerrit. *Cornelis Drebbel (1572–1633)*. Amsterdam: H.J. Paris, 1932.

Tolbert, Jane. "Censorship and Retraction: Théophraste Renaudot's *Gazette* and the Galileo Affair, 1631–1633." *Journalism History* 31, 2 (Summer 2005): 98–105.

Toulmin, Stephen. *Cosmopolis: The Hidden Agenda of Modernity*. Chicago: University of Chicago Press, 1990.

Turner, Brian S. "The Anatomy Lesson: A Note on the Merton Thesis." *Sociological Review* 38, 1 (1990): 1–18.

Turner, Gerard l'Estrange. "Three Late-Seventeenth Century Italian Telescopes, Two Signed by Paolo Belletti of Bologna." *Annali dell'Istituto e Museo di Storia della Scienza di Firenze* 9, 1 (1984): 41–64.

Vail, Jeffrey. "'The Bright Sun Was Extinguis'd': The Bologna Prophecy and Byron's 'Darkness.'" *Wordsworth Circle* 28, 3 (1997): 183–92.

Van Berkel, Klaas. "Cornelis Jacobusz. Drebbel." In *A History of Science in the Netherlands*, ed. Klaas Van Berkel, Albert Van Helden, and Lodewijk Palm. 441–43.

Van Berkel, Klaas, Albert Van Helden, and Lodewijk Palm, eds. *A History of Science in the Netherlands: Survey, Themes and Reference*. Leiden: Brill, 1999.

Van Helden, Albert. *Catalogue of Early Telescopes*. Florence: Istituto e Museo di Storia della Scienza, 1999.

———. "Galileo and the Telescope." In *The Origins of the Telescope*, ed. Albert Van Helden, Sven Dupré, Rob van Gent, and Huib Zuidervaart. Amsterdam: KNAW, 2010. 183–201.

———. *The Invention of the Telescope.* Philadelphia: American Philosophical Society, 1977.

———. "The Telescope in the Seventeenth Century." *Isis* 65 (1974): 38–58.

———. "Telescopes and Authority from Galileo to Cassini." In *Instruments*, ed. Albert Van Helden and Thomas L. Hankins. Special issue of *Osiris* ser. 2, 9 (1994): 9–29.

Van Helden, Albert, Sven Dupré, Rob van Gent, and Huib Zuidervaart, eds. *The Origins of the Telescope.* Amsterdam: KNAW, 2010.

Van Houtte, H. "Un journal manuscrit intéressant (1557–1648). Les *Avvisi* du Fonds Urbinat et d'autres Fonds de la Bibliothèque Vaticane." *Bulletin de la Commission Royale d'Histoire* 89 (1925): 359–440.

Van Strien, Kees. "Holland and the Dutch in Sterne's *Tristram Shandy.*" *Notes and Queries* (March 2007): 71–74.

Venuti, Lawrence. *The Translator's Invisibility: A History of Translation.* London: Routledge, 1995.

Vermij, Rienk. *The Calvinist Copernicans: The Reception of the New Astronomy in the Dutch Republic, 1575–1750.* Amsterdam: KNAW, 2002.

Vialardi di Sandigliano, Tomaso. "Un cortigiano e letterato piemontese del Cinquecento: Francesco Maria Vialardi." *Studi piemontesi* 34, 2 (December 2005): 299–312.

Vianello, Nereo. "Le postille al Petrarca di Galileo Galilei." *Studi di Filologia Italiana* 14 (1956): 211–433.

Virilio, Paul. *Speed and Politics*, tr. Mark Polizzotti. New York: Semiotexte, 1977.

Vittu, Jean-Pierre. "Instruments of Political Information in France." In *The Politics of Information*, ed. Brendan Dooley and Sabrina Baron. 160–78.

Vlček, Pavel. "Architecture in Prague, 1550–1650." In *Rudolf II and Prague*, ed. Eliša Fučíková et al. 345–52.

Voit, Petr. *Prazské Klementinum.* Prague: Národní knihovna v Prahe, 1990.

Wallace, William. *Prelude to Galileo: Essays on Medieval and Sixteenth-Century Sources of Galileo's Thought.* Dordrecht: Reidel, 1981.

Watt, Tessa. "Publisher, Pedlar, Pot-Poet: The Changing Character of the Broadside Trade, 1550–1640." In *Spreading the Word: The Distribution Networks of Print, 1550–1850*, ed. Robin Myers and Michael Harris. Winchester: Saint Paul's Bibliographies, 1990. 61–81.

Weber, Johannes. "The Early German Newspaper—A Medium of Contemporaneity." In *The Dissemination of News*, ed. Brendan Dooley. 69–79.

Weiss, Robert. "Henry Wotton and Orazio Lombardelli." *Review of English Studies* 19 (1943): 285–89.

Whitaker, Ewen. "Galileo's Lunar Observations and the Dating of 'Sidereus Nuncius.'" *Journal for the History of Astronomy* 9 (1978): 155–69.

Wilding, Nick. "Galileian Angels." In *Conversations with Angels: Essays Towards a History of Spiritual Communication, 1100–1700*, ed. Joad Raymond and Lauren Kassell. Houndmills: Palgrave Macmillan, 2011. 67–89.

———. *Galileo's Idol.* University of Chicago Press, forthcoming.

Wilkinson, Alexander. "'Homicides Royaux': The Assassination of the Duc and Cardinal de Guise and the Radicalization of French Public Opinion." *French History* 18, 2 (2004): 129–53.

Woolf, Daniel R. "Genre into Artifact: The Decline of the English Chronicle in the Sixteenth Century." *Sixteenth Century Journal* 19, 3 (1988): 321–54.

———. "News, History, and the Construction of the Present in Early Modern England." In *The Politics of Information*, ed. Brendan Dooley and Sabrina Baron. 80–118.

Wootton, David. *Galileo Watcher of the Skies.* New Haven, Conn.: Yale University Press, 2010.

———. "New Light on the Composition and Publication of the *Sidereus Nuncius*." *Galilaeana* 6 (2009): 61–78.

———. *Paolo Sarpi: Between Renaissance and Enlightenment.* Cambridge: Cambridge University Press, 1983.

Ziolkowski, Jan M., and Michael C. J. Putnam, eds. *The Virgilian Tradition: The First Fifteen Hundred Years.* New Haven, Conn.: Yale University Press, 2008.

INDEX

ACKNOWLEDGMENTS

This project emerged from a single sentence, rather freely translated, in Galileo's *Dialogue concerning the Two Chief World Systems.* "Why talk of publications and notoriety?" Galileo's spokesman Salviati asked during a discussion of novelty, innovation, and what we now call intellectual property. It would be difficult to overestimate the initial appeal of Salviati's rhetorical question; shorn of context, it seemed to me a plangent transcription of my own professional concerns. But my interest in the relationships between newness, newsworthiness, and the rapid development of early modern astronomy soon transcended this narrow focus, and the question eventually became the basis of a long-term project. In preparing *Evening News* for publication, if not for notoriety, I have benefitted tremendously from the support of colleagues, friends, and family.

I would like to thank, first of all, Sachiko Kusukawa for welcoming me to the History of Science ages ago, and for her warm friendship ever since. I am also grateful to Raz Chen, Ofer Gal, Stefano Gattei, Robert Goulding, Lauren Kassell, Richard Kremer, Pamela Long, Tom Settle, Nancy Siraisi, Mark Smith, Pamela Smith, Rob Van Gent, and Huib Zuidervaart for maintaining my access to this discipline. I owe special thanks to Noel Swerdlow for his introduction to the history of astronomy, and to Mario Biagioli, Massimo Bucciantini, Michele Camerota, Sven Dupré, Paula Findlen, John Heilbron, Matteo Valleriani, Nick Wilding, and David Wootton for their tremendous generosity in all things Galileian. I extend warmest thanks, as always, to Albert Van Helden for years of intellectual guidance in every aspect of the history of science, sound professional advice, and an endless supply of amusing insights.

I am also greatly indebted to a number of art historians who share my interest in early modern science, and who have accommodated intrusions from a student of literature: Horst Bredekamp, Miles Chappell, Michael

Cole, Claire Farago, David Freedberg, Michel Hochmann, Martin Kemp, Lyle Massey, Alex Marr, Steven Ostrow, Volker Remmert, Claudia Swan, Lucia Tongiorgi Tomasi, and Alessandro Tosi. Filippo de Vivo, Brendan Dooley, and Joad Raymond have been splendidly generous guides in the world of early modern journalism; Tom Mayer has been a magnanimous and patient informant about all aspects of book history. Among my confederates in literature, I owe an immense debt to my mentor John Freccero for supporting an idiosyncratic intellectual agenda with wisdom, tact, and humor. The interest and encouragement of my fellow early modernists have also been crucial: I am especially grateful to Albert Ascoli, Louise Clubb, Alison Cornish, Margreta de Grazia, Carla Freccero, Annie Jones, Vicky Kahn, Joanna Picciotto, Deanna Shemek, Carolyn Springer, and Peter Stallybrass.

Here at Princeton, I am indebted to the Renaissance Studies community, and especially to Leonard Barkan, Marina Brownlee, Jeff Dolven, Yaacob Dweck, Dan Garber, Tony Grafton, Wendy Heller, Tom Kaufmann, François Rigolot, and Nigel Smith, for helpful feedback on earlier versions of this work. Mary Harper and Carol Rigolot have encouraged this venture from the start, and have generously facilitated my contact with scholars working in adjacent areas. And I can imagine no warmer welcome than that provided, week after week, by the Program Seminar in the History of Science; I am particularly thankful to Angela Creager and Michael Gordin for their interest and collegiality.

I have been very fortunate in finding so much support at Penn Press. I owe special thanks to Caroline Hayes and to Alison Anderson for their attention to countless production details. Jerry Singerman has been a wise, loyal, and extraordinarily patient editor, and it has been a great privilege to work with him.

My greatest debt, by far, is to the usual suspects, Jim, Jimmy, and John English, whose daily conversations invariably involve notoriety, and occasionally publications. Even when *Evening News* seemed to me old news indeed, the orbit of their affection and humor always kept the project novel.